GRAVITATION AND COSMOLOGY

Related Titles from AIP Conference Proceedings

759 Material Science: 2nd Mexican Meeting on Mathematical and Experimental Physics
Edited by Roberto Olayo, March 2005, 0-7354-0244-2

757 Statistical Physics and Beyond: 2nd Mexican Meeting on Mathematical and Experimental Physics
Edited by F.J. Uribe, L. García-Colín S., E. Díaz-Herrera, March 2005, 0-7354-0242-6

751 General Relativity and Gravitational Physics, 16th SIGRAV Conference
Edited by G. Esposito et al., 0-7354-0236-1

743 The New Cosmology: Conference on Strings and Cosmology and The Mitchell Symposium on Observational Cosmology
Edited by R. E. Allen, D. V. Nanopoulos, and C. N. Pope, December 2004 , 0-7354-0227-2

736 Phi in the Sky: The Quest for Cosmological Scalar Fields
Edited by C. J. A. P. Martins, P. P. Avelino, M. S. Costa, K. Mack, M. F. Mota, and M. Parry, November 2004, 0-7354-0218-3

689 Neutrinos, Flavor Physics, and Precision Cosmology: Fourth Tropical Workshop on Particle Physics and Cosmology
Edited by J. F. Nieves and R.R. Volkas, October 2003, 0-7354-0160-8

686 The Astrophysics of Gravitational Wave Sources
Edited by J.M. Centrella, October 2003, 0-7354-0157-8

668 Cosmology and Gravitation: Xth Brazilian School of Cosmology and Gravitation; 25th Anniversary (1977-2002)
Edited by M. Novello and S. E. Perez Bergliaffa, May 2003, 0-7354-0131-4

666 The Emergence of Cosmic Structure: Thirteenth Astrophysics Conference
Edited by S.S. Holt and C.S. Reynolds, May 2003, 0-7354-0128-4

586 Relativistic Astrophysics: 20th Texas Symposium
Edited by J.C Wheeler and H. Martel, October 2001, 0-7354-0026-1

575 Astrophysical Sources for Ground-Based Gravitational Wave Detectors
Edited by J.M. Centrella, July 2001, 0-7354-0014-8

523 Gravitational Waves: Third Edoardo Amaldi Conference
Edited by S. Meshkov, June 2000, 1-56396-944-0

To learn more about these titles, or the AIP Conference Proceedings Series, please visit the webpage
http://proceedings.aip.org/proceedings

GRAVITATION AND COSMOLOGY

2nd Mexican Meeting on Mathematical and Experimental Physics

México City, México 6 - 10 September 2004

EDITORS

Alfredo Macías
Universidad Autónoma Metropolitana - Iztapalapa

Claus Läemmerzahl
ZARM, University of Bremen

Dario Nuñez
Insituto de Ciencias Nucleares
Universidad Nacional Autónoma de Mexico

SPONSORING ORGANIZATIONS
El Colegio Nacional
Universidad Autónoma Metropolitana
Consejo Nacional de Ciencia y Tecnología

AMERICAN INSTITUTE OF PHYSICS

Melville, New York, 2005

AIP CONFERENCE PROCEEDINGS ■ VOLUME 758

Editors:

Alfredo Macías
Departmento de Física
Universidad Autónoma Metropolitana - Iztapalapa
Apartado Postal 55-534
C.P. 09340, México, D.F.
México

Claus Läemmerzahl
ZARM, University of Bremen
AmFallturm
28359 Bremen, Germany

Dario Nuñez
Instituto de Ciencias Nucleares
Universidad Nacional Autónoma de México
Apartado Postal 70-543
C.P. 04510 México, D.F.
México

E-mail: amac@xanum.uam.mx
laemmerzahl@zarm.uni-bremen.de
nunez@nuclecu.unam.mx

L.C. Catalog Card No 2005922750

ISBN 0-7354-0243-4
ISSN 0094-243X
Printed in the United States of America

Contents

QUANTUM GRAVITY AND EXPERIMENTAL SEARCH

ASTROPHYSICS AND COSMOLOGY

MATHEMATICAL FORMALISM, NUMERICAL RELATIVITY AND EXACT SOLUTIONS

PREFACE TO THE VOLUME: GRAVITATION AND COSMOLOGY

The SECOND MEXICAN MEETING ON MATHEMATICAL AND EXPERIMENTAL PHYSICS was held at EL COLEGIO NACIONAL in Mexico City, Mexico, from September 6th to 10th, 2004, to set up a forum for promising and important areas of research in Physics. The main objective was to open up new lines of research for young Mexican scientists and promising graduate students, and also to provide our colleagues with a broad view of important subjects of current interest, and a forum where the corresponding results of their research could be communicated.

For his outstanding contributions to fields of higher–dimensional unification theories, like supergravity, strings and M–theory, Prof. Michael Duff was awarded the Mexican Meeting Medal 2004. For this reason, we dedicate these Proceedings to him.

These Proceedings contain lecture notes on the topics covered during the SYMPOSIUM ON GRAVITATION AND COSMOLOGY, the contributions cover the invited plenary talks, as well as the invited short talks. Plenary talks provided a pleasant and efficient way to learn about the present status of the more relevant topics of current research. In addition, short talks gave us the opportunity to acquire an idea of the research work being developed by the participants, and at the same time they facilitated the communication among them.

We would like to thank everyone who contributed to the success of the Symposium on Cosmology and Gravitation. Very special thanks are due to the invited speakers and lecturers who addressed a very interesting and high quality set of talks, and shared their deep knowledge and time with participants.

The Symposium on Cosmology and Gravitation could not have been realized without the financial support of El Colegio Nacional, and CONACyT (Mexico). We wish to thank Dr. Luis Mier y Terán, Dr. José Lema and Dr. Tomás Viveros, Principal Rector, Rector of the campus Iztapalapa, Dean at Iztapalapa of the Faculty of Basic Sciences and Engineering of the Universidad Autónoma Metropolitana, respectively, for sponsoring this international and multidisciplinary endeavor. The financial support of ICN–UNAM is gratefully acknowledged. We also thank María Eugenia López–Solache, Administrative Assistant attached to the Physics Department of the Universidad Autónoma Metropolitana, for her invaluable help and support.

We gratefully acknowledge Prof. Leopoldo García–Colín and all the staff at EL COLEGIO NACIONAL for the warm hospitality, which was extended to all participants.

We hope that these Proceedings will serve to foster the impressive growth of Cosmology and Relativity in the region, and additionally to reinforce the already existing strong ties between Mexican scientists and scientist and relativists in other parts of the world.

Alfredo Macías, Claus Lämmerzahl, and Dario Nuñez

Michael Duff

MICHAEL DUFF: BRIEF BIOGRAPHY

Michael Duff got his PhD. in theoretical physics in 1972 at Imperial College, London, under Nobel Laureate Abdus Salam. After postdoctoral fellowships in Trieste, Oxford, King's College (London), Queen Mary College (London), and Brandeis, he returned to Imperial College in 1979 on a Science Research Council Advanced Fellowship and joined the faculty there in 1980. He took leave of absence to visit the Theory Division in CERN, first in 1982, and then again as Staff Member from 1984 to 1987 when he became Senior Physicist. He has also held Visiting Professorships and Fellowships at the University of Texas, Austin, the University of California, Santa Barbara, the University of Kyoto, and the Isaac Newton Institute at the University of Cambridge. He took up his professorship at Texas A & M in 1988 and was appointed Distinguished Professor in 1992. In September 1999 he moved to the University of Michigan, where he is currently Oskar Klein Professor of Physics. In 2001, he was elected first Director of the Michigan Center for Theoretical Physics and was recently re-elected to serve until 2006.

Professor Duff's interests lie in theoretical high energy physics, especially in the quest for an all-embracing theory that would explain the origin of the Universe, its fundamental constituents and the laws that govern those constituents. His research involves quantum gravity, supergravity, Kaluza-Klein theory, superstrings, black holes, supermembranes and M-theory. In the period 1984-95, the best hope for such a "Theory of Everything" was based on Superstrings: one-dimensional objects whose modes of vibration represent the elementary particles. Superstrings live in a world with ten spacetime dimensions, six of which would have to be curled up to an unobservable small size. Recently, however, work by Professor Duff and others has indicated that superstring theory is but a small corner of a deeper theory called M-theory (where M stands for Magic, Mystery or Membrane). M-theory lives in a spacetime of eleven dimensions that describes two and five dimensional extended objects called membranes and fivebranes. The precise formulation of M–theory remains as an outstanding challenge, but it seems that quantum black holes play an important part. He is a member of Save British Science and has acted as spokesman for British Scientists Abroad, a group of expatriate scientists who are concerned about the underfunding of British Science. He is a Fellow of the American Physical Society and a Fellow of the Institute of Physics (UK). He was Carlos Graef lecturer 2003 at the Universidad Autonmoma Metropolitana-Iztapalapa.

Indeed, all his scientific papers are very clear, accurate, critical and carefully written. Assets that are reflected over and over in Mike's prominent and brilliant scientific career.

For his outstanding contributions to fields of higher–dimensional unification theories, like supergravity, strings and M–theory, Prof. Michael Duff was awarded the Mexican Meeting Medal 2004.

Alfredo Macías, Claus Lämmerzahl, and Dario Nuñez

CONTRIBUTING AUTHORS

- Miguel Alcubierre
  Instituto de Ciencias Nucleares
  Universidad Nacional Autónoma de México.
  A.P. 70-543, 04510 México D.F., México.
  E–mail: malcubi@nuclecu.unam.mx
- Giovanni Amelino–Camelia
  Dipt Fisica, Univ La Sapienza and Sez Roma1 INFN, Ple Moro 2, Rome, Italy.
  E–mail:gamelino@mail.cern.ch
- John D. Anderson
  Jet Propulsion Laboratory, California Institute of Technology,
  Pasadena, CA 91109, U.S.A.
- David Bahena
  Charles University Prague, Czech Republic.
- Argelia Bernal
  Departamento de Física, CINVESTAV–IPN.
  Apartado Postal 14–740, C.P. 07000, México, D.F., México.
- Abel Camacho
  Departamento de Física , Universidad Autónoma Metropolitana–Iztapalapa.
  P. O. Box. 55–534, México D. F., C.P. 09340 México.
  E–mail: acq@xanum.uam.mx
- Elías Castellanos
  Departamento de Física , Universidad Autónoma Metropolitana–Iztapalapa.
  P. O. Box. 55–534, México D. F., C.P. 09340 México.
- Mauricio Cataldo
  Departamento de Física, Facultad de Ciencias, Universidad del Bío–Bío, Avenida Collao 1202, Casilla 5-C, Concepción, Chile.
- Manuel G. Corona-Galindo
  Instituto Nacional de Astrofísica, Optica y Electrónica
  A. P. 256, Tonanzintla, Puebla, México
- Heinz Dehnen
  Konstanz University, Germany.
- Sergio del Campo
  Instituto de Física, Facultad de Ciencias Básicas y Matemáticas,
  Universidad Católica de Valparaíso. Avenida Brasil 2950, Valparaíso, Chile.
  Depart. de Física, Facultad de Ciencia, Universidad de Santiago de Chile.
  E–mail: sdelcamp@ucv.cl
- Hansjoerg Dittus
  ZARM, University of Bremen. 28359 Bremen, Germany.
  E–mail: dittus@zarm.uni-bremen.de

- Michael J. Duff
  Michigan Center for Theoretical Physics,
  Randall Laboratory, Department of Physics, University of Michigan,
  Ann Arbor, MI 48109-1120, USA.
  E–mail: mduff@umich.edu
- Alberto García
  Departamento de Física, CINVESTAV–IPN.
  Apartado Postal 14–740, C.P. 07000, México, D.F., México.
  E–mail: aagarcia@fis.cinvestav.mx
- Leopoldo S. García–Colín
  Departamento de Fisica, Universidad Autónoma Metropolitana–Iztapalapa.
  Apartado Postal 55–534, C.P. 09340, México, D.F., México.
  E–mail: lgcs@xanum.uam.mx
- Hector García–Compeán
  Departamento de Física, Centro de Investigación y Estudios Avanzados del IPN.
  P.O. Box 14-740, 07000, México D.F., México.
  E–mail: compean@fis.cinvestav.mx
- Alberto García Quiroz
  Departamento de Física , Universidad Autónoma Metropolitana–Iztapalapa.
  P. O. Box. 55–534, México D. F., C.P. 09340 México.
- José A. González
  Institute for Gravitational Physics and Geometry
  Penn State University, University Park, PA 16802. USA.
  E–mail: jogonzal@gravity.psu.edu
- Jaime Klapp
  Instituto Nacional de Investigaciones Nucleares.
  Ocoyoacac 52045, Estado de México, México.
  E–mail: klapp@nuclear.inin.mx
- Claus Laemmerzahl
  ZARM, University of Bremen, AmFallturm, 28359 Bremen, Germany E–mail: laemmerzahl@zarm.uni-bremen.de
- Román Linares
  Departamento de Fisica, Universidad Autónoma Metropolitana–Iztapalapa.
  Apartado Postal 55–534, C.P. 09340, México, D.F., México.
  E–mail: lirr@xanum.uam.mx
- Enrique Lozano
  Departamento de Fisica, Universidad Autónoma Metropolitana–Iztapalapa.
  Apartado Postal 55–534, C.P. 09340, México, D.F., México.
  E–mail: enrique@abaco.izt.uam.mx
- Oscar Loaiza–Brito
  Physikalisches Institut, Universitat Bonn
  Nussallee 12, D-53115 Bonn, Germany.
- Alfredo Macías
  Departamento de Fisica, Universidad Autónoma Metropolitana–Iztapalapa.

Apartado Postal 55–534, C.P. 09340, México, D.F., México.
E–mail: amac@xanum.uam.mx

- V.S. Manko
Departamento de Física, Centro de Investigación y de Estudios Avanzados del IPN.
AP 14–740, 07000 México D.F., México.
E–mail: vsmanko@fis.cinvestav.mx
- Tonatiuh Matos
Departamento de Física, Centro de Investigación y de Estudios Avanzados del IPN.
AP 14–740, 07000 México D.F., México.
E–mail: tmatos@fis.cinvestav.mx
- R. Montemayor
Instituto Balseiro and CAB
Universidad Nacional de Cuyo and CNEA
8400 Bariloche, Argentina
- Holger Müller
Physics Department, Stanford University, Stanford, CA 94305-4060, USA.
E–mail: holgerm@stanford.edu
- Michael Martin Nieto
Theoretical Division (MS-B285), Los Alamos National Laboratory,
University of California, Los Alamos, New Mexico 87545. USA.
E–mail: mmn@lanl.gov
- Ulises Nucamendi
Instituto de Física y Matemáticas
Universidad Michoacana de San Nicolás de Hidalgo
Edif. C-3, Ciudad Universitaria, Morelia, Michoacán. CP. 58040. México.
- Abdel Pérez–Lorenzana
Departamento de Física, Centro de Investigación y de Estudios Avanzados del IPN.
AP 14–740, 07000 México D.F., México.
- Hernando Quevedo
Instituto de Ciencias Nucleares
Universidad Nacional Autónoma de México.
A.P. 70-543, 04510 México D.F., México.
E–mail: quevedo@nuclecu.unam.mx
- Norma Quiroz
Departamento de Fisica, CINVESTAV. 07000 México D.F., México.
- Victor O. Rivelles
Instituto de Física
Universidade de São Paulo
Caixa Postal 66318, 05315-970, São Paulo, SP, Brazil.
E-mail: rivelles@fma.if.usp.br
- Marcelo Salgado
Instituto de Ciencias Nucleares
Universidad Nacional Autónoma de México.

A.P. 70-543, 04510 México D.F., México.
E–mail: marcelo@nuclecu.unam.mx

- Alfredo Sandoval–Villalbazo
Departamento de Física y Matemáticas, Universidad Iberoamericana
Lomas de Santa Fe 01210 México D.F., México.
E-Mail: alfredo.sandoval@uia.mx
- Eli Santos
Departamento de Física
Universidad Autónoma Metropolitana–Iztapalapa.
P.O. Box 55–534, CP 09340, México D. F., México.
- Daniel Sudarsky
Instituto de Ciencias Nucleares
Universidad Nacional Autónoma de México.
A. Postal 70–543, México D.F. 04510, México.
E–mail: sudarsky@nuclecu.unam.mx
- Slava G. Turyshev
Jet Propulsion Laboratory, California Institute of Technology,
Pasadena, CA 91109, U.S.A.
- L.F. Urrutia
Instituto de Ciencias Nucleares, UNAM.
Apartado Postal 70–543, México D.F., 04510, México.
E–mail: urrutia@nuclecu.unam.mx

QUANTUM GRAVITY AND EXPERIMENTAL SEARCH

Ten Big Questions

M. Duff

Michigan Center for Theoretical Physics, Randall Laboratory, Department of Physics, University of Michigan, Ann Arbor, MI 48109–1120, USA

Abstract. We list ten unanswered questions in fundamental theoretical physics and examine recent progress.

In the year 1900 the world-renowned mathematician David Hilbert presented twenty-three problems at the International Congress of Mathematics in Paris. These problems have inspired mathematicians throughout the last century. Indeed, Hilbert's address has had a profound impact on the direction of mathematics, reaching far beyond the original twenty-three problems themselves.

In the year 2000 the University of Michigan hosted the millennial edition of the annual international superstrings conference "Strings 2000" [2]. See Figure 2. As a piece of millennial madness, all participants were invited to help formulate the ten most important unsolved problems in fundamental physics. Each participant was allowed to submit one candidate problem for consideration. To qualify, the problem must not only have been important but also well-defined and stated in a clear way. The best ten problems were selected at the end of the conference by a selection panel consisting of David Gross, Edward Witten and myself.

In the strife-torn city of Belfast, Northern ireland, a local artist was thinking that its citizens might be less inclined to fight one another if they were to think deep thoughts. Inspired by these Top Ten Questions, he displayed ten brass plaques around the city–in pubs and restaurants and at bus stops–each one listing one of the fundamental questions in physics. See Figure 1. The locals were baffled, but intrigued.

Although the subject of this lecture was inspired the Strings 2000 competition, I shall not follow exactly the same list but instead present my own choice of what I believe are Ten Big Questions. I have been careful not to call them the ten biggest questions since no doubt everyone of you will have your own list. I should, of course emphasize that these are ten big question in the field of fundamental theoretical physics with my own idiosyncratic emphasis on M-theory. Other approaches such as loop quantum gravity and other interesting fields, such a condensed matter physics, quantum computing, biophysics etc are not treated here.

At the end of the last millennium, it became fashionable for certain pundits to declare the *End of Science*, on the grounds that all the most important scientific discoveries had already been made. To a physicist, this seems absurd because the two main pillars of twentieth century physics, namely Einstein's General Theory of Relativity and Quantum Mechanics, are mutually incompatible. At the microscopic level, general relativity fails to comply with the quantum rules which govern the behavior of the elementary particles;

CP758, *Gravitation and Cosmology: 2nd Mexican Meeting on Mathematical and Experimental Physics*
edited by A. Macías, C. Läemmerzahl and D. Nuñez

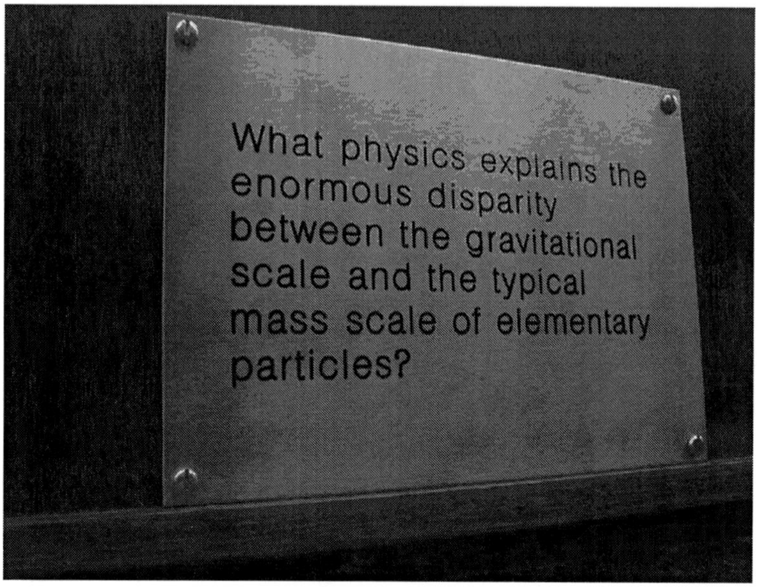

FIGURE 1. A plaque in Belfast

while on the macroscopic scale, the black holes of Einstein's theory are threatening the very foundations of quantum mechanics. Something big has to give. This augurs less the bleak future of diminishing returns predicted by the millennial Jeremiahs and more another scientific revolution.

Many physicists believe that this revolution is already under way with the theory of *Superstrings*. As their name suggests, superstrings are one-dimensional string-like objects [1]. Just like violin strings, these relativistic strings can vibrate and each mode of vibration corresponds to a different elementary particle. One strange feature of superstrings is that they live in a universe with nine space dimensions and one time dimension. Since the world around us seems to have only three space dimensions, the extra six would have to be curled up to an unobservably small size (or else rendered invisible in some other way) if the theory is to be at all realistic. This idea of extra dimensions will be an important theme in this lecture.

In 1995 superstring theory was superseded by a deeper and more profound new theory, called "M-theory" [3]. M-theory involves membrane-like extended objects with two space dimensions and five space dimensions that themselves live in a universe with eleven spacetime dimensions [4] (ten space and one time). As we shall see, it not only subsumes all of the new ideas of superstring theory but also revives older ideas on eleven-dimensional supergravity. See Figure 3. New evidence in favor of this theory is appearing daily on the internet although, as we shall see, there still much about M-theory we do not understand.

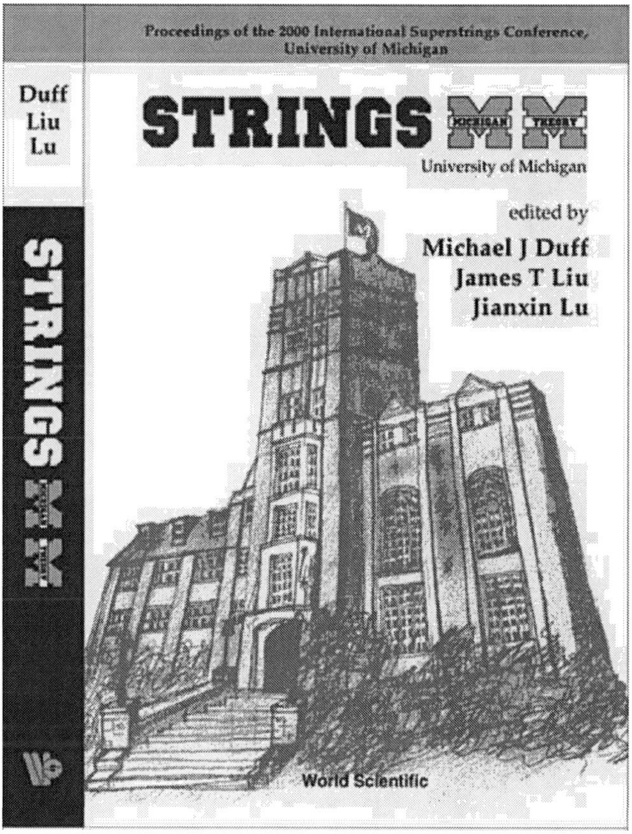

FIGURE 2. The Strings 2000 Conference.

1. CAN WE PROVE QUARK CONFINEMENT?

Theoretical physicists like to ask the big questions: How did the Universe begin? What are its fundamental constituents? What are the laws of nature that govern these constituents?

The smallest constituents of matter are, by definition, the *elementary particles*. But what is an elementary particle, exactly? How do we know when we have reached the bottom line? Well, it turns out to be easier to say what an elementary particle is not. For example, as illustrated in Figure 4, a human being (1 meter) is not elementary because he or she is made out of molecules. A molecule (10^{-8} meters) is not elementary because it is made out of atoms. An atom (10^{-10} meters) is not elementary because it is made out of electrons orbiting a nucleus which is made out of protons and neutrons (collectively called *nucleons*). Nucleons (10^{-15} meters) are not elementary because they are made out of smaller constituents called *quarks*. Finally, however, physicists believe that these quarks, together with other particles called *leptons*, of which the electron is an example,

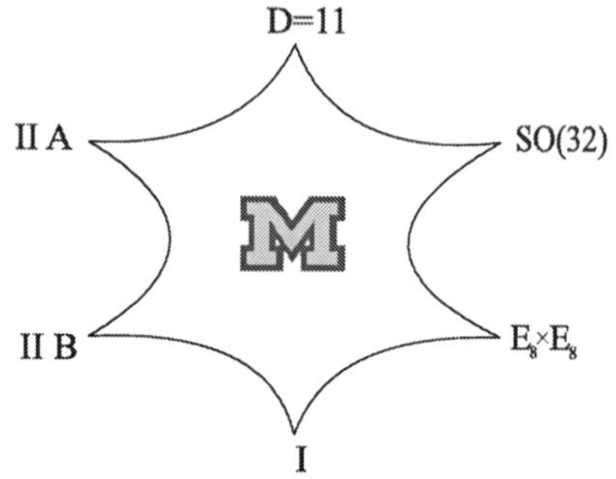

FIGURE 3. M-theory subsumes eleven-dimensional supergravity and the five ten-dimensional superstring theories.

are indeed truly elementary. They appear to have no size of their own and, as far as we can tell (which is down to about 10^{-18} meters), behave like geometrical points.

Because of this, elementary particle physics is qualitatively different from all other branches of science. In his or her everyday research, the biologist can take for granted all of chemistry; the chemist can take for granted all of atomic physics; the atomic physicist can take for granted all of nuclear physics and the nuclear physicist can take for granted all of elementary particle physics. Particle physicist are different, however, because they

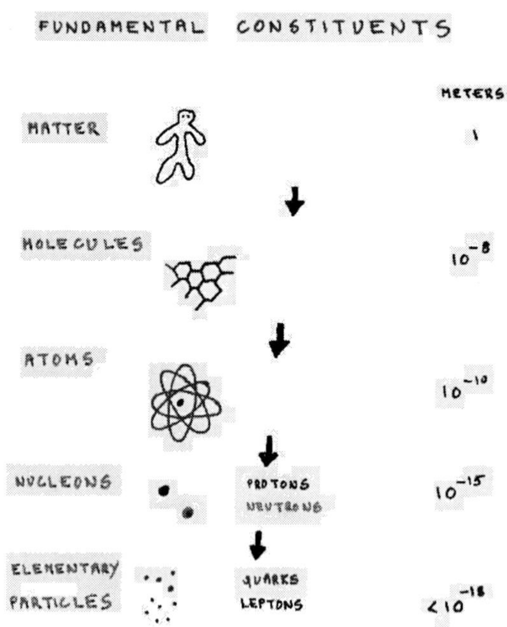

FIGURE 4. Fundamental constituents.

TABLE 1. Three families of quarks and leptons, with each quark coming in three colors (red, green and blue).

I	ν_e	e^-	u	d
II	ν_μ	μ^-	c	s
III	ν_τ	τ^-	t	b
	$Q=0$	$Q=-1$	$Q=2/3$	$Q=-1/3$

can take nothing for granted. There is no higher court of appeal. Someone once said that if you compare scientific research to a game of chess, then most scientists are trying to become masters of the game. But the particle physicist is still trying to figure out what the rules are!

Let us now take a preliminary look at the matter particles as displayed in Table 1. Theoretical consistency, borne out by experiment, requires that quarks and leptons come in families. For reasons we do not yet understand theoretically, the number of families is exactly three. Ordinary matter that we encounter on earth outside of particle accelerators is composed only of the first family, which consists of an electron neutrino with electric charge $Q=0$, an electron with $Q=-1$, an *up* quark with $Q=+2/3$ and a *down* quark with $Q=-1/3$. These particles are denoted ν_e, e, u and d respectively. All members have a non-zero rest mass except the neutrino which therefore travels at the speed of light. Apart from being heavier, the second and third families are identical replicas of the first. The second consists of a muon neutrino ν_μ, a muon μ, a *charm* quark c and a *strange* quark s. The third consists of a tau neutrino ν_τ, a tau τ, a *top* quark t and a *bottom* quark b. The six different types of quark are known as the six *flavors*. However, each quark also comes in three different *colors* labeled red, green and blue after the three primary colors. This is of course just an analogy, but it is a good one because just we can form *colorless* combinations from red, green and blue, so we can combine colored quarks to form colorless bound states. In fact, it seems to be a fact of nature that the only particles we actually observe are such colorless combinations. In particular, we never observe free quarks.

This gives rise to our first big question: Can we prove quark confinement?

It is a prediction of quantum field theory that for every particle there is an antiparticle with the same mass but opposite charge. So in addition to all the above particles, we also have their antiparticles. These are denoted by a bar, so that an anti-up quark, for example would be $\bar{u}$. Particles that were previously believed to be elementary are now known to be built out of these basic entities. For example, the proton ($Q=1$) is a *uud* combination, the neutron ($Q=0$) is a *ddu* and the positively charged pion ($Q=1$) is a $u\bar{d}$.

These fundamental building blocks of nature are held together by four fundamental forces, listed in Table 2 in order of diminishing strength. The strong nuclear force holds the protons and neutrons together inside the atomic nucleus; the electromagnetic force is responsible for all electric and magnetic phenomena and for light (light is an electromagnetic wave): the weak nuclear force is responsible for radioactivity, and lastly we have the gravitational force which is most familiar in our everyday lives. As we see, the gravitational force is incredibly weak compared to the other three forces and, for this reason, most text-books will tell you that gravity is of no consequence at the subatomic

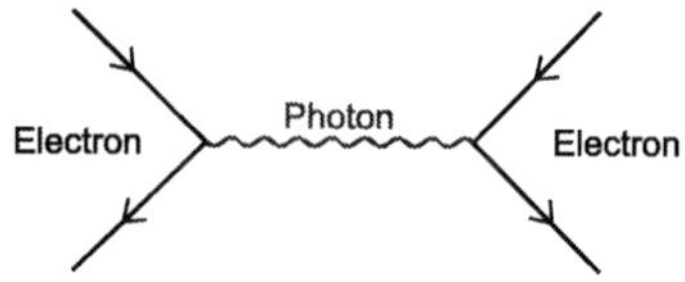

FIGURE 5. Feynman diagram representing the force between two electrons.

scale. However, I will try to convince you that most text-books are wrong: not only will gravity prove to be important but all the theoretical evidence suggests that it is actually at the root of everything else.

What exactly do we mean by a *force*? Over the years, our answer to this question has become more and more sophisticated. Nowadays, we employ quantum field theory, which is what we are led to when we try and combine quantum mechanics with special relativity. According to quantum field theory, each force is associated with a *force carrier*. For example, when an electron repels another electron, it does so by exchanging the fundamental quantum of electromagnetism, the *photon*, as in Figure 5. By the way, this picture is an example of what are called *Feynman diagrams*, named after Richard Feynman, who invented them. The study of the interactions between electrons and photons is known as *quantum electrodynamics* or *QED* and, in terms of the accuracy with which it agrees with experiment, it is the most successful theory ever invented. For example, a measure of the electron's magnetism is the g-factor. According to QED, it is given by

$$g/2 = 1.001159652190 \tag{1}$$

whereas the experimental value is

$$g/2 = 1.001159652193 \tag{2}$$

Because of this success, physicists have invented other force carriers to account for the other forces. The carriers of the strong force are known as *gluons* and they act on quarks. The study of the interactions between colored quarks and gluons is known as *quantum chromodynamics* or *QCD*. The existence of the gluon was established indirectly in 1978. The carriers of the weak force are known as W and Z bosons and, together with the photon, they act on both quarks and leptons. For example, the phenomenon of radioactivity arise from the so-called β-decay of a neutron into a proton, an electron and an antineutrino as shown by the Feynman diagram in Figure 6. This interaction is mediated by the exchange of a W boson. The W boson was discovered at CERN in 1982 and the Z boson, the following year. The collective study of the interactions of quarks and leptons with photons, W and Z bosons is known as *electro-weak theory*.

Finally, physicists have coined the word *graviton* to describe the force carrier of gravity, which acts on every other matter particle and force carrier. This is, at the moment, only a hypothetical particle and is unlikely to be detected experimentally for many years to come.

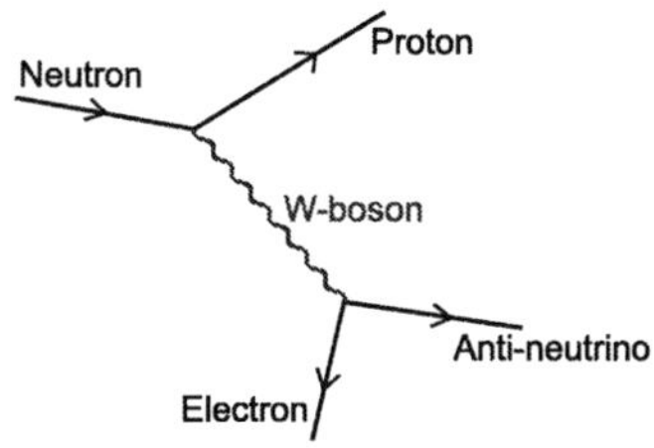

FIGURE 6. Feynman diagram representing the radioactive decay of a neutron.

2. DOES THE HIGGS BOSON EXIST?

So far, we have not discussed the origin of mass within the Standard Model. How come the neutrino is massless but the electron is not? How come the photon is massless but the W is not? It turns out that it is necessary to introduce a third kind of particle, called the *Higgs boson* after the Scottish theorist Peter Higgs, whose job it is cleverly to give masses to the particles that need masses (like the e and the W) while not giving masses to those that do not (like the ν and the γ). Thus the Higgs is neither a force carrier nor a matter particle.

Each elementary particle carries an intrinsic angular momentum or *spin*, s, which can either be an integer ($s = 0, 1, 2...$), in which case it is called a *boson*, or an odd half-integer ($s = 1/2, 3/2, 5/2...$), in which case it is called a *fermion*. The force carriers (gluons, photons, W and Z) are all bosons with $s = 1$; the matter particles (quarks and leptons) are all fermions with $s = 1/2$; the Higgs is an $s = 0$ boson. Bosons and fermions behave very differently. For example, fermions obey the *exclusion principle* of Wolfgang Pauli, which states that no two fermions can occupy the same quantum state, whereas bosons do not. They are said to obey *opposite statistics*.

The *Standard Model* is the name we give to the collection of strong, weak and electromagnetic interactions. It describes the interaction of the matter particles, force carriers and Higgs. It is consistent with every experiment that has yet been performed[1], but it also predicts other effects which have not yet been seen. In particular, there is still no direct experimental evidence for the Higgs boson. An important goal of future experiments at Fermilab in Chicago and the Large Hadron Collider at CERN in Geneva, will be to hunt for the Higgs.

This gives rise to our second big question: Does the Higgs boson exist?

3. ARE PROTONS FOREVER?

Central to the understanding of modern theories of the fundamental forces is the idea of *symmetry*: under certain changes in the way we describe the basic quantities, the

[1] Recent experiments have indicated that the neutrino has a very small but non-zero mass, but this would require only relatively mild modifications to the Standard Model.

TABLE 2. The fundamental forces.

Force	Strength	Carrier	Acts on
strong	1	gluon	quarks
electromagnetic	10^{-2}	photon	quarks and leptons
weak	10^{-5}	W and Z bosons	quarks and leptons
gravity	10^{-38}	graviton	everything

laws of physics are nevertheless seen to remain unchanged. For example, the result of an experiment should be the same whether we perform it today or tomorrow; this symmetry is called *time translation invariance*. It should also be the same before and after rotating our experimental apparatus; this symmetry is called *rotational invariance*. Both of these are examples of *spacetime symmetries*. Indeed, Einstein's general theory of relativity is based on the requirement that the laws of physics should be invariant under *any* change in the way we describe the positions of events in spacetime. In the Standard Model of the strong, weak and electromagnetic forces there are other kinds of *internal* symmetries that allow us to change the roles played by different elementary particles such as electrons and neutrinos, for example. In Grand Unified Theories, which have not yet received the same empirical support as the Standard Model, the laws remain unchanged even when we exchange the roles of the *quarks* and electrons. Thus it is that the greater the unification, the greater the symmetry required. The Standard Model symmetry replaces the three fundamental forces: strong, weak and electromagnetic, with just two: the strong and electroweak. Grand unified symmetries replace these two with just one strong-electroweak force. In fact, it is not much of an exaggeration to say that the search for the ultimate unified theory is really a search for the right symmetry.

At this stage, however, one might protest that some of these internal symmetries fly in the face of experience. After all, the electron is very different from a neutrino: the electron has a non-zero mass whereas the neutrino is massless. Similarly, the electrons which orbit the atomic nucleus are very different from the quarks out of which the protons and neutrons of the nucleus are built. Quarks feel the strong nuclear force which holds the nucleus together, whereas electrons do not. These feelings are, in a certain sense, justified: the world we live in does not exhibit the symmetries of the Standard Model nor those of Grand Unified Theories. They are what physicists call "broken symmetries". The idea is that these theories may exist in several different *phases*, just as water can exist in solid, liquid and gaseous phases. In some of these phases the symmetries are broken but in other phases, they are exact. The world we inhabit today happens to correspond to the broken-symmetric phase, but in conditions of extremely high energies or extremely high temperatures, these symmetries may be restored to their pristine form. The early stages of our universe, shortly after the Big Bang, provide just such an environment. Looking back further into the history of the universe, therefore, is also a search for greater and greater symmetry. The ultimate symmetry we are looking for may well be the symmetry with which the Universe began.

Grand Unified Theories predict the existence of magnetic monopoles and a finite lifetime for the proton. This gives rise to our third big question: Is the proton forever?

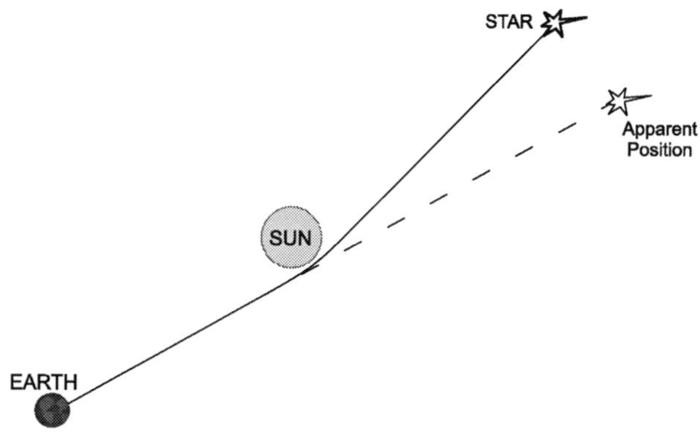

FIGURE 7. The bending of light by the Sun.

4. WHY THE ENORMOUS DISPARITY BETWEEN THE GRAVITY SCALE AND THE TYPICAL MASS SCALE OF THE ELEMENTARY PARTICLES?

Why is it so difficult to incorporate the fourth force of gravity into quantum physics? Our modern understanding of the gravitational force dates back to Einstein's 1916 theory of general relativity which asserts that the laws of physics should be the same to *all* observers, not merely those in uniform relative motion as in special relativity. According to this picture, gravity is not a *force* at all. All bodies are following straight line trajectories, but in a *curved spacetime* . (Note that it is not merely three-dimensional space which is curved, but the four-dimensional spacetime continuum.) A classic illustration of this idea is provided by the bending of light by the Sun, as shown in Figure 7. At first sight, these classical geometrical ideas seem to be at odds with the quantum picture of force-carrying gravitons. In recent times, however, we have come to appreciate that these two pictures are complementary. Indeed, the emphasis now is to explain the other three forces in geometrical terms as well.

There is a natural energy scale associated with any quantum theory of gravity. Such a theory combines three ingredients each with their own fundamental constants: Planck's constant h (quantum mechanics), the velocity of light c (special relativity) and Newton's gravitational constant G (gravity). From these we can form the so-called Planck mass $m_P = \sqrt{hc/G}$, equal to about 10^{-8} kilograms, and the Planck energy $m_P c^2$, equal to about 10^{19} GeV. (GeV is short for giga-electron-volts=10^9 electron-volts, and an electron-volt is the energy required to accelerate an electron through a potential difference of one volt.) From this we conclude that the energy at which Einstein's theory breaks down is the Planck energy. On the scale of elementary particle physics, this energy is enormous: the world's most powerful particle accelerators can currently reach energies of only 10^4 GeV.

This gives rise to our fourth big question: Why the enormous disparity between the gravity scale and the typical mass scale of the elementary particles?

TABLE 3. Behaviour of scale factor and energy density for matter, radiation and vacuum dominated universes. The solution for the scale factor is written for the case $k = 0$.

Stress Energy	w	Energy Density	Scale Factor $a(t)$
Matter	$w = 0$	$\rho \sim a^{-3}$	$a(t) \sim t^{2/3}$
Radiation	$w = \frac{1}{3}$	$\rho \sim a^{-4}$	$a(t) \sim t^{1/2}$
Vacuum (Λ)	$w = -1$	$\rho \sim \Lambda/(8\pi G)$	$a(t) \sim \exp(\sqrt{\Lambda/3}t)$

5. WHAT IS THE ORIGIN AND FATE OF THE UNIVERSE?

The standard cosmological model starts with Einstein equations coupled to a perfect fluid. See, for example, the lectures of Quevedo [6], from which we shall borrow in this section. Homogeneity and isotropy of spacetime determines the metric up to an arbitrary function of time $a(t)$, known as the scale factor, which measures the time evolution of the Universe and a discrete parameter $k = -1, 0, 1$ which determines if the Universe is open, flat or closed, respectively. The Friedmann-Robertson-Walker (FRW) metric describing the evolution of the Universe can then be written as:

$$ds^2 = -dt^2 + a^2(t) \left[\frac{dr^2}{1-kr^2} + r^2 \left(d\theta^2 + \sin^2 \theta d\phi^2 \right) \right] . \tag{3}$$

The scale factor $a(t)$ is given by solving Einstein's equations. The energy density ρ and the pressure p of the fluid satisfy an equation of state of the form $p = w\rho$. Table 3 shows the values of w for common cases corresponding to matter, radiation and vacuum domination, for which w is constant.

Einstein's equations now reduce to the Friedmann's equation:

$$H^2 = \frac{8\pi G}{3} \rho - \frac{k}{a^2} \frac{\ddot{a}}{a} = -\frac{4\pi G}{3} (\rho + 3p) , \tag{4}$$

with H the Hubble function $H \equiv \frac{\dot{a}}{a}$. These two equations imply the energy conservation equation:

$$\dot{\rho} = -3H (\rho + p) . \tag{5}$$

Equation (5) gives $\rho \sim a^{-3(1+w)}$,which then gives the solution for $a(t)$ in the Friedmann's. Table 3 shows the behaviour of ρ and $a(t)$ for typical equations of state. Notice that for the expressions for $a(t)$ we have neglected the curvature term k/a^2 from Friedmann's equation and therefore we write the solution for the flat universe case. The exact solutions for the other cases can also be found.

Under very general assumptions, based mainly on the positivity of the energy density ρ, it can be shown that the FRW ansatz necessarily implies an initial singularity, the big-bang, from which the universe starts expanding. For instance if $\rho + 3p > 0$, which is satisfied for many physical cases, the acceleration of the universe measured by $\ddot{a}$ is negative, as seen from the second Friedmann's equation. For $k = -1, 0$ the first equation tells us that, for positive energy density, the universe naturally expands forever whereas for $k = 1$ there will be a value of a for which the curvature term compensates the energy density term and $\dot{a} = 0$, after this time a decreases and the universe re-collapses.

A useful concept to introduce is the critical density

$$\rho_{\text{critical}} \equiv \frac{3H^2}{8\pi G} \,. \tag{6}$$

which, for the present time $H = H_0 \sim 65\ \mathrm{Km/s/Mpc^{-1}}$ gives $\rho_c \sim 1.7 \times 10^{-29}\mathrm{g/cm^3}$ (1 Mpc (mega-parsec) $= 3 \times 10^{22}$ meters). This allows us to define a dimensionless parameter Ω which corresponds to the ratio of the energy density of a given system to the critical density: $\Omega_i \equiv \rho_i/\rho_{\text{critical}}$, with the index i labelling the different contributions to the energy density, and the total ratio is $\Omega = \sum_i \Omega_i$. With these definitions we can write the first Friedmann equation as:

$$\Omega = 1 + \frac{k}{H^2 a^2} \,. \tag{7}$$

From this we can see the clear connection between the curvature of the spatial sections given by k and the departure from critical density given by Ω. A flat universe ($k = 0$) corresponds to a critical density ($\Omega = 1$) whereas open ($k = -1$) and closed ($k = 1$) universes correspond to $\Omega < 1$ and $\Omega > 1$ respectively.

At the atomic physics scale, the universe is cold enough for atoms to be formed and the photons are out of equilibrium, giving rise to the famous cosmic microwave background. At approximately the same time also the universe changes from being radiation dominated ($w = 1/3$) to matter dominated ($w = 0$). After this, the formation of structures such as clusters and galaxies can start, probably due to the quantum fluctuations of the early universe, leading to our present time.

The standard cosmological model has strong experimental evidence which can be summarized as follows:

- The observed expansion of the universe
- The relative abundance of the elements
- The discovery of the cosmic microwave background. In particular, the confirmation of the black body nature of the CMB is known with excellent precision, but, more importantly, the fluctuations in the temperature $\frac{\delta T}{T}$ signalling density fluctuations $\frac{\delta\rho}{\rho}$ in the early universe provide a great piece of information about the possible microscopic origin of the large scale structure formation. They also provide very strong evidence in favour of a flat universe.

Further observations and theoretical developments point also to some problems of this model which we can summarize as follows

- The big bang problem. The breakdown of classical relativity near the initial singularity
- The horizon problem. Assuming the standard expansion of the universe we receive the same information from points in the space that do not appear to be in causal contact with each other. Therefore it is actually a puzzle why the radiation is so uniform.

- The CMB anisotropies problem. The definitely observed anisotropies in the CMB are expected to be produced from physics of the early universe which is not explained in the standard model.
- The flatness problem. For Ω to be so close to one at present it had to be essentially one in the early universe with a precision of many significant figures.
- The baryogenesis problem. Combining the standard models of particle physics and cosmology we cannot explain why there seems to be an excess of matter over antimatter.
- The dark matter problem. See below.
- The monopole problem. Grand unified models typically imply the existence of topological defects such as domain walls, cosmic strings and monopoles which could have an important impact in cosmology. In particular the existence of monopoles and domain walls would over-close the universe and therefore cause new problems.
- The dark energy problem. See below.

So the fifth big question is: What are the origin and fate of the universe?

The above problems seem to require a modification of the standard model cosmology. In the 1980s the inflationary universe was proposed, offering a possible solution to the flatness, horizon and monopole problems. It was soon realised that, more importantly, it could also provide an explanation for the possible CMB anisotropies and therefore for structure formation.

The idea is that in the early universe there is a short time when the universe expanded very fast, usually an exponential expansion. If the inflationary period is long enough, it would flatten the universe quickly and would also explain why some regions could be in causal contact with each other, solving the horizon problem. Finally the fast expansion would dilute many objects, such as monopoles and other unwanted massive particles in such a way as to make them harmless for the over-closure of the universe.

Probably the most relevant property of inflation is that it can provide an explanation for the density perturbations of the CMB and therefore indirectly account for the large scale structure formation. Furthermore, most of the models of inflation imply a scale invariant, Gaussian and adiabatic spectrum which is consistent with observations. This has made inflation the standard cosmological paradigm to test present and future observations.

6. WHAT ARE DARK MATTER AND DARK ENERGY?

The survey and study of the behaviour of matter, such as rotation curves for galaxies, at many different scales, has given evidence that there should be a new kind of matter, called *dark matter*, not present in the standard model of particle physics.

Moreover, recent results form the study of high redshifted supernovae, combined with the CMB, has provided strong evidence for the fact that the universe is actually accelerating at present. This indicates that there should be a form of *dark energy* which provides $\rho + 3p < 0$ and causes the universe to accelerate. In any case this stresses the cosmological constant problem: Why is the cosmological constant almost zero? And why does it have the value it seems to have at present $\Lambda = 10^{-120} m_P^4 = (10^{-3}\mathrm{eV})^4$? Present observations point towards $\Omega = \Omega_\Lambda + \Omega_B + \Omega_{DM} = 1$ with the contribution from dark energy $\Omega_\Lambda \sim 0.73$, the contribution from dark matter $\Omega_{DM} \sim 0.23$ and the baryonic contribution a humble $\Omega_B \sim 0.04$.

Hence the sixth big question: What are dark matter and dark energy?

It has been suggested that the lightest supersymmetric particle, which would be stable, is a good candidate for dark matter. An effective cosmological constant or a time varying scalar field are the main proposals for this dark energy.

7. IS NATURE SUPERSYMMETRIC?

The goal of modern theoretical physics is to find an all-embracing *Theory of Everything* that would unite all four forces. But if current thinking about M-theory is correct, this will require three radical ideas:

1. Supersymmetry
2. Extra spacetime dimensions
3. Extended objects

M-theory, like string theory before it, relies crucially on the idea, first put forward in the early 1970s, of a spacetime *supersymmetry* which exchanges bosons and fermions. Unbroken supersymmetry would require that every elementary particle we know of would have an unknown super-partner with the same mass but obeying the opposite statistics: for each boson there is a fermion; for each fermion a boson. Spin $1/2$ quarks partner spin 0 *squarks*, spin 1 photons partner spin $1/2$ *photinos*, and so on. In the world we inhabit, of course, there are no such equal mass partners and bosons and fermions seem very different. Supersymmetry, if it exists at all, is clearly a broken symmetry and the new supersymmetric particles are so heavy that they have so far escaped detection. At sufficiently high energies, however, supersymmetry may be restored. Another challenge currently facing high-energy experimentalists at Fermilab and CERN is the search for these new supersymmetric particles. The discovery of supersymmetry would be one of the greatest experimental achievements and would completely revolutionize the way we view the physical world.

Symmetries are said to be *global* if the changes are the same throughout spacetime, and *local* if they differ from one point to another. The consequences of *local* supersymmetry are even more far-reaching: it predicts gravity. Thus if Einstein had not already discovered General Relativity, local supersymmetry would have forced us to invent it. In fact, we are forced to a *supergravity* in which the graviton, a spin 2 boson that mediates the gravitational interactions, is partnered with a spin $3/2$ *gravitino*. This is a theorist's dream because it confronts the problem from which both general relativity and grand unified theories shy away: neither takes the other's symmetries into account. Consequently, neither is able to achieve the ultimate unification and roll all four forces into one. But local supersymmetry offers just such a possibility, and it is this feature above all others which has fuelled the theorist's belief in supersymmetry in spite of thirty years without experimental support.

So all this gives rise to our seventh big question: Is nature supersymmetric?

Supergravity has an even more bizarre feature, however, it places an upper limit of *eleven* on the dimension of spacetime! There are two different ways to see this. The original explanation relies on the belief that there are no consistent ways to describe massless particles with spins greater than two, the spin of the graviton. Yet the mathematics of supersymmetry says that spins greater than two (and more than one graviton) would have to appear in dimensions greater than eleven, thus leading to a contradiction. The second explanation, to which we shall return, says the maximum dimension in which there exists supersymmetric objects such as particles, strings and membranes is eleven, where in fact we find a supermembrane. Since extra dimensions will play an important role in what follows, let us recall a little of its history.

We are used to the idea that space has three dimensions: height, length and breadth; with time providing the fourth dimension of *spacetime*. Indeed this is the picture that Einstein had in mind in 1916 when he proposed general relativity. But in the early 1920's, in their attempts to unify Einstein's gravity and Maxwell's electromagnetism, Theodor Kaluza and Oskar Klein suggested that spacetime may have a hidden fifth dimension. This Kaluza-Klein theory, and its higher dimensional generalizations, are thus tailor-made for supersymmetry.

In 1864 James Clerk Maxwell introduced the equations that describe the electromagnetic field. Albert Einstein later realized in 1905 that Maxwell's equations obey the principle of special relativity, that the laws of physics should be the same to all observers who are in uniform relative motion. In special relativity, which treats time as a fourth dimension, time t and the space coordinates x, y, z are collectively denoted x^μ where the index μ runs over $0, 1, 2, 3$. The 0 refers to time and $1, 2, 3$ to the three space coordinates (x, y, z):

$$(x^0, x^1, x^2, x^3) = (t, x, y, z) \tag{8}$$

In this modern notation, Maxwell's electromagnetic field also has four components collectively denoted $A_\mu(x)$. The field depends on its position in spacetime and so is a function of the 4 spacetime coordinates x^μ.

In 1916, Einstein introduced the principle of general relativity, that the laws of physics should be the same to *all* observers. This necessitates a gravitational field with two indices, $g_{\mu\nu}(x)$, which also has the geometrical interpretation of a metric tensor, the quantity that describes the infinitesimal distance ds between two points in four-dimensional spacetime.

$$ds^2 = g_{\mu\nu}(x) dx^\mu dx^\nu \tag{9}$$

Note that the Euclidean geometry of flat spacetime must be replaced by the Riemannian geometry of curved spacetime for which the metric tensor is itself a function of the spacetime coordinates.

In 1919, therefore, in the search for a unified theory, it was natural to attempt to combine Maxwell's electromagnetism with Einstein's gravity, since the other two nuclear forces were not as well understood. This the German-Polish mathematician Theodor Kaluza was able to do by the ingenious device of postulating a fifth dimension with coordinate θ. The five coordinates are denoted collectively x^M where the index M runs over 0, 1, 2, 3, 4.

$$(x^0, x^1, x^2, x^3, x^4) = (t, x, y, z, \theta) \tag{10}$$

He imagined a five dimensional Riemannian geometry with metric tensor $\hat{g}_{MN}(x)$ which describes the infinitesimal distance $d\hat{s}$ between two points in this five-dimensional spacetime

$$d\hat{s}^2 = \hat{g}_{MN}(x) dx^M dx^N \tag{11}$$

He then made a $4+1$ split

$$\hat{g}_{MN} = \begin{pmatrix} g_{\mu\nu} + \Phi A_\mu A_\nu & \Phi A_\mu \\ \Phi A_\nu & \Phi \end{pmatrix} \tag{12}$$

and identified $g_{\mu\nu}(x)$ with Einstein's gravitational field and $A_\mu(x)$ with Maxwell's electromagnetic field. This was all before the advent of quantum field theory, but nowadays we would refer to $g_{\mu\nu}$ as the spin 2 graviton, A_μ as the spin 1 photon and Φ as the spin 0 dilaton[2].

Of course, it is not enough to call A_μ by the name photon, we must demonstrate that it obeys the right equations and here we see the Kaluza miracle at work. If we substitute (12) into the five dimensional field equations, not only do we recover the correct Einstein equations for $g_{\mu\nu}$, but also the Maxwell equations for A_μ. So Maxwell's theory of electromagnetism was seen to be a consequence of Einstein's general relativity, provided you are willing to buy the idea of an extra spacetime dimension.

By the way, Kaluza's son, who still teaches mathematics in Germany, recalls that his father belonged to that school of theoreticians who believed that everything in nature could be derived from pure thought without the need for experiment. Consequently, he learned to swim from a textbook. He would lie on the living room couch with the book in one hand and practice his strokes with the other. He then proceeded to walk to the nearest lake and swim across it. Personally, I find it easier to believe in a fifth dimension than in the veracity of that particular story.

Attractive though Kaluza's idea was, it suffered from two obvious drawbacks. First, although the indices were allowed to range over 0, 1, 2, 3, 4, for no very good reason the dependence of the fields on the extra coordinate θ was suppressed. Secondly, if there is a fifth dimension why haven't we seen it? The resolution of both these problems was supplied by Oskar Klein in 1926. Klein insisted on treating the extra dimension seriously but assumed the fifth dimension to have circular topology so that the coordinate θ is periodic, $0 \leq \theta \leq 2\pi$. It is difficult to envisage a spacetime with this topology but a simpler two-dimensional analogy is provided by a garden hose: at large distances it looks like a line but closer inspection reveals that at every point on the line there is a little circle. So it was that Klein suggested that there is a little circle at each point in four-dimensional spacetime. See Figure 8.

---

[2] This field was considered an embarrassment in 1919, and was (inconsistently) set equal to zero. However, it was later revived and subsequently stimulated Brans-Dicke theories of gravity. The dilaton also plays a crucial role in superstring and M-theory.

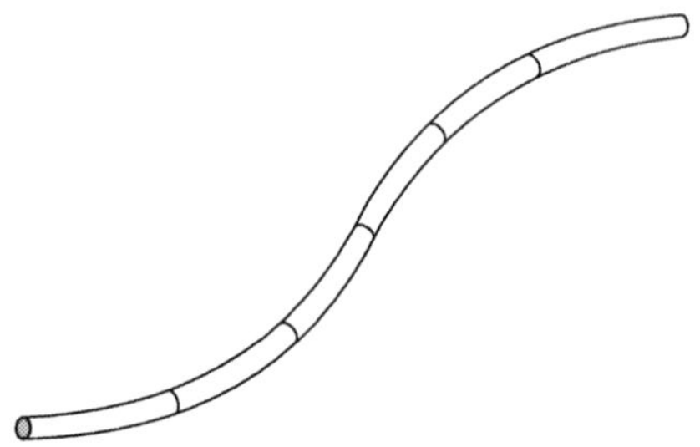

FIGURE 8. A garden hose. At large distances, it looks like a line; closer inspection reveals that at each point of the line there is a little circle.

The periodicity in θ means that the fields $\hat{g}_{MN}(x,\theta)$ may be expanded in the form

$$\hat{g}_{MN}(x,\theta) = \sum_{n=-\infty}^{n=\infty} \hat{g}_{MN(n)}(x) exp(in\theta), \tag{13}$$

The $n = 0$ modes in (13) are just Kaluza's graviton, photon and dilaton. If we now include the θ-dependence via the $n \neq 0$ modes, however, we find an infinite tower of charged, massive spin 2 particles with charges e_n given by

$$e_n = ne, \tag{14}$$

and masses m_n given by

$$m_n = |n|/R \tag{15}$$

where R is the radius of the circle and

$$e^2 = 16\pi G/R^2 \tag{16}$$

where G is Newton's constant of gravitation. Thus Klein explained (for the first time) the empirical fact that all observed particles come with an electric charge which is an integer multiple of a fundamental charge e, in other words, why electric charge is *quantized*. Of course, if we identify this fundamental unit of charge with the charge on the electron, then we are forced to take the radius of the circle to be very small: the Planck size 10^{-35} meters; much smaller than the 10^{-18} meters achievable by current particle accelerators. This satisfactorily accords with our everyday experience of living in four spacetime dimensions.

The idea for a fifth dimension came to Klein during his stay as an assistant professor at the University of Michigan, 1923-25:

> "Returning to my Ann Arbor attempts, I became immediately very eager to see how far the mentioned analogy reached, first trying to find out whether the Maxwell equations for the electromagnetic field together with Einstein's gravitational equations would fit into a formalism of five-dimensional Riemann geometry." (Oskar Klein, "From My Life in Physics".)

Klein was not at first aware of Kaluza's earlier idea but learned about it from Wolfgang Pauli to whom he had shown his manuscript. Ever the gentleman, he generously acknowledges Kaluza's prior claim, but as we have noted above, it was Klein who really took seriously the extra dimension with all its implications, and by assigning it the topology of a circle, provided the first explanation of electric charge quantization.

The Kaluza-Klein idea was forgotten for many years but was revived in the early 1980s when it was realized by Eugene Cremmer, Bernard Julia and Joel Scherk from the Ecole Normale in Paris that supergravity not only permits up to seven extra dimensions, but in fact takes its simplest and most elegant form when written in its full eleven-dimensional glory. Moreover, the kind of four-dimensional picture we end up with depends on how we curl up or *compactify* these extra dimensions: maybe seven of them would allow us to derive, a la Kaluza-Klein, the strong and weak forces as well as the electromagnetic [8]. The four dimensional theory we end up after such a compactification describes a spin 2 graviton and spin 3/2 gravitinos interacting with a collection of spin 1, spin 1/2 and spin 0 particles. This collection will depend on the choice of the compact seven-dimensional space. So the question was whether there was a choice that yielded the gluons, W and Z-bosons, photons, quarks, leptons and Higgs of the Standard Model. In the end, however, eleven dimensional supergravity fell out of favor for a couple of reasons.

First, an important feature of the real world which is incorporated into both the Standard Model and Grand Unified Theories is that Nature is *chiral*: the weak nuclear force distinguishes between right and left. However, as emphasized by Edward Witten of the Institute for Advanced Study in Princeton, among others, it is impossible via conventional Kaluza-Klein techniques to generate a chiral theory from a non-chiral one and unfortunately, eleven-dimensional supergravity, in common with any *odd*-dimensional theory, is itself non-chiral.

Secondly, despite its extra dimensions and despite its supersymmetry, eleven-dimensional supergravity is still a *quantum field theory* and runs into the problem from which all such theories suffer: the quantum mechanical probability for certain processes yields the answer *infinity*, signalling a breakdown of the theory. So it seemed in the early 1980s that we were looking for a fundamental theory which reduces to Einstein's gravity at low energies, which describes Planck mass particles and which is supersymmetric. Whatever it is, it cannot be a quantum field theory because we already know all the supersymmetric ones and they do not fit the bill.

For both these reasons, attention turned to ten-dimensional superstring theory. The idea that the fundamental stuff of the universe might not be pointlike elementary particles, but rather one-dimensional strings had been around from the early 1970s. Just like violin strings, these relativistic strings can vibrate and each elementary particle: graviton, gluon, quark and so on, is identified with a different mode of vibration. However, this means that there are *infinitely many* elementary particles. Fortunately, this does not contradict experiment because most of them, corresponding to the higher modes of vibration, will have masses of the order of the Planck mass and above and will be unobservable in the direct sense that we observe the lighter ones. Indeed, an infinite tower of Planck mass states is just what the doctor ordered for curing the non-renormalizability disease. In fact, because strings are *extended*, rather than pointlike, objects, the quantum mechanical probabilities involved in string processes are actually *finite*. Moreover, when we take the *low-energy limit* by eliminating these massive particles through their equations of motion, we recover a ten-dimensional version of supergravity which incorporates Einstein's gravity. Now ten-dimensional theories, as opposed to eleven-dimensional ones, also admit the possibility of *chirality*. The reason that everyone had still not abandoned eleven-dimensional supergravity in favor of string theory, however, was that the realistic-looking Type I string, which seemed capable of incorporating the Standard Model of particle physics, seemed to suffer from inconsistencies or *anomalies*, whereas the consistent non-chiral Type IIA and chiral Type IIB strings did not seem realistic.

Then came the September 1984 superstring revolution. First, Michael Green from Queen Mary College, London, and John Schwarz from the California Institute of Technology showed that the Type I string was free of anomalies provide the group was uniquely $SO(32)$ where O(n) stands for *orthogonal* $n \times n$ matrices. They suggested that a string theory based on the exceptional group $E_8 \times E_8$ would also have this property. Next, David Gross, Jeffrey Harvey, Emil Martinec and Ryan Rohm from Princeton University discovered a new kind of heterotic (hybrid) string theory based on just these two groups: the $E_8 \times E_8$ heterotic string and the $SO(32)$ heterotic string, thus bringing to *five* the number of consistent string theories. Thirdly, Philip Candelas from the University of Texas, Austin, Gary Horowitz and Andrew Strominger from the University of California, Santa Barbara and Witten showed that these heterotic string theories admitted a Kaluza-Klein compactification from ten dimensions down to four. The six-dimensional compact spaces belonged to a class of spaces known to the pure mathematicians as *Calabi-Yau manifolds*. The resulting four-dimensional theories resembled quasi-realistic Grand Unified Theories with chiral representations for the quarks and leptons! So at last we had a consistent quantum theory of gravity that might even explain the Standard Model of particle physics! Everyone dropped eleven-dimensional supergravity like a hot brick. The mood of the times was encapsulated by Nobel Laureate Murray Gell-Mann (inventor of the quarks) in his closing address at the 1984 Santa Fe Meeting, when he said: "Eleven Dimensional Supergravity (Ugh!)".

So ten dimensions were riding high but eleven dimensions were in the doghouse. Nevertheless, a small band of theorists clung to the idea that eleven dimensions must somehow feature in the final theory. Notwithstanding the euphoria that surrounded string theory, they allowed themselves the *liberty of doubt* about string theory as it was currently formulated and began to ask some awkward questions [9]:

- *The uniqueness problem*
  Theorists love *uniqueness*; they like to think that the ultimate *Theory of Everything* [10] will one day be singled out, not merely because all rival theories are in disagreement with experiment, but because they are mathematically inconsistent. In other words, that the universe is the way it is because it is the only possible universe. But string theories are far from unique. Already in ten dimensions there are five mathematically consistent theories as shown in Figure 3: the Type I $SO(32)$, the heterotic $SO(32)$, the heterotic $E_8 \times E_8$, the Type IIA and the Type IIB. (Type I is an *open* string in that its ends are allowed to move freely in spacetime; the remaining four are *closed strings* which form a closed loop.) If we are looking for a unique all-embracing theory, this seems like an embarrassment of riches.

The situation becomes even worse when we consider compactifying the extra six dimensions. There seem to be billions of different ways of compactifying the string from ten dimensions to four (billions of different Calabi-Yau manifolds) and hence billions of competing predictions of the real world (which is like having no predictions at all). This aspect of the uniqueness problem is called the *vacuum-degeneracy problem*. One can associate with each different phase of a physical system a *vacuum state*, so called because it is the quantum state corresponding to no real elementary particles at all. However, according to quantum field theory, this vacuum is actually buzzing with virtual particle-antiparticle pairs that are continually being created and destroyed and consequently such vacuum states carry energy. The more energetic vacua, however, should be unstable and eventually decay into a (possibly unique) stable vacuum with the least energy, and this should describe the world in which we live. Unfortunately, all these Calabi-Yau vacua have the same energy and the string seems to have no way of preferring one to the other. By focusing on the fact that strings are formulated in ten spacetime dimensions and that they unify the forces at the Planck scale, many critics of string theory fail to grasp this essential point. The problem is not so much that strings are unable to produce four-dimensional models like the Standard Model with quarks and leptons held together by gluons, W-bosons, Z bosons and photons and of the kind that can be tested experimentally in current or foreseeable accelerators. On the contrary, string theorists can dream up literally billions of them! The problem is that they have no way of discriminating between them. What is lacking is some dynamical mechanism that would explain why the theory singles out one particular Calabi-Yau manifold and hence why we live in one particular vacuum; in other words, why the world is the way it is. Either this problem will not be solved, in which case string theory will fall by the wayside like a hundred other failed theories, or else it will be solved and string theory will be put to the test experimentally. Neither string theory nor M-theory is relying for its credibility on building thousand-light-year accelerators capable of reaching the Planck energy, as some *End-of-Science* Jeremiahs have suggested.

- *The dimension problem*
  An apparently different reason for having mixed feelings about superstrings, of course, especially for those who had been pursuing Kaluza-Klein supergravity prior to the 1984 superstring revolution, was the dimensionality of spacetime. If supersymmetry permits eleven spacetime dimensions, why should the theory of everything stop at ten? Richard Feynman always used to say that, in physics, whatever is not forbidden is compulsory.
- *The membrane problem*
  This problem rose to the surface again in 1987 when Eric Bergshoeff of the University of Groningen, Ergin Sezgin, now at Texas A&M University, and Paul Townsend from the University of Cambridge discovered *The eleven-dimensional supermembrane*. This membrane has two spatial dimensions and moves in a spacetime dictated by our old friend: eleven-dimensional supergravity! It can either be a bubble-like extended object or an infinite sheet. In the same year, moreover, Paul Howe (King's College, London University), Takeo Inami (Kyoto University), Kel-

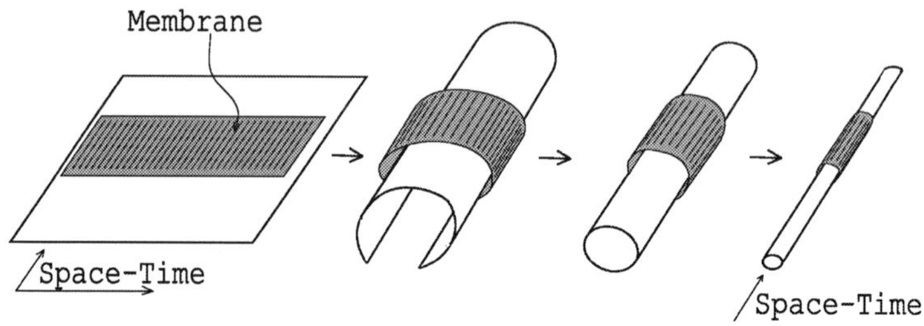

FIGURE 9. A ten-dimensional string interpreted as an eleven-dimensional membrane wrapped around the extra dimension.

logg Stelle (Imperial College) and I were then able to show that if one of the eleven dimensions is a circle, then we can wrap one of the membrane dimensions around it so that, if the radius of the circle is sufficiently small, it looks like a string in ten dimensions. See Figure 9. In fact, it yields precisely the Type *IIA* superstring. This suggested to us that maybe the eleven-dimensional theory was the more fundamental after all, though this was not a popular view among the latter-day *Flatlanders* who refused to acknowledge this extra dimension.

Membrane theory has a strange history which goes back even further than strings. The idea that the elementary particles might correspond to modes of a vibrating membrane was put forward originally in 1960 by the British Nobel Prize winning physicist Paul Dirac, a giant of twentieth century science who was also responsible for two other daring postulates: the existence of *anti-matter* and the existence of *magnetic monopoles*. (Anti-particles carry the same mass but opposite charge from particles and were discovered experimentally in the 1930s. Magnetic monopoles carry a single magnetic charge and to this day have not yet been observed. Interestingly enough, however, they provide an alternative reason why electric charge should be quantized. Once again,

$$e_n = ne, \tag{17}$$

just as Klein had said, but according to Dirac the fundamental charge e is now given by

$$e = 2\pi/g \tag{18}$$

where g is the magnetic charge. This later gave rise to speculations that Nature might be invariant under an electric-magnetic duality that exchanges electric particles and magnetic monopoles. As we shall see, both the Klein rule and the Dirac rule feature prominently in M-theory, as does electric-magnetic duality.)

When string theory came along in the 1970s, there were some attempts to revive Dirac's membrane idea but without much success. The breakthrough did not come until 1986 when James Hughes, Jun Liu and Joseph Polchinski of the University of Texas showed that, contrary to the expectations of certain string theorists, it was possible to combine the membrane idea with supersymmetry: the *supermembrane* was born.

Consequently, while all the progress in superstring theory was being made a small but enthusiastic group of theorists were posing a seemingly very different question: Once

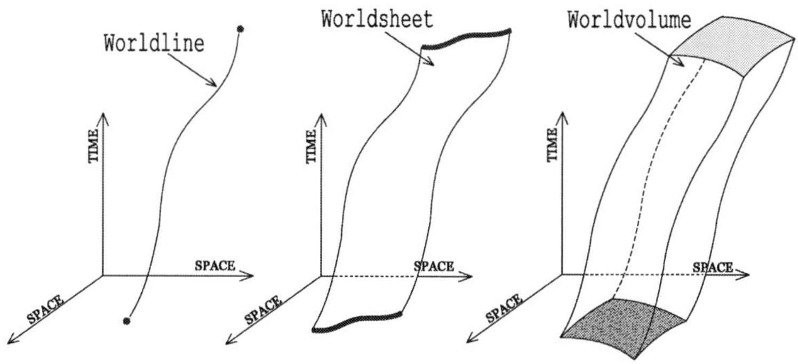

FIGURE 10. Particles, strings and membranes.

you have given up 0-dimensional particles in favor of 1-dimensional strings, why not 2-dimensional membranes or in general p-dimensional objects (inevitably dubbed *p-branes*)? Just as a 0-dimensional particle sweeps out a 1-dimensional *worldline* as it evolves in time, so a 1-dimensional string sweeps out a 2-dimensional *worldsheet* and a p-brane sweeps out a d-dimensional *worldvolume*, where $d = p+1$. See Figure 10.

Of course, there must be enough room for the p-brane to move about in spacetime, so d must be less than the number of spacetime dimensions D. In fact supersymmetry places further severe restrictions both on the dimension of the extended object and the dimension of spacetime in which it lives. One can represent these as points on a graph where we plot spacetime dimension D vertically and the p-brane dimension $d = p+1$ horizontally. This graph is called the *brane-scan*. See Figure 11. Curiously enough, the maximum spacetime dimension permitted is eleven, where Bergshoeff, Sezgin and Townsend found their 2-brane. In the early 80s Green and Schwarz had showed that spacetime supersymmetry allows classical superstrings moving in spacetime dimensions $3, 4, 6$ and 10. (Quantum considerations rule out all but the ten-dimensional case as being truly fundamental. Of course some of these ten dimensions could be curled up to a very tiny size in the way suggested by Kaluza and Klein. Ideally six would be compactified in this way so as to yield the four spacetime dimensions with which we are familiar.) It was now realized, however, that there were twelve points on the scan which fall into four sequences ending with the superstrings or 1-branes in $D = 3, 4, 6$ and 10, which were now viewed as but special cases of this more general class of supersymmetric extended object.

Other branes were later added to the scan. For example, in 1991, after Stelle and I had shown how the electric 2-brane emerges as a solution of $D = 11$ supergravity, Rahmi Guven, from University of the Bosporus, discovered an eleven-dimensional super 5-brane which acts as its magnetic dual.

Notwithstanding these and subsequent results, the supermembrane enterprise was ignored by most adherents of conventional superstring theory. Those who had worked on eleven-dimensional supergravity and then on supermembranes spent the early eighties arguing for *spacetime* dimensions greater than four, and the late eighties and early nineties arguing for *worldvolume* dimensions greater than two. The latter struggle was

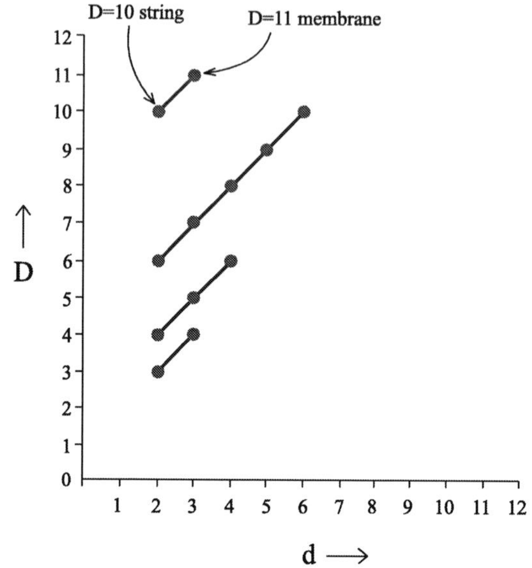

FIGURE 11. The branescan, circa 1987.

by far the more bitter.

The five superstrings problem and the dimension problem were suddenly solved simultaneously by Witten in his, by now famous, talk at the University of Southern California in February 1995. Witten put forward a convincing case that the distinction we used to draw between the five consistent string theories was merely an artifact of our approximation scheme and that when looked at exactly, there was really only one theory, which subsumed all the others. Moreover this theory was a supersymmetric theory in eleven dimensions! In fact, when viewed at distances much larger than the Planck length, it is approximated by our old friend eleven-dimensional supergravity! In the coming months, literally hundreds of papers appeared on the internet confirming that, whatever this eleven-dimensional theory may be, it certainly involves supermembranes in an important way. Consequently Witten dubbed it *M-theory* "where M stands for Magic, Mystery or Membrane according to taste".

Even the chiral $E_8 \times E_8$ string, which according to Witten's earlier theorem could never come from eleven-dimensions, was given an eleven-dimensional explanation by Petr Horava (Princeton University) and Witten. The no-go theorem is evaded by compactifying not on a circle (which has no ends), but on a line-segment (which has two ends). It is ironic that having driven the last nail into the coffin of eleven-dimensions (and having driven Gell-Mann to utter "Ugh!"), Witten was the one to pull the nail out again! He went on to argue that if the size of this one-dimensional space is large compared to the six-dimensional space, then our world is approximately five-dimensional. This may have important consequences for confronting M-theory with experiment.

What do we now know with M-theory that we did not know with old-fashioned string theory? Here are a few examples, references to which may be found in [4].

- Electric-magnetic duality in $D = 4$ is a consequence of membrane/fivebrane duality in $D = 11$. Moreover, the Klein charge quantization rule and the Dirac charge quantization rule play dual roles. They are thus complementary rather than contradictory.
- *Exact* electric-magnetic duality, first proposed for maximally supersymmetric quark-gluon theories, has been extended to *effective* duality by Nathan Seiberg (Princeton) and Witten to less supersymmetric theories: the so-called Seiberg-Witten theory. This has been very successful in providing the first proofs of quark confinement (albeit in the as-yet-unphysical super QCD) and in generating new pure mathematics on the topology of four-manifolds. Seiberg-Witten theory and other dualities of Seiberg may, in their turn, be derived from M-theory.
- Indeed, it seems possible that all supersymmetric quantum field theories with any symmetry, and their spontaneous symmetry breaking, admit a geometrical interpretation within M-theory as the worldvolume fields that propagate on the common intersection of stacks of p-branes wrapped around various holes of the compactified dimensions, with the Higgs field given by the brane separations.
- New insights into the uniqueness problem. See section 9.
- Another interconnection was uncovered by Polchinski who realized that the Type *II* super *p*-branes carrying a certain kind of charge may be identified with the so-called Dirichlet-branes (or *D*-branes, for short) that he had studied some years ago by looking at strings with unusual boundary conditions. Dirichlet was a German mathematician who first introduced such boundary conditions. These *D*-branes are just the surfaces on which open strings can end. This *D*-brane technology has opened up a whole new chapter in the history of supermembranes.
- A by-product of these membrane breakthroughs has been an appreciation of the role played by black holes in particle physics and string theory. In fact they can be regarded as black branes wrapped around the compactified dimensions. These black holes are tiny (10^{-35} meters) objects; not the multi-million solar mass objects that are gobbling up galaxies. However, the same physics applies to both and there are strong hints that M-theory may even clear up many of the apparent paradoxes of quantum black holes raised by Stephen Hawking in Cambridge. Ever since the 1970's, when Hawking used macroscopic arguments to predict that black holes have an entropy equal to one quarter the area of their event horizon, a microscopic explanation has been lacking. But treating black holes as wrapped *p*-branes, together with the realization that Type II branes have a dual interpretation as Dirichlet branes, allowed Andrew Strominger and Cumrun Vafa of Harvard to make the first microscopic prediction in complete agreement with Hawking. The fact that M-theory is clearing up some long standing problems in quantum gravity, such as the black hole entropy problem gives us confidence that we are on the right track.
- It is known that the strengths of the four forces change with distance. In supersymmetric extensions of the Standard Model, one finds that the strengths of the

strong, weak and electromagnetic forces all meet at about at very short distances entirely consistent with the idea of Grand Unification. The strength of the gravitational force also almost meets the other three, but not quite. This near miss has been a source of great interest, but also frustration. However, in a universe of the kind envisioned by Witten, spacetime is approximately a narrow five dimensional layer bounded by four-dimensional walls[3]. The particles of the Standard Model live on the walls but gravity lives in the five-dimensional bulk. As a result, it is possible to choose the size of this fifth dimension so that all four forces meet at this common scale. Note that this is much larger than the Planck length, so gravitational effects may be much closer than we previously thought; a result that would have all kinds of cosmological consequences. Indeed, this has lead to a revival of a variation on the Kaluza-Klein theme whereby our universe is a 3-brane embedded in a higher-dimensional spacetime. Thus the strong, weak and electromagnetic forces might be confined to the worldvolume of the brane while gravity propagates in the bulk. It has recently been suggested that, in such schemes, the extra dimension might be much larger than 10^{-35} meters and may even be a large as a millimeter [11]. In yet another variation due to Lisa Randall of Harvard and Raman Sundrum of Johns Hopkins, the extra fifth dimension is infinite. It has even been suggested that the Big Bang was nothing but the collison of two such 3-branes.

- Another interesting offshoot of D-branes has been provided by Juan Maldacena at the Institute for Advanced Study in Princeton, who has looked at M-theory curled up to a particular 5-dimensional spacetime called anti de Sitter space, a negatively curved saddle-shaped spacetime named after the Dutch mathematician Willem de Sitter. According to Maldacena this 5-dimensional world, including all its gravitational interactions, may be completely described by a non-gravitational theory that resides on its 4-dimensional boundary. This holds promise not only of a deeper understanding of M-theory, but may also throw light on theories that live on the boundary such as the 4-dimensional quark theories that govern the strong nuclear interactions. If this so-called "holographic" picture is correct, and it is looking increasingly plausible, then the universe is like the wall of PlatoÕs cave and we are the shadows projected on it.

Thus branes are no longer the ugly-ducklings of string theory, but now play center stage in theoretical physics: as the microscopic constituents of M-theory, as the higher-dimensional progenitors of black holes and as entire universes in their own right.

---

[3] If this picture is correct, we really need only five dimensions with the other six going along for the ride. Why should Nature behave like this? The only good answer to this question I could find is in Mother Goose's Nursery Rhymes:

Nature requires five,
Custom allows seven,
Idleness takes Nine,
And Wickedness Eleven.

8. WHAT IS THE RESOLUTION OF THE BLACK HOLE INFORMATION PARADOX?

Black holes behave as if they had an entropy proportional to the area of their horizon and the discovery of Hawking that black holes radiate thermally accorded beautifully with the requirements of thermodynamics But the calculation of this radiation raised a sharp paradox. If the hole evaporates away then an initial pure state evolves to a mixed state, leading Hawking to postulate that in the presence of gravity we must generalize quantum mechanics so that systems are described not by wavefunctions but by density matrices.

As we have seen, however, the recent progress in string and M-theory suggests very strongly that black holes behave just like any other composite particle in quantum mechanics: their entropy can be attributed to the degeneracy of their internal states, and Hawking emission is a unitary process where the de-excitation of an internal state leads to emission of a quantum. In particular the entropy of extremal and near extremal black holes have been reproduced exactly from a count of microstates. If we accept these suggestions from string and M-theory, then we apparently solve the information problem: the information in the matter thrown into the hole is recovered in the radiation that emerges, and there is no need for a modification of quantum mechanics. So what was wrong with Hawking's original calculation? In a recent unpublished talk by Hawking himself, it is claimed that there is no information loss after all. Everything is still very uncertain. What is certain is that the debate will continue.

Hence our eighth big question: What is the resolution of the black hole information paradox?

9. IS OUR UNIVERSE UNIQUE?

In string theory, the vacuum degeneracy problems arises because there are billions of Calabi-Yau vacua which are distinct according to classical topology. Like higher-dimensional Swiss cheeses, each can have different number of p-dimensional holes. This results in many different kinds of four-dimensional Standard-like Models with different symmetries, numbers of families and different choices of quarks and leptons. Moreover, M-theory introduces new effects which allow many more possibilities, making the degeneracy problem apparently even worse. However, most (if not all) of these manifolds are in fact smoothly connected in M-theory by shrinking the p-branes that can wrap around the p-dimensional holes in the manifold and which appear as black holes in spacetime. As the wrapped-brane volume shrinks to zero, the black holes become massless and effect a smooth transition from one Calabi-Yau manifold to another. Although this does not yet cure the vacuum degeneracy problem, it puts it in a different light. The question is no longer why we live in one topology rather than another but why we live in one particular corner of the unique topology. This may well have a simpler explanation. A fuller discussion may be found in Brian Greene's book [1].

In any event we are still left with the ninth big question: Is our universe unique? Or, as Einstein put it: Did God have any choice?

The traditional view is that the laws of nature are unique and that all parameters are in principle calculable. So we will one day understand why, for example, the cosmological constant has the value that it does, why there are three families of quarks and leptons and why spacetime appears to have three space and one time dimension.

The anthropic view, in its strongest form, is that all these parameters are tailored to yield intelligent beings like us. Few scientists adhere to this extreme viewpoint. However, many subscribe to a weaker version that some of the parameters are not calculable but accidents of nature. For example the value of the cosmological constant may be no more fundamental than the size of the earth's orbit around the sun. The best we can do is to use our existence as a way of placing constraints on its value.

A third perspective that is currently popular in the string and M-theory community is the *Landscape*. We just happen to live one universe out of the many, possibly infinite, number of universes that make up the multiverse. This seems, at first, to be an admission of defeat. How can we possibly make any predictions? In fact, proponents of the landsape point of view claim that, although it falls short of the ideal traditional view of a unique universe, it is not without predictive power. In the vast landscape, we cannot hope to find the state that describes our universe but we can shift our interest to statistics of these states. Nor can we hope to predict all the feature of nature with arbitrary precision, but perhaps we can find correlations among properties of the states. It does seem necessary to adopt the weak anthropic principle for some quantities such as the cosmological constant. Disappointing? Perhaps, but still worth exploring. After all, maybe this is as good as it gets.

10. DOES M-THEORY DESCRIBE NATURE?

So are we quarks, strings, branes or what?

New York Times, Tuesday September 22, 1998

I hope to have convinced you that the answer involves "branes or what", but there is still no final answer to the exact definition of M-theory, although several different proposals have been made. Is M-theory to be regarded literally as membrane theory? In other words should we attempt to "quantize" the eleven dimensional membrane in some, as yet unknown way? Personally, I think the jury is still out on whether this is the right thing to do.

M-theory has sometimes been called the *Second Superstring Revolution*, but I feel this is really a misnomer. It certainly involves new ideas every bit as significant as those of the 1984 string revolution, but its reliance upon supermembranes and eleven dimensions makes it is sufficiently different from traditional string theory to warrant its own name. One cannot deny the tremendous historical influence of superstrings on our current perspectives. Indeed, it is the pillar upon which our belief in a quantum consistent M-theory rests. In my opinion, however, the focus on one-dimensional objects moving in a ten-dimensional spacetime, that prevailed during the period 1984-1995, will ultimately be seen to be a small corner of *M*-theory [3]. Indeed, future historians may well judge this era as a time when theorists were like boys playing by the sea shore, and diverting themselves with the smoother pebbles or prettier shells of ten-dimensional superstrings while the great ocean of eleven-dimensional M-theory lay all undiscovered before them.

ACKNOWLEDGMENTS

I would like to thank my Mexican colleagues for the great honor of bestowing on me this Gold Medal, especially Alfredo Macias for his hospitality.

REFERENCES

1. Brian Greene, *The Elegant Universe*, W. W. Norton, 1998.
2. *Strings 2000*, Proceedings of the 2000 International Superstrings Conference, University of Michigan, Ann Arbor, 10 - 15 July 2000, edited by Michael J Duff, James T Liu & Jianxin Lu (World Scientific 2001). Also published in I. J. M. P. **16** (2001) 641.
3. Michael Duff, *The theory formerly known as strings*, Scientific American 278 (1998) 54-59 (February).
4. M. J. Duff, *The world in eleven dimensions: supergravity, supermembranes and M-theory*, (IOP Publishing, 1999).
5. *2001: A Spacetime Odyssey*, Proceedings of the Inaugural Conference of the Michigan Center for Theoretical Physics, University of Michigan, Ann Arbor, 20-25 May 2001, edited by Michael J. Duff & James T. Liu (World Scientific, to appear).
6. F. Quevedo, *Lectures on string / brane cosmology*, Class. Quant. Grav. **19**, 5721 (2002) hep-th/0210292.
7. O. Klein. *From my life in physics*, in The Oskar Klein Memorial Lectures, Vol. 1, ed. G. Ekspong (World Scientific, 1991), p. 85.
8. Daniel Z. Freedman and Peter van Nieuwenhuizen, *The Hidden Dimensions of Spacetime*, Scientific American 252 (1985) 62.
9. M. J. Duff, *Not the Standard Superstring Review*, Plenary talk at the *Institute of Physics Conference on Nuclear Physics and Particle Physics*, University of Birmingham, April 1987. CERN Preprint TH.4818/87 (1987). Also published in *Proceedings of the XVth International School on Subnuclear Physics*, Erice, August 1987 (Ed. Zichichi, Plenum Press, 1991).
10. Steven Weinberg, *Dreams of a final theory*, Pantheon,1992.
11. N. Arkani-Hamed, S. Dimopoulos and G. Dvali, *The universe's unseen dimensions*, Scientific American, August 2000, 62.

The Search for Quantum Gravity Signals

G. Amelino–Camelia*, C. Lämmerzahl†, A. Macias** and H. Müller‡

*Dipt Fisica, Univ La Sapienza and Sez Roma1 INFN, Ple Moro 2, Rome, Italy
†ZARM, University of Bremen, Am Fallturm, 28359 Bremen, Germany
**Departamento de Física, Universidad Autónoma Metropolitana-Iztapalapa, Apartado Postal 55-534, C.P. 09340 México, D.F., Mexico
‡Physics Department, Stanford University, Stanford, CA 94305-4060, USA

Abstract. We give an overview of ongoing searches for effects motivated by the study of the quantum-gravity problem. We describe in greater detail approaches which have not been covered in recent "Quantum Gravity Phenomenology" reviews. In particular, we outline a new framework for describing Lorentz invariance violation in the Maxwell sector. We also discuss the general strategy on the experimental side as well as on the theoretical side for a search for quantum gravity effects. The role of test theories, kinematical and dymamical, in this general context is emphasized. The present status of controlled laboratory experiments is described, and we also summarize some key results obtained on the basis of astrophysical observations.

1. INTRODUCTION AND PRELIMINARY REMARKS

1.1. The search for quantum gravity

Our present description of the laws of physics may be characterized as obtained from two types of constituents. The first type of constituent are theoretical frameworks which apply to *all* physical phenomena at *any* instant. These "universal" or "frame" theories are Quantum Theory (all matter is of microscopic origin), Special and General Relativity; SR and GR, (all kinds of matter locally have to obey the principles of Lorentz symmetry and behave in the same way in gravitational fields), and statistical mechanics which is a method to deal with all kinds of systems for a large number of degrees of freedom. The second type of constituent is nonuniversal and pertains to the description of the four presently-known interactions: the electromagnetic, the weak, the strong and the gravitational. The first three interactions are all described within a single formalism, in terms of a gauge theory. So far only gravity has not been successfully included into that scheme. One reason for that might be that gravity appears on both sides: it is an interaction but it is at the same time also a universal theory. Universal theories like relativity and gravity are geometric in origin and do not rely on the particular physical system under consideration, whereas a description in terms of a particular interaction heavily makes use of the particular particle content. Therefore, gravity plays a distinguished role which may be the reason for the difficulty encountered in attempting to unify the other interactions with gravity and attempting to quantize gravity, see Table 1.

The most pressing problem for present-day theoretical physics is the unification of

CP758, *Gravitation and Cosmology: 2nd Mexican Meeting on Mathematical and Experimental Physics*
edited by A. Macías, C. Läemmerzahl and D. Nuñez

TABLE 1. The structure of present physical theories.

Frame theories	**Interactions**
Quantum theory Special Relativity General Relativity Statistical mechanics	Electrodynamics Gravity Weak interaction Strong interaction
Problem	**Wish**
Incompatibility of quantum theory and General Relativity	Unification of all interactions

quantum theory with gravity, the so-called "quantum-gravity problem". The standard scheme of quantization has been proven to lead to inconsistencies when applied to gravity. In particular, the emerging theory would not be perturbatively renormalizable. Various ideas have been explored in alternative to the standard quantization procedure, and some of the most popular approaches are based on string theory, canonical/loop quantum gravity, or non–commutative geometry. It is hoped that these quantum-gravity approaches may also provide a scheme for the unification of the four interactions, something that string theory already comes very close to doing.

A general feature of all these quantum-gravity scenarios is the appearance of new effects, often mediated by new fields, and in particular some of these effects appear to require a modification of some of the fundamental principles which SR and GR are based on.

1.2. Implications of a new quantum gravity theory

The possibility of solving the quantum-gravity problem is often described as intellectually exciting but of mere academic interest. However, the fact that a significant sample of quantum-gravity approaches appears to lead to modifications of some aspects of SR and GR suggests that the implications may go well beyond the academic interest. This point applies in particular to the area of modern metrology, the definition, preparation and transport of physical units. Since various atomic clocks on Earth are located at different height and geographical positions, it is clear that the uniqueness of the definition of time in terms of the TAI, the international atomic time, relies on the validity of SR and GR, see Fig.1.

Another key objective of modern metrology is to base all units on distinguished quantum effects. The reason for that is the precision achieved and the universal reproducibility which is based on the uniqueness of quantum mechanics. Beside the second and the meter, where this already has been done, one can base the unit of the electrical resistance, the Ohm, on the quantum Hall effect using $R_H = K/n$ where $n \in \mathbb{N}$ and $K = h/e^2$ is the von–Klitzing constant, and the unit of the electrical voltage on the Josephson effect via $U_J = n\nu/K_J$ where $K_J = 2e/h$ is the Josephson constant and ν a given frequency, see Fig.2. These quantum definitions of units heavily rely on the validity of basic dynamical equations like the Maxwell, the Schrödinger, and the Dirac equation. If one of these equations was to be modified then some definitions of units would also

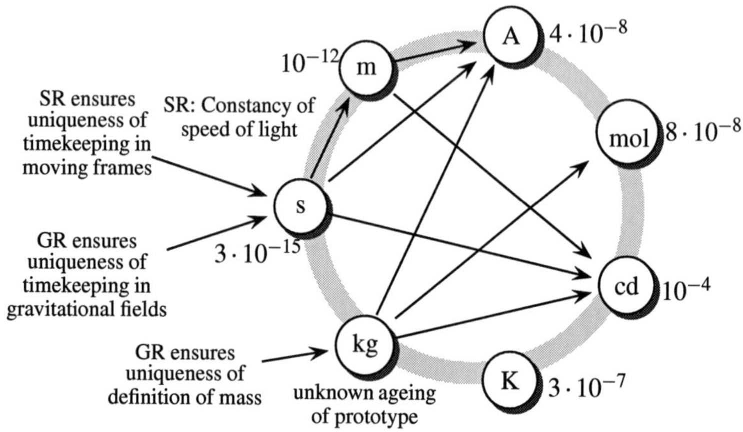

FIGURE 1. The SI–units (s = second, m = meter, A = Ampere, mol = mole, cd = candela, K = Kelvin, kg = kilogram) and their interdependences [1]. The numbers indicate the stability of the corresponding unit. The uniqueness of the transport of the definition of the second and of the meter depends on the validity of SR and GR. A replacement of the mechanical definition of the Ampere or other quantities also requires conventional quantum theory and Maxwell theory and thus the validity of the SR and GR. All units, but the Kelvin, are thus influenced by relativity.

be affected. Therefore, *any high precision test of Lorentz invariance is also a test of the modern metrological scheme.*

Moreover, if it turns out that the Maxwell and Dirac equations have to be modified then this may complicate future high precision navigation.

1.3. The magnitude of quantum gravity effects

Since the typical laboratory energies are of the order of 1 eV and the characteristic quantum-gravity energy scale is expected to be of the order of the Planck energy which is about 10^{28} eV, the quantum-gravity effects in laboratory experiments are likely to come in at the order of 10^{-28} (or lower) which appears to be well beyond the reach of laboratory experiments, in spite of the many high-precision devices which are becoming available for searching for new effects. As stressed in several recent "quantum gravity phenomenology" reviews [2, 3, 4, 5] this suggests that the relevant phenomenology should rely on contexts of interest in astrophysics, rather than use controlled laboratory experiments. However, there may nevertheless be an important role for laboratory experiments in this phenomenology, especially considering the following observations:

- The arguments that suggest that the characteristic quantum-gravity energy scale should be of order 10^{28} eV must at present be viewed as inconclusive. One really needs the correct theory, which is still not established, in order to reliably estimate this scale. It may well be that the characteristic quantum-gravity energy scale is actually much lower than 10^{28} eV. For example, in scenarios with "large extra

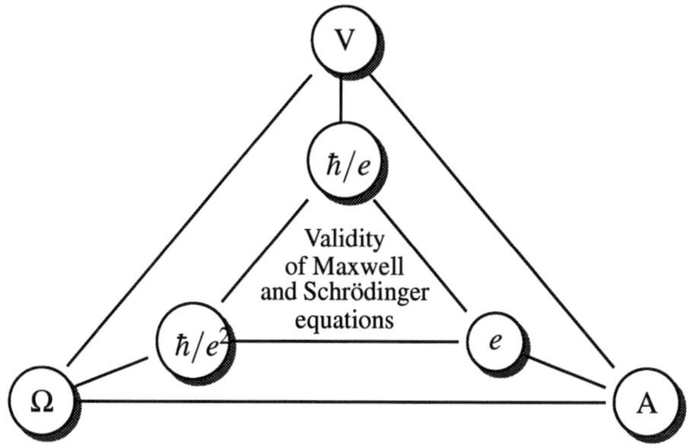

FIGURE 2. Electrical units of the voltage V (Volt), electrical resistance Ω (Ohm), and electrical current A (Ampere) based on distinguished quantum effects. The experimental effects are interpreted in terms of the ordinary Maxwell and Schrödinger equation.

dimensions" the quantum-gravity effects, including deviations from Newton's law, would be accessible at much lower energies (see e.g. [6]). Similarly, in some string-theory-motivated "dilaton scenarios" [7, 8] the Universality of Free Fall (UFF) would be violated already at the 10^{-13} level, and the PPN parameter γ, which in ordinary Einsteinian gravity is exactly 1, might be different from unity by up to 10^{-5}.

- Even within the assumption that the quantum-gravity effects actually do originate at the Planck scale, one can find some (however rare) cases in which the small 10^{-28} effect is effectively "amplified" [2] by some large ordinary-physics number that characterizes the laboratory setup. For example, there has been considerable work on the possibility that quantum-gravity effects might significantly affect some properties of the neutral-kaon system [9, 10] and in those analyses the effects are amplified by the fact that the neutral-kaon system hosts the peculiarly small mass difference between long-lived and the short-lived kaons $M_{L,S}/|M_L - M_S| \sim 10^{14}$. Similarly, in studies of quantum-gravity-induced interferometric noise [11] one can exploit the fact that some scenarios predict noise that increases as the frequency of analysis decreases, and therefore, searching for fundamental noises in high precision long–term stable devices (like optical resonators) may give new access to this domain of quantum gravity effects [12].
  In an analogous way, the sensitivity of some clock–comparison experiments is amplified by a factor of $h\nu/m_p c^2 \sim 10^{18}$ where ν is the clock frequency used and m_p the proton mass, see Sec.4.3.6.
- In some rare cases the laboratory experiment actually does have the required 10^{-28} sensitivity, as illustrated by the analysis reported in Ref. [13], which studies the implications of quantum-gravity-induced wave dispersion for the most advanced modern interferometers (the interferometers whose primary intended use is the search of classical gravity waves).

- The fact that low-energy data suggest that the electroweak and strong interactions unify at the GUT energy scale of $\sim 10^{16}$ GeV may encourage us to conjecture that also the gravitational interaction would be of the same strength as the other interactions at that scale. This would mean that the characteristic energy scale of quantum gravity should be some three orders of magnitude smaller than the Planck energy (see e.g. [14]).

It appears therefore that we should complement the astrophysics searches with a wide program of laboratory/controlled experiments searching for the effects predicted by some quantum-gravity approaches. Although in most cases the potential sensitivity of astrophysics searches will be higher, the laboratory experiments have the advantage of providing access to a much larger variety of potential phenomena, and for the development of the relevant phenomenology it would be an extraordinary achievement to be able to analyze a new effect within a controlled/repeatable laboratory experiment with the possibility of a systematic modification of boundary and initial conditions, rather than relying on the "one chance" observations available in astrophysics.

1.4. Main quantum gravity schemes

As mentioned, the most popular approaches to the quantum-gravity problem are based on string theory, canonical/loop quantum gravity, or non–commutative geometry, see, e.g., [15]. For different reasons the development of phenomenology for each of these three ideas is still at a rather early stage of development. But several hints for experimentalists have already emerged.

- In canonical/loop quantum gravity the key difficulty is the fact that the techniques for obtaining the classical limit of the theory have not yet been developed. Since our phenomenology will usually be structured as a search of corrections to the classical effects, this is a very serious issue. However, several authors [16, 17, 18, 19], guided by the intuition from working with some candidate quasiclassical states, made analyses that led to the expectation that Lorentz symmetry should be broken in Loop Quantum Gravity, and as a result the Maxwell and Dirac equations should include extra terms of higher derivatives. But clearly these violations from Lorentz symmetry still cannot be viewed as a "prediction" of Loop Quantum Gravity because of the heuristic nature of the underlying arguments, and indeed there are some authors (see, *e.g.* Ref. [20]) who have presented arguments in favour of exact Lorentz symmetry for Loop Quantum Gravity.
- String theory schemes in any dimensions always predict new fields which couple in different ways to the various matter sectors. As a result a large variety of effects are *allowed* by string theory, including effects leading to violation of Lorentz invariance, violation of the UFF and violation of the Universality of the Gravitational Redshift (UGR) is encountered. From the viewpoint of the phenomenologist a disappointing aspect of string theory is that it appears that it cannot be falsified on the basis of these effects. String theory may predict many new low-energy effects, but it can also be easily tuned to avoid all of them. But of course it is still

interesting to look for these new effects, and indeed a rich phenomenology is being developed. In particular, there is considerable work on a string-theory-motivated dilaton scenario [7] with deviations from the UFF at the order 10^{-13} and deviations of the PPN parameters γ and β at the order 10^{-5} and 10^{-9}, respectively. Another large phenomenological effort is being devoted to a string-theory-motivated general framework, the Standard Model Extension (SME) [21], for the description of violations of Lorentz invariance that are codified in constant parameters in still linear and Lagrangian based generalized Maxwell and Dirac equations.

- Much of the quantum-gravity work based on noncommutative geometry has focused on the hypothesis that the correct quantum gravity should have as flat-spacetime limit a noncommutative version of Minkowski spacetime, with spacetime coordinate noncommutativity of the type $[x_\mu, x_\nu] = i\theta_{\mu\nu} + i\zeta^\sigma_{\mu,\nu} x_\sigma$ for some appropriate choice of the coordinate-independent $\theta_{\mu\nu}$ and $\zeta^\sigma_{\mu\nu}$. In a noncommutative spacetime there is an absolute limitation on the localization of a spacetime point (event), and as a result the relevant theories are inevitably nonlocal. In some frameworks it is possible to give an effective commutative-spacetime description of this nonlocality, in which in particular – in terms of partial differential equations – higher order derivatives may occur. Another rather generic feature of these noncommutative versions of Minkowski spacetime is the emergence of anomalous dispersion relations. The simplest and most studied example of such an anomalous dispersion relation is

$$m^2 \simeq E^2 - \vec{p}^2 + \eta \vec{p}^2 \frac{E}{E_P} + O\left(\frac{E^4}{E_P^2}\right), \tag{1}$$

where E_P is the Planck scale and η is a numerical factor (expected to be of order 1) describing the strength of the quantum gravity modification ($E_\mathrm{QG} \equiv E_\mathrm{P}/\eta$ is the characteristic scale of the modification of the dispersion relation).

	string theory	canonical/loop quantum gravity	non–commutative geometry
geometrical characteristics	starts from fixed (flat or deSitter) background	dynamical	starts from fixed background
particle characteristics	depends on particle model	model independent	model independent
violations of	Lorentz invariance	Lorentz invariance? (may depend on boundary conditions)	Lorentz invariance
	Universality of Free Fall	Universality of Free Fall? (not yet calculated)	Universality of Free Fall? (not yet calculated)

1.5. Methods of phenomenological generalizations of dynamical equations

Since it was not covered extensively in recent quantum-gravity-phenomenology reviews [2, 3, 4, 5] we will here devote particular attention to the study of phenomenological generalizations of dynamical equations like the Dirac or the Maxwell equations. As always in quantum gravity phenomenology, these studies must be structured in terms of some test theories, bridging the gap between the rich (but often untreatable) formalisms used in the study of the full quantum gravity problem and the language which is appropriate to describe experiments. The structure of these test theories should be obtained by calculating some low energy approximation of the full quantum gravity scenario. This has been attempted for the "Liouville String" scenario [22, 23] as well as for the loop quantum gravity approach [16, 17, 18, 19]. But since various technical issues remain to be understood about the relevant approximation schemes, the phenomenology is being developed on the basis of rather general test theories describing modifications of the Dirac and the Maxwell equations.

One approach, which is being pursued mostly as starting point of the SME [24, 21, 25], adopts some general Lagrangian which is still quadratic in the field strengths or in the fermionic fields and requires further building principles like conservation of energy-momentum, Lorentz-covariance, conventional quantization, Hermiticity, microcausality, positivity of energy, gauge invariance, and power-counting renormalizability[1]. The parameters of the SME are just additional interactions with constant fields[2]. The main advantages of this approach are the mathematical consistency and physical interpretability of the new theory in conventional terms and the fact that it provides the most conservative modification of established theories. But of course this may also turn into a disadvantage, if it eventually turns out that the correct quantum gravity requires more novel features, such as effects that are not described by power-counting-renormalizable terms. And indeed some authors have chosen to look beyond the SME setup, considering Planck-scale-suppressed effects which are in fact not described by power-counting-renormalizable terms (see, *e.g.*, Refs. [28, 16, 29]).

Of course, it is also possible to renounce to the assumption of a Lagrangian generating the dynamical equations, and in fact there is a rich phenomenology being developed introducing the generalizations directly at the level of the dynamical equations. This is of course more general than the Lagrangian approach (see [30, 31] for examples dealing with generalized Dirac and Maxwell equation). In the case of the Dirac equation [30] one is led to effects like the violation of the UFF, of the UGR and of Local Lorentz Invariance (the effects violating Local Lorentz Invariance have been also obtained later

[1] We denote by Standard Model Extension (SME) the model described with this name in Refs.[21, 25]. Readers should however notice that in the recent literature this same model is sometimes referred to as "minimal SME", and instead with "SME" one denotes a generalization which allows for nonrenormalizable terms, while insisting on coordinate-independence, availability of a Lagrangian, and availability of a field-theoretic description. Within this terminology, the model we refer to as "SME" is thus called the "minimal SME".

[2] This kind of generalizations in the photonic sector have been introduced and discussed earlier by Ni [26] and Haugan and Kauffmann [27].

on in [25]). More effects than in the SME are encountered for the generalized Maxwell equation discussed in [31]. Charge conservation, which automatically comes out from the Lagragian approach, can be violated in models generalizing the field equations [31]. In addition, in [31] also more Lorentz Invariance violating parameters than in the SME have been found. However, particular care must be taken for the mathematical consistency of the formalism if one "by hand" generalizes the dynamical equations. This will be automatically secured if one employs a constructive axiomatic scheme in order to derive equations, as shown for example in Refs. [32, 33, 34] within a derivation of the generalized Dirac equation in terms of fundamental properties of the dynamics of fields.

1.6. Kinematical test theories for Lorentz invariance

Instead of modifying the dynamical equations, some authors have preferred to introduce new-physics effects at the level of kinematics. In particular, there is significant work on kinematical test theories for Lorentz invariance. Later on in these notes we will comment on some recent proposals of this type which have been motivated rather directly by some approaches to the quantum-gravity problem. We here want to comment on a predecessor, the Robertson–Mansouri–Sexl (RMS) theory, which was introduced by Robertson [35] and Mansouri and Sexl [36], where a modification of Lorentz–transformations are the basis for, e.g., anomalous effects of light propagation what can be tested. Compared with the dynamical approach the kinematical approach is more powerful since it is independent of the specific particle model under consideration: it discusses the transformation properties of observed quantities under transformations between inertial systems. One may however be uncomfortable with the fact that this RMS model assumes that in one frame light propagates isotropic. This requires to single out a preferred frame, which is usually taken to be the frame given by the cosmological radiation background. But this preferred role for the cosmological radiation background, which is after all a classical-physics feature, is not necessarily compelling. And one might ask how this programme should proceed if it turns out that, for example, there is also a stochastic gravitational wave background, which selects a different frame with respect to the one natural for the radiation background. Should one then make a choice? This choice would influence significantly the interpretation of the experiments. Some authors are also uncomfortable with the RMS scheme because of its lack of generality with respect to certain issues. In particular, it is not obvious that one should assume that in the preferred frame light propagates isotropically, something that might be violated in a Finslerian space–time, and the RMS scheme also does not allow a description of possible violations of Lorentz invariance leading to birefringence effects.

1.7. The scheme of exploring new physics with generalized Maxwell and Dirac equations

As mentioned in this notes we will devote particular attention to the possibility of exploring new physics with generalized Maxwell and Dirac equations. There is a long history in the discussion of violations of Lorentz invariance and of the UFF (also called Weak Equivalence Principle) due to some modifications of the Maxwell and Dirac equations.

Very early considerations are based on Mach's principle and led to considerations of a hypothetical anomalous inertial mass term in the Schrödinger equation [37] which subsequently has been tested with very high precision by Hughes [38] and Drever [39], see also Sec.4.3.6. Later discussions have been performed by Liebscher and Bleyer [40, 41]. The most general Dirac equation based on basic properties like unique evolution, superposition principle, finite propagation speed and probability conservation has been discussed in [33]. Based on this, a generalized Pauli equation has been derived [30] which is most appropriate to confront this generalized theory with experiments. Since the spin–$\frac{1}{2}$–sector of the SME is contained in this general scheme, also there a generalized Pauli equation can be derived [42].

Modifications of the Maxwell equations in view of a discussion of violations of Lorentz invariance and of the UFF has been first introduced by Ni [43, 26] and Haugan and Kauffmann [27]. The same sort of modifications can be found in the SME [25, 44]. More general modifications which also include charge non–conservation have been introduced in [31].

For exploring new physics in these frameworks, at first one discusses and explores isolated effects for the Dirac and the Maxwell equation. These are propagation phenomena of electromagnetic and matter waves and, in the case of electrodynamics, the static solutions for a point–like charge and magnetic moment. In a next step, combined effects which appear in electromagnetically bound systems have to be calculated, see, e.g., Sec.4.3.3. These combined effects govern the physics of atoms, molecules, and solids which serve as realizations of clocks, and of rulers.

1.8. What is a violation of Lorentz invariance?

A key focus for these notes are studies of possible departures from Lorentz symmetry. For this phenomenology a key question is how we may interpret or how we may proceed if we are detecting a signal which appears to violate Lorentz invariance. After all the attempts at a conventional-physics explanation have failed, and it is therefore certain that one is dealing with new physics, it would be nice to have some criteria for establishing what constitutes a robust basis for accepting or recognizing a violation of Lorentz invariance.

Of course in science one never establishes how things "are", but one rather finds models that reproduce all observations. Therefore any interpretation will always be subject to further scrutiny. Still it is useful to keep in mind that some checks of consistency with the hypothesis of violation of Lorentz symmetry can be made. In particular, it is (at

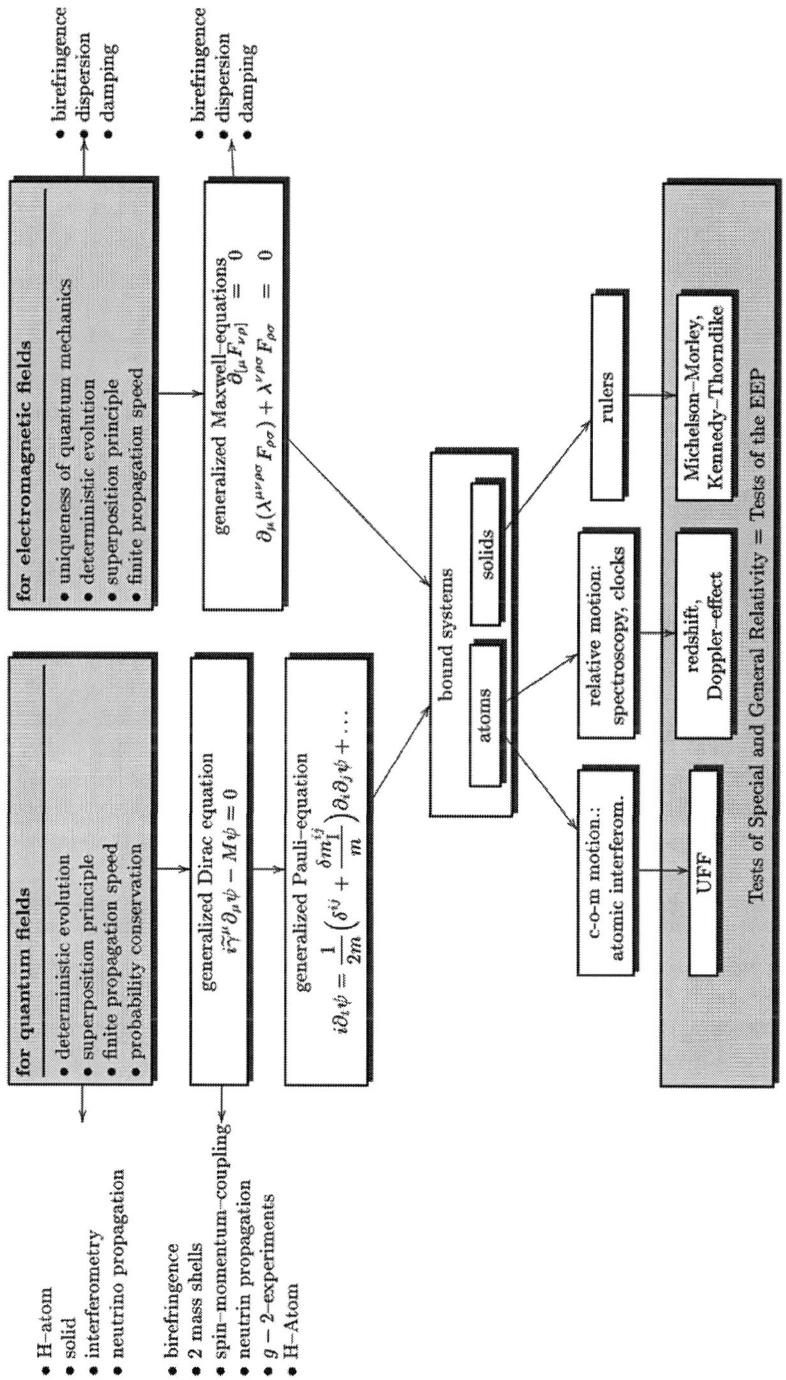

FIGURE 3. Schematics of the implications of modified Dirac and Maxwell equations.

least in principle) relatively easy to distinguish between the hypothesis of a violation of Lorentz symmetry and the hypothesis of the discovery of a new interaction. If the effect signals a new interaction it should be possible to shield it (like the electromagnetic interaction), transform it away (like metrical gravity) and/or identify a cause of the effect. A violation of Lorentz invariance should instead be universal and it should not be possible to transform it away.

1.9. Main directions in searches for new physics

At present there are at least six main directions in the search for possible quantum gravity effects:

Search for orientation dependent effects. The effects one is looking for depend on the orientation of the laboratory. Early experiments of this kind are Michelson–Morley (see Sec.4.3.4) and Hughes–Drever experiments (see Sec.4.3.6). For power-counting-renormalizable effects the results of all experiments of this kind can be related to the SME, which gives the most general parametrization of power-counting-renormalizable anisotropy effects in various basic equations. Presently, for photons the anisotropy is limited by 10^{-15} and for nuclear matter by 10^{-30}.

Search for a violation of the Universality of Free Fall. Within the string-theory-inspired dilaton scenarios there is a distinguished prediction of a violation of the UFF at the 10^{-13} level [7]. Presently, the UFF is confirmed at the level $5 \cdot 10^{-13}$. Also anomalous spin couplings are considered (see, e.g., [45] for a short review), which are beyond the standard coupling to the gravitational field. Other works considering possible modifications of the UFF for various particles are [46, 47].

Search for a violation of the Universality of the Gravitational Redshift. String theory may also host deviations from the UGR [48]. In fact, it has been shown that violations and UFF and of UGR are strongly related [49, 50].
Presently, the UGR is confirmed by clock–clock comparison (atomic fountain clock vs. H–maser) at the $1.5 \cdot 10^{-5}$–level [51].

Search for a modified dispersion relation. Searches of modifications of the dispersion relation of the form (1), and its generalizations, constitute another direction of experimental quantum gravity efforts. This type of effect is beyond the SME, since it could not be described by power-counting-renormalizable terms. A possible manifestation of the modified dispersion relations is an energy dependent propagation velocity which should lead to different time–of–arrivals of the same event on a distant star when looked at it in different frequency channels, see e.g. [52]. Another effect is a frequency dependent position of interference fringes [13].

Search for space–time fluctuations. Search for fundamental space–time fluctuations [11] and associated fundamental decoherence effects in quantum systems [53, 54] have been also considered extensively in association with various classes of experiments [55, 12]. There has also been work on the implications of space–time fluctuations for the spreading of signals, which causes a sharp signal to turn into a fuzzy signal upon relatively long propagation times [2, 56].

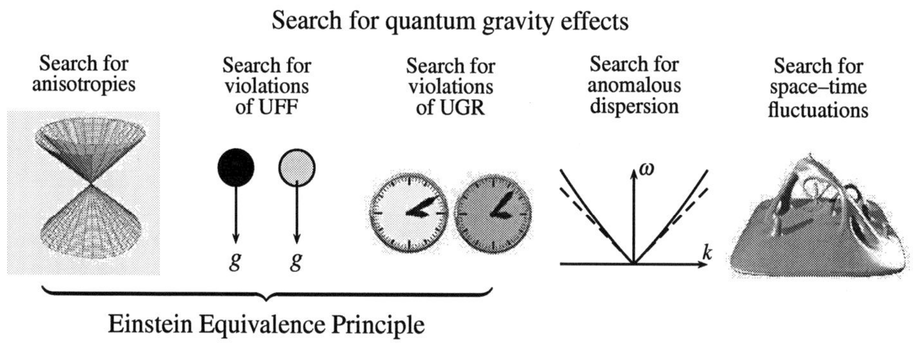

FIGURE 4. The main directions and characteristics of searches for quantum gravity induced effects: anisotropy effects, violation of UFF and UGR, modified dispersion and space–time fluctuations. A search for violations of isotropy, UFF and UGR amounts to a search of a violation of Einstein's Equivalence Principle.

Search for departures from *CPT* symmetry Since the *CPT* theorem is based on locality and Lorentz covariance, it is not surprising that the quantum-gravity approaches that involve some nonlocality and/or departures from Lorentz symmetry provide motivation for searches of departures from *CPT* symmetry. Noteworthy limits have been obtained using some analyses of the neutral-kaon system (see Ref. [57] and references therein) and other limits are now being pursued in the context of neutrino physics [10]. Indeed, it has been shown that *CPT* violation implies a violation of Lorentz invariance [58].

2. QUANTUM GRAVITY PREDICTIONS

2.1. String theory

The low-energy physics of String theory, as presently understood, could be very rich, with the presence of new fields and the possibility of a variety of new effects. This may provide the basis for a large phenomenological effort, even though one should keep in mind that the new effects are not genuine "predictions" of String theory: String theory could make room for these effects but it could equally well suppress them all. It appears at present not possible to falsify String theory on the basis of low-energy phenomenology, but it would nonetheless be very exciting if any of the new effects that String theory may host was actually found.

Lorentz invariance. A possible violation of Lorentz invariance, due to spontaneous symmetry breaking, has been considered [59]. Though requirements like renormalizability and gauge invariance rule out spontaneous symmetry breaking of the Lorentz

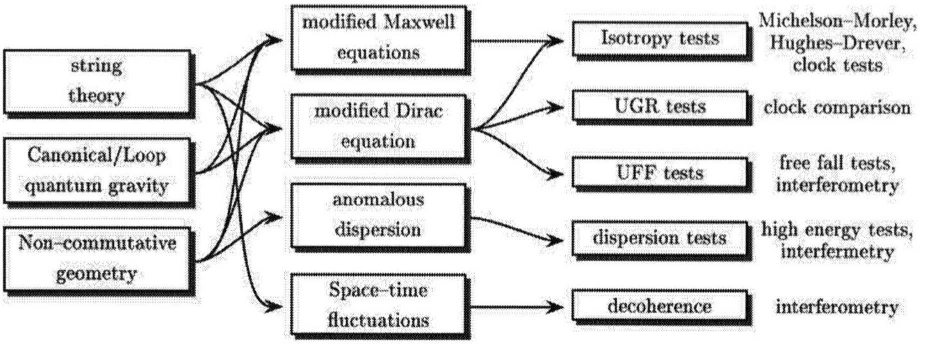

FIGURE 5. Quantum gravity, effective intermediate (test) theories and the experimental search.

group in ordinary field theories it might happen for string theories. Lorentz symmetry breaking may arise from tensor–tensor–scalar couplings that are allowed because strings are extended objects which contain an infinite number of particle modes.

Universality of Free Fall. The UFF is the most basic principle underlying GR. It is responsible for the possibility to geometrize the gravitational interaction and, thus, for the present understanding of space and time. However, in some string-inspired dilaton scenarios [60, 61, 7, 8] is has been claimed that the UFF might be violated, in terms of the Eötvös parameter, at the 10^{-13} level. Similar predictions are made by Wetterich [62] from a quintessence model where the dynamics of a cosmological field introduced to explain the dark energy, yields a violation of the UFF at the 10^{-14} level. Also in the "Liouville noncritical string" scenario [63] there are predictions for a violation of UFF due to the interaction of particles with the space–time foam which, however, are too small to be of today's experimental interest.

Universality of Gravitational Redshift. The same string-theory scalar field, which may couple to different matter fields in a different way thus leading to a violation of UFF, also leads to the effect that clocks based on different physical principles behave differently in the gravitational field and, thus, violate the UGR. This has been shown in [64, 48]. The relation between violations of UFF and UGR has been discussed in [49, 50]. Also quintessence models predict a violation of UGR [65].

Anomalous dispersion. Since, in principle, the effective equations for the electromagnetic field and also for the Dirac field contain higher order derivatives, anomalous dispersion relations of the structure given in Eq.(1) could also come out from string theory [25]. However, in most cases they are neglected in the string-theory context since it is believed that they are less significant than the modifications, e.g., of the Dirac matrices in the case of the Dirac equation or of the constitutive relation in electrodynamics.

Space–time fluctuations and decoherence. In the most popular "critical superstring" formulation of String Theory there has been so far no direct argument in favour of spacetime fluctuations and decoherence. However, in the "Liouville noncritical String" approach [66] space–time fluctuations and decoherence have been considered. For decoherence, possible implications for the neutral-kaon system have been investigated (see, *e.g.*, Ref. [67] and references therein), while a direct investigation of spacetime fluctuations has been attempted by Percival and coworkers [53, 54].

2.2. Loop quantum gravity

As mentioned, in canonical/loop quantum gravity the key difficulty is the fact that the techniques for obtaining the classical limit of the theory have not yet been developed. In a certain sense one has an otherwise attractive candidate for quantum gravity, which however has not been shown to actually contain classical gravity as a limit. As a way to get some intuition for the type of effects that the theory might predict, when fully understood, several authors [16, 17, 18, 19] have proposed to start with the exploration of the properties of some candidate quasiclassical states.

Lorentz invariance. By introducing the so–called weave states, which are candidate quasiclassical states of the space–time geometry, in the range between pure quantum geometry and classical geometry, one can motivate effective equations for the propagation of spinors and of electromagnetic fields [17, 18, 19, 68]. The resulting effective equations possess terms which violate Lorentz invariance, and also terms of higher order derivative. As mentioned there are some critiques on this approach, notably by Kozameh [20] who claimed that the breaking of LI came in by a particular choice of boundary conditions. Recently, it was also observed [69] that a plausible path toward the Loop-Quantum-Gravity classical limit should lead to a situation in which Lorentz symmetry is neither preserved nor broken, it would be deformed, in the sense we discuss more carefully later in our remarks on noncommutative spacetimes.

Universality of Free Fall and Unversality of Gravitational Redshift. To our knowledge there has been no analysis so far suggesting violations of the UFF and UGR in Loop Quantum Gravity. This is not surprising in light of the mentioned "classical limit" problem: one should first verify UFF and UGR in the classical limit and then look for possible quantum corrections.

Anomalous dispersion. A modification of the energy-momentum dispersion relation is rather typical (though not inevitable) when Lorentz symmetry is broken. The arguments [17, 18, 19, 68] that suggest breaking of Lorentz symmetry in Loop Quantum Gravity do indeed provide motivation for modifications of the dispersion relation, since they lead to modified Maxwell and Dirac equations with higher order derivatives, suggesting a modified dispersion relation of the structure (1).

Space–time fluctuations and decoherence. There has not been much discussion of Space–time fluctuations and decoherence in Loop Quantum Gravity. However, a recent proposal [70] for a reformulation of Loop Quantum Gravity does lead to decoherence and the need to set up the analysis within a density-matrix framework.

2.3. Non–commutative geometry

The formalism of noncommutative geometry appears to have rather wide (though not completely general) applicability in the study of the quantum-gravity problem. In particular, a large number of studies has been devoted to the hypothesis that the correct quantum gravity might admit a regime which could be based on a non-commutative version of Minkowski spacetime. In some cases non-commutative spacetimes prove useful at an effective-theory level (for example, in certain string theory pictures [71] spacetime noncommutativity provides an effective theory description of the physics of strings in presence of a corresponding external background), while other quantum-gravity approaches (see, *e.g.*, Ref. [72]) explore the possibility that a noncommutativity might be needed for the correct fundamental description of spacetime.

The most studied noncommutative versions of Minkowski spacetime all fall within the parametrization ($\mu = 0,1,2,3$)

$$[x_\mu, x_\nu] = i\theta_{\mu\nu} + i\zeta^{\sigma}_{\mu\nu} x_\sigma \,, \tag{2}$$

with coordinate-independent $\theta_{\mu\nu}$ and $\zeta^{\sigma}_{\mu\nu}$. The choice of $\theta_{\mu\nu}$ and $\zeta^{\sigma}_{\mu\nu}$ is the key theoretical input and the aspect to be determined experimentally.

Lorentz invariance. An intense research effort has been devoted to the implications of noncommutativity for the classical Poincaré symmetries of Minkowski spacetime. This is in fact one of the few aspects of the relevant theories which can be analyzed rather exhaustively, and it leads to some ideas for experimental testing. For the simplest noncommutative versions of Minkowski spacetime, the canonical noncommutative spacetimes characterized by coordinate noncommutativity of type

$$[x_\mu, x_\nu] = i\theta_{\mu\nu} \tag{3}$$

with coordinate-independent $\theta_{\mu\nu}$, a full understanding has been matured, and in particular it has been established that the Lorentz-sector symmetries are broken [71, 73, 74] (in a sense analogous to the popular mechanism of spontaneous symmetry breaking) by this type of noncommutativity.

At the next level of complexity [75], the one of Lie-algebra noncommutative spacetimes

$$[x_\mu, x_\nu] = i\zeta^{\sigma}_{\mu\nu} x_\sigma \,, \tag{4}$$

the description of symmetries is in some cases more demanding at the technical level. The most studied of these Lie-algebra versions of Minkowski spacetime is "κ-Minkowski [76, 77] ($l, m = 1,2,3$)

$$[x_m, t] = \frac{i}{\kappa} x_m \,, \quad [x_m, x_l] = 0 \,. \tag{5}$$

κ-Minkowski is a Lie-algebra spacetime that clearly enjoys classical space-rotation symmetry; moreover, at least in a Hopf-algebra sense (see, *e.g.*, Ref. [78, 79]), κ-Minkowski is invariant under "noncommutative translations". It is particularly noteworthy that in

κ-Minkowski boost transformations are necessarily modified [78, 79, 80]. A first hint of this comes from the necessity of a deformed law of composition of momenta, encoded in the so-called coproduct (a standard structure for a Hopf algebra). As a result in this spacetime one expects departures from classical Lorentz symmetry but of a type that does not lead to the emergence of a preferred class of inertial observers (the "Relativity Principle" is still upheld). One usually refers to these cases as "deformations of Lorentz symmetry" (in alternative to the more familiar mechanisms that may lead to broken Lorentz symmetry).

Unversality of Free Fall and Universality of Gravitational Redshift. Concerning the UFF and the UGR these approaches based on noncommutative versions of Minkowski spacetime are insufficient for the derivation of rigorous predictions, since indeed they focus on the Minkowski limit. They are motivated by the idea of some form of "quantization" of gravity, but at present they only incorporate this quantization in the noncommutativity of spacetime and gravitational effects are not yet described (although some attempts of generalization are under way).

Anomalous dispersion. The fact that noncommutative versions of Minkowski spacetime typically lead to departures from classical Lorentz symmetry (broken Lorentz symmetry in most cases, and, as mentioned, "deformed" Lorentz symmetry in some special cases like κ-Minkowski) in turn leads to the emergence of departures from the special-relativistic energy-momentum ("dispersion") relation.

For the cases affected by IR/UV (infrared/ultraviolet) mixing the phenomenology based on modified dispersion relations must be conducted cautiously, since the IR/UV mixing may lead to infrared singularities, which are still not fully understood (or at least there is still a lively debate on how to handle them in phenomenology). We do not to comment on this possible phenomenology in these notes, but refer to Refs. [81, 74, 82].

When there is no singularity in the infrared, and assuming that the onset of the new effects is characterized by the Planck length, one will typically find (and indeed one finds in the examples which have been analyzed) dispersion relations that fall within the class

$$m^2 \simeq E^2 - \vec{p}^2 + \eta p^2 \left(\frac{E^n}{E_P^n} \right) + O\left(\frac{E^{n+3}}{E_P^{n+1}} \right), \tag{6}$$

where η and n are model-dependent parameters which should be determined experimentally. Not only are η and n model dependent, but within a given model they may also take different values for different particles. In particular, in the analysis [73] of some noncommutative spacetimes one finds birefringence (different polarizations of light propagate at different speeds). The fact that some noncommutative spacetimes host this particle dependence suggests that they might also give rise to departures from the UFF and the UGR, but, as mentioned, the relevant formalisms are still at too early a stage of development for producing detailed predictions concerning the UFF and the UGR.

Space–time fluctuations. One of the original motivations for introducing the formalism of noncommutative geometry was the one of formalizing the limitations on the localization of spacetime points (events) at the quantum-gravity (Planck-scale) level, which are suggested by a variety of arguments combining general relativity and quantum mechanics (see, *e.g.*, Refs. [83, 72, 84, 85, 86, 87, 88]). Unfortunately the analysis of the relevant formalisms has not yet been developed to the point of a phenomenologically useful description of this localization limits. The noncommutativity of the coordinates of course implies some associated limits on the combined measurement of pairs of coordinates, but phenomenology would need a more advanced level of description, one in which one could for example derive the implications of noncommutative geometry for the activities of an interferometer. Clearly if our interferometers are operating in a fuzzy (noncommutative) spacetime, rather than a sharp classical spacetime, one should expect that at some point the accuracy of the interferometer would be affected. This is indeed the expectation of those working on noncommutative spacetimes, but it has proven so far too hard to attempt to derive from the noncommutative geometry the relevant physical effects.

3. THE TEST THEORIES

The study of the quantum-gravity problem has provided motivation for the study of a rather large number of new effects. In light of the nature of the indications that are coming from the theory side it appears necessary to structure the phenomenological efforts according to a few alternative strategies.

In some cases a given quantum picture of spacetime or a given quantum-gravity scenario definitely predicts a specific effect, *i.e.* we can identify in the formalism an effect which, if not found, could falsify the theoretical picture. Then there is scope for a sharply aimed phenomenological effort, characterized only by the few parameters that the specific theoretical picture involves. A situation of this type is maturing in the study of certain noncommutative spacetimes, perhaps most notably the κ-Minkowski spacetime, whose single length parameter (here denoted by λ) could soon be constrained at a beyond-Planckian level.

In other cases, especially in ambitious attempts at a complete solution of the quantum-gravity problem, such as string theory and loop quantum gravity, one is dealing with an extremely complex and rich formalism which appears to have the potentiality of giving rise to a plethora of new effects, even though one is not able (at least at present) to establish that a certain new effect is definitely present in the relevant theoretical framework. This type of frameworks provide motivation for a general approach for the description of deviations from standard theories. Since even the most basic principles of present-day physics are not necessary "safe" in these new theories, one should be prepared for modifications of the fundamental equations describing physics. The most fundamental equations are the equations governing the standard model, that is, the gauge theory of the electroweak and strong interaction and the Einstein field equation, and the phenomenology should explore all candidate modifications of these equations.

Concerning the target sensitivity that this phenomenology should set for itself, obviously the Planck energy scale can provide some tentative guidance, but it appears that the phenomenology should be prepared for the possibility that another, possibly lower, scale might characterize at least some quantum-gravity effects. For those new effects, as in the example of some modified dispersion relations mentioned earlier, whose description automatically brings about a to-be-determined energy scale, it is of course of primary importance from a quantum-gravity perspective to find ways to test the hypothesis that the relevant energy scale is indeed the Planck scale. But even a lower sensitivity may turn out to be sufficient to uncover a quantum-gravity effect.

This large phenomenological effort, which as a whole should consider a large number of possible effects (and parametrizations of the effects) and a large range of possible magnitudes for these effects, must of course be structured in terms of some reference test theories. Reference to some widely-adopted test theories of course facilitates the comparison of sensitivities achieved by different experiments, and the test theories can also provide a language that bridges the gap between experiments and (usually rather formal) theory work on quantum gravity. But even in setting up such test theories a few alternative strategies should be explored simultaneously. A key issue is whether the new effects end up taking the form of new terms in the old type of formalisms or require even a new formalism. Usually, even when a new formalism is required for the full new theory, one can approximately incorporate the new effects in the old formalism. And on the basis of this experience it is natural to set up some test theories that indeed describe the new effects via some new terms in the old type of equations. But the phenomenology should also keep in mind that there have been cases in the history of physics where the description of the new physics really demanded a new formalism. For example, in order to introduce the new effects of special relativity in quantum mechanics one was tempted to formulate a "relativistic quantum mechanics" (introducing new structures in the old formalism) but it turned out to be necessary to invent relativistic quantum field theory. We now understand that there is a limit of relativistic quantum field theory that can be described approximately in terms of a "relativistic quantum mechanics", but the nature of the limiting procedure by which the old formalism emerges can only be established *a posteriori*, when the full theory has been developed and understood. The study of the quantum-gravity problem has confronted us with new features like the mentioned IR/UV mixing, which for example should act as a warning against assuming *a priori* that a naive low-energy limit of quantum gravity should give us back our present theories. Another example is the deformed law of composition of momenta in κ-Minkowski spacetime, which is incompatible with the standard field-theoretic setup and can only be switched off when all other κ-Minkowski modifications of Lorentz symmetry are switched off. Even at very low energies, if one keeps the κ-Minkowski dispersion relation (which could find room in the standard field-theory setup), but neglects the modification of the law of composition of momenta (which is not compatible with the standard field-theory setup), inconsistent results are obtained, including some significant *violations* of Lorentz symmetry (whereas κ-Minkowski should only involve a *deformation* of Lorentz symmetry).

So alongside the test theories that describe the new effects still in the old language, we should also develop some test theories that are prepared for the need of a new formalism. In practice the second type of test theories will need to simply avoid using too much of our present formalism. For example, as stressed here later in this Section, some phenomenology of Planck-scale departures from Lorentz symmetry focuses on pure kinematics only, in order to avoid the assumption that dynamics should be described within our present formalisms. An example of the opposite type is the SME framework, where one considers a plethora of new effects, but all still codified according our present rules, including power-counting-renormalizability.

In this Section we give a partial overview of test theories that can be used in quantum-gravity phenomenology. We describe in some detail, since it was not covered in detail in other recent quantum-gravity-phenomenology reviews [2, 3, 4, 5], a phenomenological scheme for the modification of the Maxwell and the Dirac equation. And we describe more briefly, since pedagogical descriptions are already available in Refs. [2, 3, 4, 5], some other test theories that are relevant for the search of anomalous dispersion and of quantum-spacetime fluctuations. In any case, since the present theories are well proven by all current experiments, it is reasonable to introduce tests theories through small modifications of the present theories.

A further important aspect of test theories is that, from an experimenter's point of view, detailed test theories also allow 'bookkeeping' of the possible effects that can be bounded by different experiments: All, even very different, effects are related to the same set of parameters. Only by the use of test theories one is able to 'compare' different experiments.

3.1. The generalized Maxwell equation

3.1.1. The model

A very general model for generalized Maxwell equations is based on the assumption that the homogeneous Maxwell equations $dF = 0$ are still true. This can be based on a definition of the electromagnetic field strength and the charge based on the Aharonov–Bohm effect [31]. In our approach the principles to formulate generalized inhomogenous Maxwell equations are:

- linear in the field strength,
- first order in the differentiation, and
- small deviation from the standard Maxwell equations.

Therefore, the generalized Maxwell equation which we are going to discuss and to confront with experiments is

$$\lambda^{\mu\nu\rho\sigma}\partial_\nu F_{\rho\sigma} + \chi^{\mu\rho\sigma}F_{\rho\sigma} = 4\pi j^\mu \,. \tag{7}$$

The requirement of this equations to describe a small deviation from the standard theory leads to $\lambda^{\mu\nu\rho\sigma} = \eta^{\mu[\rho}\eta^{\sigma]\nu} + \chi^{\mu\nu\rho\sigma}$ where $\eta^{\mu\nu}$ is the Minkowski metric diag(+ − − −) and, in that frame, all components of $\chi^{\mu\nu\rho\sigma}$ and $\chi^{\mu\rho\sigma}$ are small compared to unity. Therefore, all effects are calculated to first order in these quantities only.

In our approach, the constitutive tensor $\lambda^{\mu\nu\rho\sigma}$ and, thus, the tensor $\chi^{\mu\nu\rho\sigma}$ are assumed to possess the symmetry $\lambda^{\mu\nu\rho\sigma} = \lambda^{\mu\nu[\rho\sigma]}$ only (in the SME the constitutive tensor possesses the symmetry of the Riemann tensor [44]). The decomposition of this constitutive tensor and of $\chi^{\mu\rho\sigma}$ reads (see Ref. [31] for the definition of the various irreducible parts)

$$
\begin{aligned}
\chi^{\alpha\beta\mu\nu} &= {}^{(1)}W^{\alpha\beta\mu\nu} + \frac{1}{2}\varepsilon^{\mu\nu}{}_{\rho\sigma}\eta^{\rho[\alpha}\Psi^{\beta]\sigma} + \frac{1}{12}X\varepsilon^{\alpha\beta\mu\nu} - \eta^{\mu[\alpha}\Phi^{\beta]\nu} + \eta^{\nu[\alpha}\Phi^{\beta]\mu} \\
&\quad + \eta^{\mu[\alpha}W_{\rm a}^{\beta]\nu} - \eta^{\nu[\alpha}W_{\rm a}^{\beta]\mu} - \frac{1}{6}W\eta^{\alpha[\mu}\eta^{|\beta|\nu]} + {}^{(1)}Z^{\alpha\beta\mu\nu} \\
&\quad + \frac{1}{2}\varepsilon^{\mu\nu}{}_{\rho\sigma}\eta^{\rho(\alpha}\Upsilon^{\beta)\sigma} + \frac{1}{3}\left(2\eta^{\mu(\alpha}\Delta^{\beta)\nu} - 2\eta^{\nu(\alpha}\Delta^{\beta)\mu} - \eta^{\alpha\beta}\Delta^{\mu\nu}\right) \\
&\quad + \frac{1}{4}\eta^{\alpha\beta}Z^{\mu\nu} + \frac{1}{2}\left(\eta^{\mu(\alpha}\Xi^{\beta)\nu} - \eta^{\nu(\alpha}\Xi^{\beta)\mu}\right) \qquad (8) \\
\chi^{\mu\rho\sigma} &= {}^{(1)}\chi^{\mu\rho\sigma} + \varepsilon^{\mu\nu\rho\sigma}a_\nu + t^{[\rho}\eta^{\sigma]\mu}\,. \qquad (9)
\end{aligned}
$$

A 3+1 decomposition of the generalized Maxwell equations gives $(i,j=1,2,3)$

$$
\begin{aligned}
4\pi\rho &= \nabla\cdot E + \chi^{000i}\dot{E}_i + \chi^{00ij}\dot{B}_{ij} + \chi^{0i\rho\sigma}\partial_i F_{\rho\sigma} + \chi^{0\rho\sigma}F_{\rho\sigma} \qquad (10) \\
4\pi j^i &= \dot{E}_i - (\nabla\times B)^i + \chi^{i00j}\dot{E}_j + \chi^{i0jk}\dot{B}_{jk} + \chi^{ij\rho\sigma}\partial_j F_{\rho\sigma} + \chi^{i\rho\sigma}F_{\rho\sigma}\,, \qquad (11)
\end{aligned}
$$

where $E_i = F_{0i}$ and $B_i = \frac{1}{2}\varepsilon_{ijk}F_{jk}$. Using the homogeneous Maxwell equations, the time derivative of B can be replaced by a spatial derivative of the electric field. The appearance of the term χ^{000i} makes both equations to dynamical equations for the electric field rendering them to be an overdetermined system. Therefore we have to require the vanishing of the coefficient χ^{000i}. Since this should be true for any chosen frame of reference, we have to require $\chi^{(\mu\nu\rho)\sigma} = 0$, what is identical to ${}^{(1)}Z_{(\alpha\beta\mu)\nu} = 0$, $\Xi_{\mu\nu} = 0$, and $\Delta_{\mu\nu} = -\frac{3}{4}Z_{\mu\nu}$.

Due to the vanishing of these irreducible parts we get $\partial_\mu(\lambda^{\mu\nu\rho\sigma}\partial_\nu F_{\rho\sigma}) = 0$ so that only $\chi^{\mu\rho\sigma}$ may lead to charge non–conservation:

$$
4\pi\partial_\alpha j^\alpha = \chi^{\alpha\mu\nu}\partial_\alpha F_{\mu\nu} = {}^{(1)}\chi^{\alpha\mu\nu}\partial_\alpha F_{\mu\nu} + \eta^{\alpha\mu}t^\nu\partial_\alpha F_{\mu\nu}\,. \qquad (12)
$$

The validity of the homogenous Maxwell equations allows to require the vanishing of $\lambda^{\mu[\nu\rho\sigma]}$ without any physical consequences. This leads to $X = 0$, $Z^{\mu\nu} = W_{\rm a}^{\mu\nu}$, and $\Upsilon^{\mu\nu} = \frac{1}{2}\Psi^{\mu\nu}$.

3.1.2. Radiation effects

In order to discuss the effects due to the anomalous terms in radiation phenomena, we derive the wave equation

$$
0 = \ddot{E}_i - \Delta E_i + (\nabla(\nabla\cdot E))^i + \chi^{i\mu\nu j}\partial_\mu\partial_\nu E_j + 2\chi^{i0j}\dot{E}_i + \chi^{ikj}\partial_k E_j \qquad (13)
$$

and make a plane wave ansatz $E = E^0 e^{-i(k\cdot x - \omega t)}$ which results in equations for the amplidude and its derivative:

$$
\begin{aligned}
0 &= \left((\omega^2 - k^2)\delta_{ij} + k_i k_j + 2\chi^{i\mu j\nu} k_\mu k_\nu\right) E_j^0 . && (14)\\
0 &= -2\omega \dot{E}_i^0 - 2k\cdot\nabla E_i^0 + k_j \partial_i E_j^0 + k_i \partial_j E_j^0 \\
&\quad + 2\chi^{i\mu\nu j}\left(k_\mu \partial_\nu E_j^0 + k_\nu \partial_\mu E_j^0\right) + 2\omega\chi^{i0j} E_j^0 - k_k \chi^{ikj} E_j^0 && (15)\\
0 &= \ddot{E}_i^0 - \Delta E_i^0 + \partial_i \partial_j E_j^0 + 2\chi^{i\mu\nu j} \partial_\mu \partial_\nu E_j^0 . && (16)
\end{aligned}
$$

The first equation gives the dispersion relation

$$
\omega = \left(1 + \rho(n) \pm \sqrt{\sigma^2(n) - \rho^2(n)}\right) |k| \tag{17}
$$

with $\rho(n) = -\frac{1}{2}\Phi^{\mu\nu} n_\mu n_\nu$ and $\sigma^2(n) = \frac{1}{2}\eta_{\alpha\gamma}\eta_{\beta\delta}\chi^{\alpha\mu\nu\beta}\chi^{\delta\rho\sigma\gamma} n_\mu n_\nu n_\rho n_\sigma$, where $n_\mu = k_\mu/\omega = (1, k/|k|)$. This generalizes results in Ref. [89, 44].

The $\pm$ sign in front of the square root indicates a hypothetical birefringence. If no birefringence is observed, then ${}^{(1)}W^{\mu\nu\rho\sigma} = 0$ and $\Psi^{\mu\nu} = 0$. For the interpretation of this result, see [90]. It has been shown [44] that from astrophysical observations birefringence can be excluded at the level of 10^{-32}. The remaining $\rho(n)$ leads to an anisotropic speed of light which has been excluded in laboratory experiments [91] at the 10^{-15} level. Vanishing anisotropy implies $\Phi^{\mu\nu} = 0$. Furthermore, from (15) a propagation equation for the amplitude can be derived

$$
v^\mu \partial_\mu E_i^0 = -\left(2\omega\chi^{i0j} - k_k \chi^{ikj}\right) E_j^0 , \tag{18}
$$

where v^μ is the group velocity of the light ray. This is directly related to the charge non–conservation parameter $\chi^{\mu\rho\sigma}$. Since no precession of the polarization has been inferred from astrophysical observation [92] the tensor $\chi^{\mu\rho\sigma}$ vanishes at the order 10^{-42} GeV.

From all these requirements, the remaining generalized Maxwell equations are (we let aside the factor W since this can be absorbed into a redefinition of the electric charge and current)

$$
4\pi j^\alpha = \partial_\nu F^{\mu\nu} - \frac{1}{2}\eta^{\alpha\mu} Z^{\nu\beta}\partial_\beta F_{\mu\nu} + \frac{3}{2}\eta^{\beta\mu} Z^{\nu\alpha}\partial_\beta F_{\mu\nu} + \frac{1}{2}\eta^{\alpha\beta} Z^{\mu\nu}\partial_\beta F_{\mu\nu} . \tag{19}
$$

In the SME approach the $Z^{\mu\nu}$ are absent. Therefore, in our approach it is not possible to establish Lorentz invariance by radiation experiments only. Further experiments are needed.

3.1.3. Electromagnetostatics

The 3+1 decomposition of the above equations gives

$$\frac{\rho}{\varepsilon_0} = -\nabla \cdot E - \hat{\zeta} \cdot (\nabla \times E) - c\zeta \cdot (\nabla \times B) \tag{20}$$

$$\mu_0 j = \frac{1}{c^2}\dot{E} - \frac{1}{c^2}\hat{\zeta} \times \dot{E} - (\hat{\zeta} \cdot \nabla)B + \nabla \times B - \frac{1}{c}\nabla(\zeta \cdot E) + \frac{1}{c}\zeta(\nabla \cdot E), \tag{21}$$

where we used SI units and defined $\zeta^i := \frac{3}{2}Z^{0i}$ and $\hat{\zeta}_i := \frac{3}{4}\varepsilon_{ijk}Z^{jk}$.

The generalized Maxwell equations for a point charge at the origin are given by (20,21) with $\rho = q\delta(r)$ and $j = 0$. With $E = \nabla\phi$ and $B = \nabla \times A$ and the gauge $\nabla \cdot A = 0$ we get to first order in the perturbations

$$\phi = \frac{1}{4\pi\varepsilon_0}\frac{q}{r}, \qquad A = \frac{q\zeta}{4\pi\varepsilon_0 cr}. \tag{22}$$

This gives a magnetic field

$$B = \frac{q}{4\pi\varepsilon_0 c}\frac{\zeta \times r}{r^3}. \tag{23}$$

Therefore, our model includes the feature that a point charge also creates a magnetic field.

If the source of the Maxwell equations is a magnetic moment m localized at the origin, then the Maxwell equations are (20,21) with $\rho = 0$ and $j = m \times \nabla\delta(r)$. We assume again a static situation and get to first order

$$A = \frac{\mu_0}{4\pi}\frac{m \times r}{r^3}, \qquad \phi = \frac{\mu_0 c}{4\pi}\frac{(\zeta \times m) \cdot r}{r^3}. \tag{24}$$

A magnetic moment also creates an electric field of an electrical dipole with dipole moment $d = \mu_0 c\varepsilon_0 \zeta \times m$. This feature is "dual" to the previous case.

For the new Lorentz invariance violating parameters ζ and $\hat{\zeta}$ there seem to exist no experimental results. However, we expect strong bounds on the ζ^i components from measurements with SQUIDs and from atomic spectroscopy. With SQUIDs weak magnetic fields of down to 10^{-14} T can be measured. If we assume that a measurement with SQUIDs of a magnetic field from a point charge does not lead to any magnetic field larger than the SQUID sensitivity, then, from $|\lambda\zeta/(2\pi\varepsilon_0 c\rho)| \leq 10^{-14}$ T for a line charge density $\lambda = 0.01$ C/m at a distance of 1 cm, we get the estimate $|\zeta| \leq 2.7 \cdot 10^{-17}$. We strongly encourage experimentalists to carry out such an experiment.

The ζ^i will also lead to a hyperfine splitting, additional to the usual one. We get for the interaction Hamiltonian of an electron in the magnetic field (23) of the nucleus

$$H_\zeta \;=\; \mu_{\mathrm{el}} \cdot \frac{q\zeta \times r}{4\pi\varepsilon_0 cr^3}. \tag{25}$$

If we choose the z–axis in direction of ζ, then the corresponding energy shift $\Delta E_{nlm} = \langle \psi_{nlm} \mid H_\zeta \mid \psi_{nlm} \rangle$ leads, e.g., to

$$\Delta E_{210} = -\frac{q\zeta_z \mu_x}{48\pi\varepsilon_0 ca^2}. \tag{26}$$

With $\mu_z = e\hbar/m_e$ this yields $\Delta E_{210} = \zeta_z 1.8 \cdot 10^{-2}$ eV. The state of the art of high precision measurements of energy levels is of the order $\Delta E/E \approx 10^{-15}$. Since the measured energy levels are still well described within the standard theory one gets for energies of about 10 eV at best an estimate $|\zeta_z| \leq 10^{-14}$ which, however, is not as good as a direct measurement discussed above might yield.

The parameter $\hat{\zeta}$ gives rise to small deviations from the unperturbed quantities only and, thus, cannot be measured such precisely.

3.2. The generalized Dirac equation

3.2.1. The model

The Dirac equation can be derived from the principles of (i) unique evolution, (ii) superposition principle, (iii) locality, (iv) conservation of probablility, and (iv) Lorentz invariance. The requirements (i) – (iii) leads to a linear system of partial differential equations of first order, requirement (iv) ensures the hyperbolicity of these equation. Using (i) to (iv) one gets

$$0 = i\gamma^\mu \partial_\mu \psi - M\psi \tag{27}$$

where the matrices γ^μ in general do not fulfill a Clifford algebra, that is, there is no underlying Riemannian metric. As a consequence, the characteristic surface of this equation is given, as in the case of the Maxwell equation, by a forth order equation $0 = \det(\gamma^\mu k_\mu)$ and shows birefringence and anisotropy. The same phenomenon we encounter with the mass shell equation $0 = \det(\gamma^a p_a + M)$.

Also models with higher order derivatives as they appear in, e.g., in the effective equations for spin–$\frac{1}{2}$ particles within loop quantum gravity, have been considered [93].

3.2.2. Propagation of Dirac particles

There are two quasi–classical phenomena which one can derive from this generalized Dirac equation [30]: the equation of motion for the position in a nonrelativistic WKB aproximation

$$\ddot{a}^i = -\left(\delta^{ij} + \alpha^{ij} + \bar{\alpha}^{ij}_k S^k\right) \partial_j U + \beta_{kl} \delta^{ij} \partial_j U^{kl} \tag{28}$$

where U is the Newtonian potential, U^{ij} the Newtonian gravitational potential tensor [94], S^i the spin and α^{ij}, $\bar{\alpha}^{ij}_k$, and β_{ij} are anomalous terms connected with the $X^{\mu\nu}$ and an anomalous coupling to U^{ij}, see [30]. The corresponding equation for the precession of the spin is

$$\dot{S}^i = \varepsilon^{ijk}\Omega_j S_k \qquad \text{with} \qquad \Omega_i = \tau_i^{kl} p_k p_l + \tau_i^j p_j + \tau_i , \tag{29}$$

where again the anomalous parameters τ_k^{ij}, τ_j^i, and τ_i are connected with the $X^{\mu\nu}$ and the anomalous coupling to U^{ij}. In both cases the dynamics is influenced by anomalous parameters. The acceleration can be probed, e.g., by atom interferometry, the spin precession by the corresponding spin precession experiments performed, e.g., in high energy experiments.

3.2.3. Spectroscopy

For quantum particles also spectroscopy can give valuable information about the underlying dynamics of the fields. The anomalous parameters give rise to additional energy shifts and splitting of spectral lines. As an example, we mention the energy shifts occuring the Hughes–Drever like experiments (see Sec.4.3.6) where a valence particle (a valence electron or a valence proton in the atomic's nucleus) leads to a splitting of the Zeeman lines [30] which, among others, yields the best estimates on an hypothetical anomalous inertial mass of the proton and on a hypothetical space-time torsion [95].

Further effects like a modification of the spreading of a wave packet have been discussed in [96].

3.3. Generalized gravitational field

In physics, geometry cannot be separated from the motion of physical objects. That something can be "geometrized" means that the dynamics of a certain object is independent from properties characterizing the object. In GR, the geometrization comes from the independence of the path of structureless particles from the mass and decomposition, the only characteristics of a structureless particle. The path is determined by its initial position and velocity only and, thus, given by an equation of the form $\ddot{x}^\mu + H^\mu(x,\dot{x}) = \alpha\dot{x}^\mu$. Then the motion is not related to the particular particles and, thus, can be assigned to an underlying geometry. The geometry underlying this geometrization of the path dynamics is called "path structure". This very general structure becomes more familiar when we assume that at each space-time point there is a frame so that the equation of motion for all particles reduces to $\ddot{x}^\mu = 0$ (Einstein's elevator). Then the equation of motion reads $\ddot{x}^\mu + H^\mu_{\rho\sigma}(x)\dot{x}^\rho\dot{x}^\sigma = \alpha\dot{x}^\mu$. The compatibility of this structure with the light cone structure then leads to a Weylian space–time which can be reduced to the ordinary Riemannian space-time by imposing the additional requiremet of the non–occurence of the second clock effect [97, 98].

The next step in establishing a theory for gravity then is to determine the equations which give the metric in terms of the matter content in the universe. In a particular case we get Einstein's field equations. A general parametrizations for a wide range of possible field equations for the metric is given by the PPN formalism, see e.g. [94]. With the most important PPN parameters β and γ, the metric is given by

$$
\begin{aligned}
g_{00} &= 1-2U+2\beta U^2 && (30)\\
g_{0i} &= V_i && (31)\\
g_{ij} &= -(1+2\gamma U)\,\delta_{ij} && (32)
\end{aligned}
$$

where U is thr Newtonian potential and V_i is a gravitational field connected with a matter current. In this representation the metric is very well adapted for a confrontion with experimental data, e.g., from light deflection, redshift, gravitational time delay, perihelion shift, and data from binary systems. Einstein's field equations are characterized by $\beta=\gamma=1$. Recently, it has been proposed in a string theory inspired dilaton model that β and γ may differ from unity by 10^{-9} and 10^{-5}, respectively. The latter one is very close to the recent measurement by the Cassini mission [99] which gave $|\gamma-1|\leq 2\cdot 10^{-5}$.

3.4. Anomalous dispersion

As mentioned various quantum–gravity scenarios lead us to considering Planck-scale modified dispersion relations of the general type (6). The derivation of the modified dispersion relation can take a rather different path in different approaches. For example in κ-Minkowski spacetime, where one has the support of the deformed (Hopf) algebra, the dispersion relation is obtained rather simply, through a Casimir of the spacetime-symmetry algebra, just like the classical dispersion relation is obtained through the mass Casimir of the Poincaré (Lie) algebra. But in some scenarios the argument may be rather involved, as shown by the case of Loop Quantum Gravity, in which we are still unable to perform a full rigorous derivation of the dispersion relation. This would require a sort of "Minkowski vacuum" for Loop Quantum Gravity, which is still unknown. One therefore relies on some states which are not the vacuum, but may reproduce some of the characteristics of the vacuum. For example, some authors [16, 17] have performed the analysis using the so-called Loop-Quantum-Gravity "weave states".

A key point for the development of test theories based on this modified dispersion relations concerns the possibility of assuming that a field-theoretic formulation be admissible. Some authors [29, 100] have relied on the field-theoretic setup, in spite of the fact that it is inevitably nonrenormalizable (within the field-theoretic setup Eq. (6) corresponds to nonrenormalizable dimension-5 operators). Other authors, perhaps the majority, have been concerned not only with the nonrenormalizability of the corresponding field theory, but also with the possibility that the relevant quantum-gravity scenarios might have to be based rather significantly on decoherence, something that field theory by construction cannot accommodate. These other authors have preferred avoiding to commit to a formulation of dynamics and are basing their phenomenology exclusively on kinematics.

The pure-kinematics test theories only rely on the form of the dispersion relation and the form of the law of energy-momentum conservation, which is usually assumed to be unmodified.

A formal description of a field-theory-based test theory hosting the modified dispersion relations (6) is discussed in Refs. [29, 101]. For example, for a pure abelian gauge theory one introduces the extra term

$$\frac{1}{E_P} n^a F_{ad} n \cdot \nabla (n_b \tilde{F}^{bd}) , \tag{33}$$

where n_a is a background four-vector that codified the breakup of Lorentz symmetry.

3.5. Space–time fuzziness and decoherence

The quantum-gravity literature on spacetime fuzziness (or "spacetime foam") originates from a rich collection of arguments [83, 72, 84, 85, 86, 87, 88], combining general relativity and quantum mechanics, which suggest that in the Planck-scale regime there should be some absolute limitations on the measurability of distances. It turns out to be most convenient [5] to characterize operatively this spacetime fuzziness as an irreducible (fundamental) Planck-scale contribution to the noise levels in the readout of interferometers. Interferometer noise can in principle be reduced to zero in classical physics. Ordinary quantum properties of matter already introduce an irreducible noise contribution. Spacetime fuzziness would introduce another irreducible source of noise, reflecting the fact that the distances involved in the experiment would be inherently unsharp in a foamy spacetime picture.

While, as mentioned in the preceding sections, nearly all approaches to the quantum-gravity problem predict some limitations on the accuracy of localization, and therefore predict some spacetime fuzziness, one is usually unable to rigorously derive from first principles a detailed description of the physical consequences of this fuzziness, such as the mentioned interferometric noise. A phenomenology is being developed nonetheless, exploiting [5, 102, 103, 104] the fact that the role of the Planck scale in the needed formulas has only a limited number of options, as a result of the constraints introduced by dimensional analysis.

The key input needed for this phenomenology turns out to be the power spectrum $\rho(f)$ of the Planck-scale-induced strain noise [102]. Combining some intuition about the stochastic-like features of spacetime fuzziness and the dependence on the Planck length one can rather easily reach a model of $\rho(f)$. For example if the effects depend linearly on the Planck length L_p ($\equiv 1/E_p$) and the underlying phenomena are of random-walk type one is inevitably led to

$$\rho_h \sim L_p f^{-2} \Lambda^{-2} \equiv \zeta L_p f^{-2} L^{-2} , \tag{34}$$

The proportionality to the square of the inverse of the frequency is a direct result of the assumption of random-walk-type processes. The length scale Λ is needed on the basis of the dimensional analysis of the equation, and is to be treated as a free parameter to be constrained experimentally. One may choose to make reference to a dimensionless parameter, ζ, which may be used to express the ratio of Λ with the length L of the arms of an interferometer or of an optical resonator.

Other hypothesis [5, 102, 103, 104] about the stochastic-like features of the underlying Planck-scale processes lead to other forms of f dependence (and L_p dependence) of the strain noise power spectrum. In general one should find ($\alpha > -1$)

$$\rho_h \sim \zeta_{\alpha\beta} L_p^{1+\alpha} f^{-2-\beta} L^{-2-\alpha-\beta} . \tag{35}$$

There is therefore some interest in attempting [5, 102, 103, 104, 12] to improve limits on the parameters $\zeta_{\alpha\beta}$.

The idea of spacetime fuzziness also motivates some research work on Planck-scale-induced decoherence. It is in fact rather plausible that spacetime fuzziness might affect the time evolution of quantum-mechanical states in such a way that, for example, a pure state might evolve into a mixed state. A Planck-scale decoherence scenario which has been extensively studied, for what concerns the phenomenological aspects, is the one of Ref. [57] (and references therein) which is motivated by the "Liouville Strings" approach [66, 67] and describes the decoherence effects within a density-matrix formalism.

4. CONTROLLED LABORATORY EXPERIMENTS

The realization that, as a result of the large astrophysical propagation distances and (in some cases) the large energies of the particles involved, in a few astrophysics contexts the analysis of quantum-gravity effects can achieve Planck-scale sensitivity has generated a large interest over the last few years, as documented in Refs. [2, 3, 4, 5] and references therein. However, the realm of astrophysics is also affected by some key limitations, which encourage us to complement this phenomenology with controlled/laboratory experiments. In astrophysics we are just limited to the role of "observers" rather than having the opportunity to devise, set up and control a measurement procedure. In interpreting astrophysical observations one is faced with the problem that the source as well as the space in between (that might contain gasses, other electromagnetic fields etc.) are not under control of the experimenter, and this will, in the majority of cases, lead only to model-dependent results. Moreover, as mentioned above, the study of the quantum-gravity problem can motivate the search of a variety of new effects, which we would like to study one by one, whereas in astrophysics we must take what Nature offers, and this is not going to cover all aspects of the phenomenology of interest.

In this Section we discuss a few significant plans for a quantum-gravity-phenomenology with controlled/laboratory experiments. The description of examples of astrophysics studies is postponed to the next section, and will be structured more succinctly, since several recent reviews (*e.g.*, Refs. [2, 3, 4, 5]) have already covered in detail the approach based on astrophysics.

4.1. Universality of Free Fall

One of the first physical laws stated and subject to tests is the UFF. It began with the free fall tests of Galileo and its tests using a tilted plane. The best tests today use torsion balances and confirm the UFF at the level of $5 \cdot 10^{-13}$ [105]. The best free fall tests reach the level of 10^{-10} [106, 107, 108]. Future tests to be performed in space on dedicated satellite missions, namely MICROSCOPE and STEP, should be able to tests the UFF to a level of 10^{-15} and 10^{-18} respectively [109, 110]. Therefore, MICROSCOPE as well as STEP are able to cover predictions from some quantum gravity motivated predictions.

4.2. Universality of the Gravitational Redshift

In these tests the ticking rates of two clocks is compared while the clocks change their position in a gravitational field. As in tests of the UFF, all pairs of clocks have to be considered. Presently, the best tests has been reported in [51] where two atomic clocks, an atomic fountain clock and a hydrogen maser, are subject to the changing gravitational field of the Sun. The two clocks show the same redshift to an accuracy of $1.5 \cdot 10^{-5}$. Other tests compared atomic clocks and clocks based on microwave and optical resonators and gave in both cases the same redshift within $2 \cdot 10^{-2}$ [111, 112].

4.3. Lorentz invariance

4.3.1. Accurate, clean, and comprehensive tests

Test of the principle of (local) Lorentz invariance (LI) are an extremely rich field for the search of fundamental physics. One reason is the universal validity of LI as a framework for all theories of nature. A violation of LI would thus manifest itself in virtually all branches of physics, and this allows to exploit the highest precision measurements as tests of LI. In test theories, the possible violations of LI are encoded into a set of parameters. The effects of these parameters in various experiments can be calculated (at least in principle), so that results on the parameters can thus be obtained from experiments. These theories have been advanced to a level where the outcome of all these experiments can be compared. This also answers the question whether two experiments test different aspects of relativity or whether they are different methods for observing the same effect.

Ideally, the experimental verification of LI should be *accurate, clean,* and *comprehensive.* Accurate refers to the ability to detect very small violations and clean refers to how well the hypothetical effect sought in experiments is understood and referred to a set of parameters in the test theory. This is a requirement to both the experiment and the underlying theory. It ensures that a violation of relativity does not accidentally cancel. Comprehensiveness denotes the degree of completeness to which the theoretically conceivable violations of LI are excluded.

This summary wants to present the most important experimental techniques that have been used for bounding violations of LI. The variety of methods makes this an interesting and challenging subject for experimental physicists. As a general observation, tight limits on spin-dependent parameters can be obtained by comparing similar particles having different polarizations. The spin-independent effects have to be measured by comparing to independent standards, e.g., a macroscopic body as a length standard. Another common feature of most experiments is that they seek for a Lorentz-violating dependence of an effect as a function of the orientation of the apparatus or the velocity of the laboratory frame in space. For a laboratory located on Earth, the relevant rotations and velocities can be given by Earth's orbit as well as rotation.

4.3.2. Applications and every day tests

LI has found many applications in fundamental and applied research, in technology and, thus, even in every day life. One striking example on how every day these may be are the accelerated electrons in cathode ray tubes used in color television sets: Their mass is increased by about 5% relative to their rest mass. For a less mundane example, the accurate quantum field theoretical prediction of the electron's anomalous magnetic moment $g = 2.002\ldots$ to twelve significant digits [113] (maybe the most accurate prediction of a 'complicated' number in any science) also is a confirmation of LI.

In the global positioning system, a receiver determines its position by comparing timing signals received from accurate reference clocks located in satellites. The positions of the satellites are known, and the difference in the timing signals gives the relative distances that allows the receiver to determine its spatial coordinates and the local time. However, since the reference clocks are moving with respect to the receiver, their rates are subject to time dilation (and also to the redshift due to Earth's gravity). Not taking into account these effects would lead to a position error of many kilometers [114].

There are also technologies whose very operating principle is a relativistic effect. Certain microwave oscillators ('gyrotrons') [115] and the free electron laser [116] use the relativistic Doppler effect to generate short-wavelength radiation. In a free electron laser, a beam of electrons is directed along an arrangement of periodically poled magnets, where the period p is of the order of a centimeter. The moving electrons have a kinetic energy $E \gg m_e c^2$, and thus see a Lorentz-contracted period $pm_e c^2/E$. The corresponding oscillating magnetic field in the electron's rest frame thus has a frequency $\nu_e = cE/(pm_e c^2)$ and forces the electrons to an oscillatory motion that leads to the emission of electromagnetic waves with a frequency ν_e. The Lorentz transformations back to the laboratory frame concentrate these waves into a tight cone in the forward direction of the electron beam, with a blue-shifted frequency $\nu_{\rm lab} = \nu_e E/(m_e c^2) = cE^2/(pm_e^2 c^4)$. Thus, highly concentrated radiation and a small wavelength $\lambda = p(m_e c^2/E)^2$ are achieved. The two Lorentz transformations give the quadratic dependence on $1/E$.

Particle accelerators are maybe the most prominent example of a technology where relativistic effects are built in. The velocity-dependence of the relativistic mass, for example, is taken into account in the construction of the machines, and, most importantly, the understanding of the high-energy particle reactions provides maybe the richest field for the application of relativistic (quantum field) theories. These have not only given rise to an enormous output of fundamental physics discoveries, but the apparent correctness of relativistic quantum field theory is also a confirmation of relativity itself.

While the success of relativistic physics in fundamental and applied research gives us confidence into the theory, the 'every-day tests' provide no high-precision confirmation: On the one hand, machines like particle accelerators, microwave oscillators etc. are constructed in such a way that small tolerances do not lead to malfunction, and this also means that small errors in the underlying theory may not affect functionality. Also, the 'every day tests' are usually not clean: In accelerator experiments, for example, an incomplete understanding of the hadronic processes might cover a tiny violation of relativity. For these reasons (and for achieving comprehensiveness), LI has been tested in dedicated experiments since its inception.

4.3.3. Macroscopic matter effects

Lorentz violation affects the properties of macroscopic bodies through a modification of its microscopic constituents. These modifications, in particular the change of the geometry, have been studied for the interpretation of experiments. As an important result, within the SME matter effects do not cancel the sensitivity in interferometer or cavity tests of Lorentz invariance. For a model which may lead to a cancellation see [117]. Instead, they enhance the sensitivity for Lorentz violation in electrodynamics, but only slightly for cavity materials presently in use. Moreover, the theory concerning the Dirac sector [118] allowed to derive the first experimental limits on some of the electron coefficients $c_{\mu\nu}$, at a level of 10^{-14}. The theories summarized here constitute a complete description of all SME effects that influence experiment using vacuum-filled cavities. These experiments are thus particularly clean tests of Lorentz invariance.

Photon sector. Within the electrodynamic sector of the SME, Lorentz violation leads to a modified velocity of light $c = c_0 + \delta c$ and also to a modified Coulomb potential of a point charge e, $\Phi(\vec{x}) = e^2/(4\pi|\vec{x}|) + V$, where [44]

$$V = \frac{e^2}{8\pi}\frac{x\cdot\kappa_{DE}\cdot x}{|x|^3}. \tag{36}$$

The influence of this on solids can be treated for ionic crystals [119], which in the simplest case (e.g., NaCl) consists of a lattice of ions with opposite charges. The lattice is formed by the balance between attractive Coulomb forces and a quantum mechanical repulsion due to the overlap of the ionic orbitals. Perturbative calculations show that the change in the repulsive potential due to Lorentz violation is negligible. For estimating the influence of the modification of the Coulomb potential V, the force generated by the modified Coulomb potential is summed over all ions and equated to the elastic force associated with a distortion of the lattice. This leads to the relative change of the length

$$\frac{\delta L_z}{L} = A\left[(2\sigma - 3\tau_{\|})(\kappa_{DE})_{\|} - 3\tau_{\perp}(\kappa_{DE})_{\perp}\right] \tag{37}$$

with σ and τ being constants obtained from summing the Coulomb potential in analogy to the Madelung constants. $(\kappa_{DE})_{\|} = (\kappa_{DE})_{zz}$, $(\kappa_{DE})_{\perp} = (\kappa_{DE})_{xx} + (\kappa_{DE})_{yy}$ and

$$A = -\frac{1}{2}\frac{e^2\nu_1\nu_2}{8\pi E_Y}\frac{N_m N_A \rho}{Ma}. \tag{38}$$

TABLE 2. Length change coefficients.

Material	$a_{\parallel}$	$a_{\perp}$	Material	$a_{\parallel}$	$a_{\perp}$
NaCl	-0.28	0.10	LiF	-1.06	0.37
sapphire	-0.03	0.01	quartz	-0.11	0.04

ν_1 and ν_2 are the number of valence charges for the atoms, and E_Y is the elastic modulus. The contributions of the length change and the change in the speed of light give the total frequency change of a cavity filled with vacuum due to Lorentz violation in electrodynamics

$$\frac{\delta \nu_{\mathrm{cav}}}{\nu_{\mathrm{cav}}} = -a_{\parallel} \hat{N} \kappa_{DE} \hat{N} - \left(\frac{1}{2} + a_{\perp}\right) \left[\hat{E}^* \kappa_{DE} \hat{E} + (\hat{N} \times \hat{E}^*) \kappa_{DE} (\hat{N} \times \hat{E})\right] . \qquad (39)$$

Here, $a_{\parallel} = A(2\sigma - 3\tau_{\parallel})$ and $a_{\perp} = -3A\tau_{\perp}$. This has been simplified by noting that astrophysical tests lead to $(\kappa_{HB}) = -(\kappa_{DE})$. For practical materials sapphire and quartz the length change effect is negligible (see Table 2). For future experiments using resonators made of other materials, however, the influence might be stronger and *enhances* the sensitivity.

The model can be extended to include the additional non-Lagrangian coefficients of the most general Maxwell equations that are linear and first order in the derivatives [31, 120]. It is found that the non-SME terms do not additionally modify the geometry of crystals.

Fermionic sector. Here, the starting point is the nonrelativistic hamiltonian $h = h_0 + \delta h$ of a free electron in the SME, as described in [42]. Most of the terms contained therein (Tab. 6) have zero expectation value within the rest frame of a solid. The only term that doesn't drop out for non-spin-polarized materials (spin-polarized materials can also be treated[118]) is the modification of the kinetic energy of the electron, $\delta h = E'_{jk} p_j p_k$. Since the electrons inside crystals have a nonzero expectation value $\langle p_i p_j \rangle$, which is a function of the geometry of the lattice, Lorentz violation will cause a geometry change ('strain') of the crystal.

Strain is conventionally expressed by the strain tensor e_{ij}. For $i = j$, it represents the relative change of length in x_i-direction, and for $i \neq j$, it represents the change of the right angle between lines originally pointing in x_i and x_j direction. A general linear relationship between the Lorentz violating quantity E'_{jk} and strain is given by

$$e_{dc} = \mathscr{B}_{dcpj} E'_{pj}. \qquad (40)$$

with a 'sensitivity tensor' $\mathscr{B}_{dcpj}$ that has to be determined from a model of the crystal. $\mathscr{B}_{dcpj}$ can be taken as symmetric in the first and last index pair; symmetry under exchange of these pairs will hold only for some simple crystal geometries, like cubic. Thus, the tensor has at most 36 independent elements. To calculate the sensitivity tensor, the electronic states are described by Bloch wave functions to determine $\langle p_i p_j \rangle$; the corresponding strain is calculated using elasticity theory. As a result, the sensitivity

TABLE 3. Sensitivity coefficients.

Mat.	$\mathscr{B}_{11}$	$\mathscr{B}_{12}$	$\mathscr{B}_{13}$	$\mathscr{B}_{14}$	$\mathscr{B}_{31}$	$\mathscr{B}_{33}$	$\mathscr{B}_{41}$	$\mathscr{B}_{44}$
Au	24.13	-11.06						12.34
Al_2O_3	3.58	-1.05	-0.53	0.014	-0.57	3.14	0.004	5.08
Nb	6.80	-2.40						17.9
fused quartz	2.64	-0.32						3.95

tensor $\mathscr{B}_{dcjp} = \mu_{dcmp}\kappa_{mj} + \mu_{dcmj}\kappa_{mp}$, where

$$\kappa_{mj} = \frac{N_{e,u}\hbar^2}{m|\det(l_{ij})|} k_{ml} k_{jk} \overline{\xi}_{lk}, \tag{41}$$

can be calculated. $N_{e,u}$ is the number of valence electrons per unit cell, $|\det(l_{ij})|$ is the volume of the unit cell expressed by the determinant of the matrix of the primitive direct lattice vectors, k_{ml} is the matrix containing the primitive reciprocal lattice vectors, and μ_{dcmp} the elastic compliance constants. The symmetric 3×3 matrix $\overline{\xi}_{lk}$ is given by the Fourier coefficients of the Bloch wave functions. Its six parameters are unknown at this stage and can, e.g., be determined from a simple model that leads to $\overline{\xi}_{lk} \sim \delta_{lk}$. To eliminate these unknowns, an alternative method is used to calculate the strain for the simple case of isotropic Lorentz violation $E'_{jk} \sim \delta_{jk}$, and the result is compared to Eq. (40) [118]. This yields six equations from which κ_{ab} (that depends solely on material properties) can be determined and re-inserted into Eq. (40):

$$\mathscr{B}_{dcjp} = \mu_{dcmp}\lambda_{aamj} + \mu_{dcmj}\lambda_{aamp}. \tag{42}$$

Note that the theory now needs no assumptions that go beyond the use of Bloch states. Thus, it is very generally applicable and accurate.

For convenience, we arrange the independent elements of e_{ab} into a six-vectors $\mathbf{e} = (e_{xx}, e_{yy}, e_{zz}, e_{yz}, e_{zx}, e_{xy})$ and express Eq. (40) as $\mathbf{e} = \mathscr{B}\cdot\mathbf{E}'$, where $\mathscr{B}$ is a 6×6 'sensitivity matrix'. Tab. 3 gives values for the materials presently used for cavities. Gold is included as an example for a material with exceptional sensitivity.

The influence of the electron terms $c_{\mu\nu}$ on hydrogen molecules have also been calculated [121]. Here, an explicit wave-function can be obtained from first principles, and Lorentz-violating changes in the frequencies of electronic and (ro-) vibrational transitions, as well as the bond length, have been obtained. This allows new tests of Lorentz symmetry that use molecules. A change in the index of refracion of dielectric media is also connected to violation of LI [122]. It has to be taken into account for certain cavity experiments.

4.3.4. Cavity experiments

Polarization experiments, see Sec.5, comparing light with light, cannot measure all 19 coefficients of the ordinary constitutive tensor $(k_F)_{\kappa\lambda\mu\nu}$, whereas cavity and interferometer experiments, comparing light with matter, can. Moreover, cavity experiments can measure coefficients for Lorentz violation in the electron's equation of motion that have not been measured by other methods. Cavity experiments have been developed out of the well known interferometer experiments originally performed by Michelson, Morley [123, 124] as well as Kennedy, Thorndike [125, 126], and others.

The basic principle is to measure the time of flight of light rays that transverse a path, which is usually defined by a pair of mirrors and a spacer. A change connected to the orientation or the velocity of the apparatus in space would indicate Lorentz violation. Interferometers indicate such a change by a shift in fringes formed by the interference of two beams. In cavities, however, interferometry between multiple beams is used. Each transverses the cavity a large number of times that is limited by losses such as imperfect mirrors. Today, about 10^5 reflections are used, and $\sim 10^6$ seem technically possible. Thus, a large effective distance of propagation is reached in a compact apparatus, that can be much better shielded from temperature fluctuations and (seismic) vibrations, and is less sensitive to gravitational bending[3].

In cavity experiments, one measures the resonance frequencies

$$\omega = 2\pi \frac{mc}{2nL} \tag{43}$$

(m is a constant mode number, c is velocity of light parallel to the cavity axis, n the index of refraction of the medium, and L the cavity length) of a cavity, defined by the boundary conditions for standing waves. A Lorentz-violating shift in c and/or L and n connected to a rotation or boost of the apparatus would lead to a measurable shift in ω. The shift in c is, of course, due to Lorentz violation in the photonic sector. The shift in L has been determined above.

The first experiment to make full use of the potential that reference cavities offer for precision experiments was performed by Brillet and Hall [127] (see Tab. 4). A laser was stabilized to a resonance of a fused quartz cavity (actually, ultra-low expansion (ULE) glass ceramics). Both the cavity as well as the laser were rotating on a platform. Their frequency was compared to a stationary methane (CH_4) frequency standard. The data showed frequency variations of a few parts in 10^{13} that where ascribed to a slight tilt of the rotation axis with respect to Earth's gravity. After a transformation of the signal into the sidereal frame, however, this effects becomes oscillatory and can thus be separated from any signals of cosmic origin. The experiment was performed before the developement of an elaborate theoretical framework. A detailed re-analysis of the published result within the SME is, unfortunately, not possible, since only a single signal component is given. For deriving detailed SME results, it would be desirable to have at least 10 independent components.

In a smilar setup, a non-rotating quartz cavity (ULE) was compared against a CH_4 standard by Hils and Hall [128]. The signal period was given by Earth's rotation. On that time-scale, however, the ULE cavity showed a significant drift that limited the accuracy of that experiment. The experiment was interpreted in the RMS test theory (Sec.1.6)

[3] In the large interferometers dedicated to the search for gravitational waves, shielding against seismic vibrations with frequencies above a couple of Hz is effectuated by pendulum-like mounting of the interferometer end mirrors. Thus, however, the dc position of the mirrors is much less stable than in the rigid cavities described here.

TABLE 4. Overview of recent cavity experiments (WGR = whispering-gallery resonator). The notation $1-2(4)$ in the column for κ_{e-} indicates that the experiment provides 4 independent limits on parameters of $\tilde{\kappa}_{e-}$, the lowest individual being about 1, the highest about 2 parts in 10^{15} (absolute mean plus standard error). If no number in brackets is given, all matrix elements have been limited. The notation ~ 5 indicates that, although a detailed analysis within the SME was not performed, the experiment should limit at least one of the parameters at the indicated level.

Ref.	type	$\tilde{\kappa}_{e-}$ $\times 10^{15}$	$\tilde{\kappa}_{o+}$ $\times 10^{11}$	c_{AB} $\times 10^{15}$
[127] (1979)	rotat. quartz cavity/CH_4	~ 6	~ 10	
[128] (1980)	quartz cavity/CH_4			
[112] (2000)	sapphire cavity/iodine			
[129] (2003)	Nb microwave cavities	140-430 (4)	200 (1)	
[91] (2003)	sapphire cavities	4-18 (4)	3-28	
[118] (2003)	reanalysis of [127, 91]	4-18 (4)	3-28	2-100 (3)
[130] (2004)	sapphire WGR/Cs clock	2-8 (4)	3-5	

TABLE 5. Limits from recent cavity experiments.

Parameter	[91, 118]	[91, 130, 122]
$(\tilde{\kappa}_{e-})^{XY}/10^{-15}$	1.7 ± 2.6	-1.7 ± 1.6
$(\tilde{\kappa}_{e-})^{XZ}/10^{-15}$	-6.3 ± 12.4	-4.0 ± 3.3
$(\tilde{\kappa}_{e-})^{YZ}/10^{-15}$	3.6 ± 9.0	0.5 ± 2.52
$((\tilde{\kappa}_{e-})^{XX}-(\tilde{\kappa}_{e-})^{YY})/10^{-15}$	8.9 ± 4.9	2.8 ± 3.3
$(\tilde{\kappa}_{o+})^{XY}/10^{-11}$	14 ± 14	-1.8 ± 1.5
$(\tilde{\kappa}_{o+})^{XZ}/10^{-11}$	-1.2 ± 2.6	-1.4 ± 2.3
$(\tilde{\kappa}_{o+})^{YZ}/10^{-11}$	0.1 ± 2.7	2.7 ± 2.2
$c^e_{(YZ)}/10^{-15}$		0.21 ± 0.46
$c^e_{(XZ)}/10^{-15}$		-0.16 ± 0.63
$c^e_{(XY)}/10^{-15}$	$\lesssim 1$	0.76 ± 0.35
$(c^e_{XX}-c^e_{YY})/10^{-15}$	$\lesssim 8$	1.2 ± 0.64
$\lvert c^e_{XX}+c^e_{YY}-2c^e_{ZZ}-0.25(\tilde{\kappa}_{e-})^{ZZ}\rvert/10^{-12}$	$\lesssim 1$	$\lesssim 1$

and provided the best limits of 6.6×10^{-5} on the velocity-dependence parameter at the time. As above, a detailed analysis within the experiment from the published data is not feasible. This experiment was improved by a factor of about three by Braxmaier *et al.* [112], who used a cavity made from crystalline sapphire, operated at the temperature of liquid helium (4 K). The crystalline material showed a remarkable absence of creep, i.e., no systematic long term drifts of the cavity length. Thus, it was possible to search for a signal with 1 year period, given by Earth's orbit.

The first experiment to determine simultaneous limits on the photonic SME coefficients was published in 2003 [129]. Two superconducting microwave cavities made of Niobium (Nb) were used to give separate limits on four components of $\tilde{\kappa}_{e-}$ and one limit on $\tilde{\kappa}_{o+}$ (Tab. 4). Although the accuracy of this experiment was lower than that of older experiments, it determined simultaneous limits on several parameters, thus representing progress towards a comprehensive verification of LI.

The experiment of Müller *et al.* [91, 131], improved both the accuracy as well as the coverage of parameter space. It determined four elements of κ_{e-} to a level of about 10^{-15}. The three elements of $\tilde{\kappa}_{o+}$ enter the signal suppressed by the velocity of Earth's orbit, $\beta \sim 10^{-4}$, and where thus limited to about 10^{-11} (Tab. 5). That one of these seven parameters was different from zero at 1.8σ does *not* mean that Lorentz violation was detected. Rather, the probability of a mean value to be larger than the standard error is 23% even if the true value is zero, so one or two such instances are to be expected if the error bars are realistic. The experiment compared two cryogenic sapphire resonators pointing in orthogonal directions over a period of 399 days. This made it possible to independently measure Fourier signal components separated by as little as 1/1 year in frequency space, and thus allowed the experiment to give separate limits on the elements of $\tilde{\kappa}_{o+}$.

The experiment of Wolf *et al.* used a slightly different setup, in which resonances of a cryogenically cooled microwave whispering gallery resonator (WGR) are used. The WGR is a cylinder made from crystalline saphirre, in which the microwave mode travels along the perimeter, guided by total internal reflection. Also this experiment was operated for a sufficiently long time to state separate limits on the elements of $\tilde{\kappa}_{o+}$. In this experiment, the results for the three parameters are significant at about 2σ. As explained above, however, this is to be expected from standard statistics even if all parameters are zero, and thus not an indication of Lorentz violation.

The theory of the influence of fermionic LI violation in the cavity geometry [118] made it possible to state the first limits on three electron parameters contained in $c_{\mu\nu}$. Since this influence is material dependent, it can be separately measured by perfoming experiments with different cavity materials, here the experiments of Brillet and Hall [127] and Müller *et al.* (Tab. 5). Theoretically, all components of $c_{\mu\nu}$ can be limited by this method, if sufficiently detailed experimental results are available. As explained above, this is unfortunately not the case for the older experiments.

For a complete determination of the photon and electron coefficients, the experiment of Müller *et al.* has to be compared to an experiment that gives as many signal components, but uses dissimilar cavities. The experiment by Wolf *et al.* [130] could be used for this purpose. However, in the whispering-gallery type cavity, 98% of the electromagnetic field energy are confined within a refractive material. Thus, a shift in the index of refraction also connected to Lorentz violation alters the Lorentz-violation signal. This has been taken into account in [121], leading to more comprehensive and more accurate limits (Tab. 5). The additional hypothetical shift in the index of refraction of sapphire contributes to the higher accuracy.

4.3.5. Doppler shift experiments

To measure the relativistic Doppler shift, the frequency of a moving oscillator is measured as a function of the velocity with respect to the detector. Such measurements can be easily described within the RMS framework, where a parameter α describes the strength of the second order Doppler shift. $\alpha = -1/2$ is predicted by SR. In the SME and other dynamical test theories, the analysis is more complicated, since it must contain the nature of the moving clock. The most precise test yet gives $|\alpha_{MS} + 1/2| \lesssim 1.8 \times 10^{-7}$

[132] by spectroscopy on lithium ions moving within a storage ring. The accuracy of this result is limited by the knowledge of the rest-frame transition frequency of lithium.

Other methods to measure the Doppler shift include γ-ray spectroscopy using a rotating Mössbauer absorber [133]. While the resolution of the frequency measurement is much higher, the velocity is limited to maybe 100 m/s due to centrifugal forces in the rotor, making these experiment less accurate.

In the SME and other dynamical test theories, the interpretation of such experiments is different: Here, the physics of the clock and the detector is described by dynamical equations, from which the received frequency can be calculated. A departure from the relativistic Doppler shift would arise from the extra terms in the dynamical equations. Thus, the analyis must contain the nature of the moving clock. Although this was already discussed [134], at the time of this writing no such analysis has been performed yet.

4.3.6. Clock-comparison experiments

The basic idea is to compare the oscillation frequencies of two physical systems ("clocks") and look for any change with the orientation or velocity of the clocks in space. The highest precision limits are achieved by comparing the frequencies of dissimilar clocks as they rotate with the Earth.

If the influence of the Lorentz-violating parameters on the frequencies of the clock can be calculated (by use of perturbation theory), definite bounds can be obtained by a comparison to the outcome of the experiment. However, the parameters of the SME are so manifold that a calculation of the individual influences is generally difficult, especially if nuclear energy levels are involved. In these cases, the problem has been simplified by use of simplified models for nuclear matter and assuming that all but a few parameters for Lorentz violation are zero [42]. This gives insight into how many of the parameters can be bounded at which level of accuracy. Further assuming no cancellations between the parameters, limits on many parameters for Lorentz violation that lead to spin-dependent effects for the electron, the proton, and the neutron could be obtained (Tab. 6).

In the tests originally performed by Hughes [38] and Drever [39] at the beginning of the sixties, the clocks are based on transitions in atoms or ions between states characterized by the orientation of a electronic or nuclear spin (hyperfine or Zeeman transitions). If the rotation–invariance of the energy $E = mc^2$ of the nucleus was violated, the energy levels of the states involved would depend on the orientation of the clock in space. This would cause a modulation of the clock frequency ν_1 with its orientation, that can be detected by comparing it to the frequency ν_2 of a dissimilar second clock (practically, a clock based on another atomic species). A particular feature of these experiments is that the magnitude of the frequency change sought has to be compared to the rest frame energy m of the investigated fermion (i.e., $m \sim 1$ GeV for protons), but not to the absolute frequency ν of the clocks. It is therefore advantageous to use a very low frequency. If that is in the kHz range, where the photon energy is about $h\nu = 10^{-12}$ eV, the relative change $\delta\nu/\nu$ in the frequency is 10^{18} times larger than $\delta m/m$. Hence the enormous resolution that is possible in clock-comparsion experiments.

TABLE 6. Combinations of Lagrangian parameters and the current order-of magnitude bounds (expressed as the decimal logarithm), obtained by using assumptions described in the text (from [42]). $c_{(XY)} = \frac{1}{2}(c_{XY} + c_{YX})$.

parameter	proton GeV	neutron GeV	electron GeV
$\tilde{b}_J = b_J - md_{J0} + \frac{1}{2}\varepsilon_{JKL}(g_{KL0} - H_{KL})$	-27	-30	-27
$\tilde{c}_Q = m(c_{XX} + c_{YY} - 2c_{ZZ})$			
$\tilde{c}_{Q,J} = 2mc_{(JZ)}$ $(J = X, Y)$		-25	
$\tilde{c}_- = m(c_{XX} - c_{YY})$		-27	
$\tilde{c}_{XY} = 2mc_{(XY)}$	-27		
$\tilde{d}_J = 2md_{(J0)} - \frac{1}{2}(md_{J0} + \frac{1}{2}\varepsilon_{JKL}H_{KL})$	-25	-28	-22
$\tilde{g}_{D,J} = m\varepsilon_{JKL}(g_{K0L} + \frac{1}{2}g_{KL0}) - b_J$	-25	-28	-22
$\tilde{g}_Q = m(g_{X0X} + g_{Y0Y} - 2g_{Z0Z})$			
$\tilde{g}_{Q,J} = m(g_{J0Z} + g_{Z0J})$			
$\tilde{g}_- = m(g_{X0X} - g_{Y0Y})$			
$\tilde{g}_{XY} = m(g_{X0Y} + g_{Y0X})$			

The leading systematic effect in these experiments are magnetic fields, which directly couple to the spin. While excellent magnetic shielding is thus a prerequisite for performing tests of spin-dependent effects, the most recent experiment [135] uses an additional trick to reduce their influence. A vapour cell is filled with two species of noble gas (^{129}Xe and ^{3}He). Population inversion for both is achieved by coupling to optically pumped Rb vapour. Thus, maser oscillations of both species of noble gas at frequencies in the kHz range (see Tab. 7) are achieved. The point is that both masers depend on the same magnetic field, since they are spatially perfectly overlapped. Thus, the frequency of either can be used as a sensitive probe for stabilizing the magnetic field, while the other serves as a probe for Lorentz violation. Additionally, due to the simple structure of ^{3}He nuclei, the experiment obtains limits that do not rely on the above assumptions that have been made in order to simplify the theory. The frequency change of the He maser $\delta\nu_J = -3.5\tilde{b}^n_J + 0.012\tilde{d}^n_J - 0.012\tilde{g}^n_{D,J}$ due to Lorentz violation could be calculated. Because of the relatively small numerical factors, the influence of the parameters $\tilde{b}^n_J, \tilde{g}^n_{D,J}$ can be neglected, and the experiment results in a limit $\tilde{b}^n_J = (6.4 \pm 5.4) \times 10^{-32}$ GeV [135, 136]. Space–based experiments of this type are also planned for the near future [137, 138].

A similar trick is used in 'co-magnetometer' experiments [143]. Since the spin-dependent Lorentz violation fields couple to matter in a similar way as magnetic fields, magnetometers (based on the precession frequency of a spin) are sensitive to these terms. The relative sensitivity to Lorentz violation and magnetic background fields may be different for different types of atoms, so simultaneous operation of two spatially overlapped magnetometers can be used to eliminate the sensitivity to the magnetic feld.

TABLE 7. Overview of clock–comparison experiments with sensitivity to neutron and proton mass anisotropy δm. The quoted accuracy applies for the measurement of ν_1.

Reference	clock 1	ν_1	clock 2	ν_2	$\delta m/m_{n,p}$
[139]	^{9}Be	303 MHz	^{1}H	23.91401 GHz	10^{-26}
[140]	^{201}Hg	5.5 Hz	^{199}Hg	15 Hz	10^{-28}
[141]	^{21}Ne	1.000 kHz	^{3}He	9.650 kHz	10^{-28}
[142]	^{199}Hg	4.0321 Hz	^{133}Cs	1.858 kHz	10^{-29}
[135]*	^{3}He	1.7 kHz	^{129}Xe	4.9 kHz	5×10^{-31}

* see also [136]

4.3.7. Other tests

Limits on spin dependent electron coefficients have been obtained by torsion balances with spin polarized solids (i.e., permanent magnets). One experiment yielded $\tilde{b}^e_Z \simeq (2.7\pm1.6)\times10^{-25}m_e$ [144, 145]; in a similar experiment [146], $((\tilde{b}^e_X)^2+(\tilde{b}^e_Y)^2)^{1/2} \leq 6.0\times10^{-26}m_e$ and $|\tilde{b}^e_Z| \leq 1.4\times10^{-25}m_e$ have been found.

Hydrogen spectroscopy can prospectively limit linear combinations of $\tilde{b}^e_J, \tilde{b}^p_J$, to about 10^{-27} GeV [147]. Due to the simplicity of Hydrogen, the sensitivity of these tests is accurately calculable, making this a clean test. Comparing the frequencies of hydrogen masers, [148] find $|\tilde{b}^p_J+\tilde{b}^e_J| \lesssim 2\times10^{-28}$ GeV.

4.4. Anomalous dispersion in interferometry

The study of anomalous dispersion relations, although usually motivated on the basis of a break up of Lorentz symmetry, deserves to be discussed separately, if nothing else because of the huge interest that has been attracted by this phenomenology in the recent literature (see, *e.g.*, Refs. [2, 3, 4, 5]). Anomalous dispersion is of course a prototypical propagation effect, and its study would immediately suggest resorting to astrophysics contexts, where the propagation distances can be gigantic. We here want to stress that however there is at least one class of controlled/laboratory experiments where extremely high sensitivity to anomalous dispersion can be achieved. This has been stressed in Ref. [13] on the basis of the observation that conceivably, if we manage to operate them at high sensitivity using two different-wavelength beams, modern intereferometers (of the type used for gravity-wave searches) might achieve [13] sensitivity to $\eta \sim 1$, for the $n=1$ case in Eq. (6).

4.5. Space–time fluctuations and decoherence

Concerning the models of Planck-scale-induced strain noise in interferometry described in Section 3.5 the quality of experimental limits is improving quickly. For example, already available data constrain ζ to be much smaller than 1 in the random-walk scenario of Eq. 34. This is achieved both using large "free-mirror" laser-light interferometers [102, 103, 104, 12], such as TAMA (soon improving with LIGO and VIRGO), and using small-size laser-light interferometers whose mirrors are rigidly connected [12].

The fact that we are still unable to derive a quantitative description of spacetime fuzziness from the various quantum-gravity proposals does not allow us to have a good intuition for the significance of these limits. And, while theory work attempts to clarify the situation, it appears necessary to attempt to push the limits as far as possible.

For what concerns the density-matrix Planck-scale decoherence formalism mentioned in Section 3.5 the best limits were obtained using data from the CPLEAR neutral-kaon experiment. The neutral-kaon system is very sensitive to new physics because of the delicate balance of conventional-physics scales that governs its peculiar features. The density matrix formalism of Ref. [57] involves several dimensionless parameters which one may guess to be all obtained as roughly the ratio between the kaon mass and the Planck scale. For some of the parameters, CPLEAR data already exclude this possibility experimentally.

5. OBSERVATIONS IN ASTROPHYSICS

As mentioned, the realization that tests of some quantum-gravity effects with extremely high sensitivity could be based on observations in astrophysics has generated over these past few years a rather large interest. In turn this has resulted in the production of quite a few general reviews of the subject, of which Refs. [2, 3, 4, 5] are just a representative small sample. We refer the reader to these reviews (and references therein) for a detailed discussion of the relevant phenomenological proposal. Still, in order to render our own review somewhat self-contained, in this Section we comment briefly on some of the most studied methods.

5.1. SME astrophysics

There has been a sizeable effort of constraining SME parameters using astrophysical observations of radiation. Of course the interpretation of the data on radiation emitted by distant sources as tests of LI is faced with the problem that the source as well as the space in between (that might contain gasses, other electromagnetic fields etc.) are not under control of the experimenter, and the analysis inevitably acquires a certain level of model dependence.

The distance of the source can only be inferred from studying other radiation, and therefore the analysis must always compare radiation with radiation, which in some cases renders the analysis insensitive to spin-independent effects. But of course the extremely long propagation distance is very valuable for constraints on other types of effects.

The SME 4-vector $(k_{AF})^\kappa$ from the photon sector (Tab. 8) (that is also expected to vanish for theoretical reasons [149]) has been ruled out by birefringence measurements on radiation emitted by distant radio galaxies [92]. The idea is that $k_{AF} \neq 0$ leads to a

TABLE 8. Effects of the SME parameters of the photonic sector on the phase velocity of light. # denotes the number of degrees of freedom contained in each symbol. $\beta = v/c$ is the velocity of the experiment relative to the frame of reference in which the parameters are defined.

Parameters	#	makes c dependent on	limit	Ref.
$\vec{k}_{AF}$	4	circular polarization	10^{-42} GeV	[92]
$(\tilde{\kappa}_{e+})^{jk}$	5	linear polarization	10^{-32}	[150, 44]
$(\tilde{\kappa}_{o-})^{jk}$	5	linear polarization	10^{-32}	[150, 44]
$(\tilde{\kappa}_{e+})^{jk}$	5	direction of propagation	10^{-15}	[91, 130]
$(\tilde{\kappa}_{o+})^{jk}$	3	direction of propagation, $\mathscr{O}(\beta)$	10^{-11}	[91, 130]
$\tilde{\kappa}_{tr}$	1	motion of experiment $\mathscr{O}(\beta^2)$		

splitting of photons in two circularly polarized modes with different phase velocities, according to a dispersion relation

$$|k| = \omega \mp \frac{1}{2}\left((k_{AF})_0 - |k_{AF}|\cos\theta\right). \tag{44}$$

Here, $k = (\omega, k)$ denotes the wave vector and k_{AF} the spatial components of k_{AF}. The angle θ is between the direction of k_{AF} and k. The $+$ and $-$ signs correspond to right- and left-handed circularly polarized waves, respectively. As in the Faraday effect, this causes a rotation of the polarization of a linearly polarized plane wave by an angle $\delta\phi = ((k_{AF})_0 - |k_{AF}|\cos\theta)L/2$, where L denotes the distance travelled and θ is the angle between the diection of propagation and k. The radio galaxies are expected to emit radiation that is polarized either parallel or perpendicular to their observed elongation. A rotation of the polarization would cause a difference between the observed polarization and elongation; if no difference is observed, an upper limit on k_{AF} can be derived. Studying 160 sources, [92] one obtains $|k_{AF}| = \sqrt{(k_{AF})^\alpha (k_{AF})_\alpha} \lesssim 10^{-42}$ GeV.

The SME matrices $(\tilde{\kappa}_{e+})^{jk}$ and $(\tilde{\kappa}_{o-})^{jk}$ from the photon sector (Tab. 8) have also been restricted by polarization measurements. A difference of the phase velocity Δv_p of the two polarization modes travelling over a distance L would lead to a change in the relative phase

$$\Delta\phi \approx 2\pi \Delta v_p \frac{L}{\lambda}, \tag{45}$$

where λ is the wavelength. This modifies the polarization, e.g., light that is initially linearly polarized would in general become elliptically polarized. The large factor L/λ makes the experiment very sensitive. A change of the polarization state at the receiver occurs as a consequence of either a change in L or λ. For a astrophysical source, L is fixed, but one can measure the dependence on λ (making the assumption that the emitted polarization is relatively constant over a range of wavelengths). From data on 16 sources at 0.04 Gpc$\leq L \leq$3.53 Gpc, a limit of 2×10^{-32} (90% confidence level) on all elements of $(\tilde{\kappa}_{e+})^{jk}$ and $(\tilde{\kappa}_{o-})^{jk}$ is obtained [150, 44].

Similar analyses of synchrotron radiation from the Crab nebula and other sources are reported by [151]. As stressed in the next subsection, they interpret the result in terms of a modified dispersion relation of the form (6). Since their analysis uses birefringence methods, it should be possible to interpret the same result in terms of the SME: The wavelength of the $\sim 0.3\,\mathrm{MeV}$ radiation is $\sim 4\times 10^{-12}\,\mathrm{m}$, and the distance is assumed to be $L = 0.5\,\mathrm{GPc}$=$1.5\times 10^{25}\,\mathrm{m}$ or about $10^{37}\lambda$. Since the upper limit on polarization changes with λ is of the order of one radian, an analysis of these results in terms of the SME will likely lead to a limit of the order of 10^{-37} on at least one component of either $(\tilde{\kappa}_{e+})^{jk}$ or $(\tilde{\kappa}_{o-})^{jk}$.

Another kind of astrophysical limit is derived from the following idea: If Lorentz symmetry is not exact, the limiting velocity for different particles does not need to be equal. A 'fast' particle with a higher limiting velocity may then emit Cherenkov radiation that consists of particles with lower limiting velocities. Thus, its velocity would eventually be reduced to below the threshold for Cherenkov radiation in a time that is short compared to astrophysical timescales, so particles above a certain energy would not exist in cosmic radiation. From observations of cosmic radiation with energies of up to 3×10^{11} GeV, the maximum velocity of several particles can be constrained to deviate no more than a few parts in $10^{-20}\ldots 10^{-24}$ from c [152].

5.2. Anomalous dispersion

A special role within the phenomenology inspired by the possibility of Lorentz-symmetry violations is played by effects of anomalous dispersion. The fact that (as also stressed earlier in these notes) several approaches to the quantum-gravity problem motivate the study of Planck-scale-modified dispersion relations, has led, over the last few years, to a very large effort on the phenomenology side.

The first popular idea for searches of Planck-scale anomalous dispersion was based [28] on properties of gamma-ray bursts. These bursts travel cosmological distances and contain a rich time structure, and as a result they can be used to set rather significant limits on the wavelength dependence of the speed of photons which would follow from a dispersion relation of the type (6). The present best limits obtained using this strategy constrain (see, *e.g.*, Refs. [52, 153, 154]) the parameter $\eta < 300$ for the case $n = 1$ in (6). And planned telescopes should provide an improvement of a few orders of magnitude within a few years [155, 156].

For the case in which the modified dispersion relation is assumed to hold within a field-theoretic setup very stringent (beyond-Plankian) limits can be obtained from an analysis of synchrotron radiation from the Crab nebula [100]. However, the field-theoretic setup in the photon sector imposes birefringence and invites one to consider a particle/helicity dependence of η, leading to many parameters. Only one of these parameters can be constrained using the Crab-nebula synchrotron-radiation analysis [157, 100].

The case of the Crab-nebula synchrotron radiation is interesting because, as stressed already in the previous subsection, this same set of data can be valuable both for the SME and for the phenomenology based on (6). Clearly the physics described by the SME matrices is different from the energy-dependent effects [28] associated with (6). In principle, simultaneous limits on both effects could be obtained by studying the energy-dependence of the phase-shift. This illustrates that some observation is sometimes interpreted in two theoretical frameworks to produce results that appear to be unrelated, while the actual observation is the same.

Still concerning the hypothesis of a description of birefringence based on (6) a limit of order $|\eta| < 2 \cdot 10^{-4}$, for $n = 1$, can be inferred [158] using observations of polarized light from distant galaxies.

Perhaps the most exciting opportunity for this phenomenology based on (6) is provided by the study of ultrahigh-energy cosmic rays. Before reaching our observatories ultrahigh-energy cosmic rays travel gigantic (cosmological) distances and our expectations concerning the structure of the cosmic-ray spectrum depend strongly on the interactions of cosmic rays with photons in the cosmic microwave background, which in particular should produce a cut off in the spectrum around $5 \cdot 10^{19}$ eV. A modification of the dispersion relation could lead [159, 160, 161, 162, 163] to a modification of this cut off prediction. The Auger cosmic-ray observatory, now starting to take data, has the sensitivity needed to probe very small values of η for $n = 1$, and even in the case $n = 2$ Auger could still be sensitive to η of order 1.

5.3. Spacetime fuzziness and decoherence

Some proposals concerning the use of observations in astrophysics to explore the quantum-gravity ideas of spacetime fuzziness and decoherence have also been made. In Ref. [164] it is argued that evidence of a good phase coherence of light from extragalactic sources could be used to constrain spacetime-fuzziness scenarios. This proposal opens a valuable phenomenological window, but it might be necessary to seek improved analyses, taking into account the comments made in Ref. [165] concerning the quantum-gravity aspects, and in Ref. [166] concerning the astrophysics aspects.

Concerning decoherence certain aspects of neutrino astrophysics may play a role [10] somewhat similar to the one played by laboratory studies of the neutral-kaon system.

6. SUMMARY AND OUTLOOK

We have given a rather general overview of the phenomenology work which finds motivation in the study of the quantum-gravity problem. The field has grown so wide that we shall not here argue that we gave a complete overview, but we did set as one of our primary goals the one of illustrating some alternative strategies which are being pursued for this phenomenology. Within this spectrum of research programmes one finds on one side a phenomenology of the type illustrated by studies of the modified dispersion relation (6), with only a minimal parametrization, a nearly exclusive focus on the hypothesis that the onset of the new effects should be characterized by a scale which is rather close to the Planck scale, and the expectation that the new effects should not be describable in terms of renormalizable field-theory operators. And on the opposite side the strategy involves very general parametrizations, a phenomenology which looks for the new effects even at scales much different from the Planck scale, but assuming throughout that the present formalisms can accommodate the new effects, so that one is led for example to the SME field theory and/or to the type of modified Maxwell and Dirac equations which we discussed. To our knowledge there are no other reviews covering this wide spectrum of approaches, and we thought it might be beneficial to provide one. In spite of the differences in strategy there are a number of common challenges for all these approaches, besides the obvious common goal of contributing to the search of a solution for the quantum-gravity problem.

Another key goal for our work was to stress that this candidate effects that have been discussed in the quantum-gravity literature are not merely of academic interest, as it is sometimes assumed. We believe that the example of metrology, which we discussed in Subsection 1.2, should act as a clear warning of the potential (even technological!) implications of these studies.

We expect that this field will prove to be among the most exciting of the next 10 or 20 years, since over this period we can clearly see several opportunities for sharp improvement of the present sensitivities toward the relevant effects. A good example is provided by the Auger cosmic-ray data which will surely amuse us for the next couple of years, and are going to provide a much clearer picture of the high-energy cosmic-ray spectrum. Similarly, the next generation of gamma-ray telescopes, such as GLAST, will provide a huge improvement for the relevant gamma-ray studies.

While many Lorentz-violating effects have been bounded, some to impressive levels of precision, improvements in terms of accuracy, cleanliness and comprehensiveness are expected from improved versions of the old experiments and from experiments of new kinds. New turntable versions of the cavity tests are performed at several locations. The shorter timescale of the rotation (minutes compared to 24 h for Earth's rotation) allows to collect much more data in a given time and to exploit the time scale of the optimum sensitivity of the experiment. This should allow two to three orders of magnitude improvements, if the systematic effects associated with active rotation (like the bending of the resonators caused by a slight tilt of the rotation axis relative to Earth's gravity) can be controlled. As an alternative idea, cavity experiments could also search for birefringence in isotropic materials induced by Lorentz violation. Such experiments could perform the same tests as conventional Michelson-Morley type experiments, but would be immune against a broad range of systematic effects, including length changes of the cavity [122].

A very interesting challenge is to put some of these experiments into space. This would have several advantages. The absence of seismic vibrations is important for experiments involving macroscopic matter. Experiments in space also offer the possibility to choose the time scale of rotation and orbit of the satellite which provide the necessary modulation of the Lorentz violation in the frame of the experiment. On Earth, both time scales (24 h and 365 d) are relatively long. In space, these would be reduced to minutes and hours without introducing the systematic effects of turntable experiments on Earth. Several proposals [138, 167, 168] have been made for experiments on the International Space Station ISS and dedicated satellites.

New types of terrestrial experiments are suggested by new theoretical work. For example, the relatively low accuracy of the bounds on the photon parameters $\tilde{\kappa}_{0+}$ can be improved by exploring the static limit of Lorentz-violating Maxwell equations [31, 169]. Such experiments can also bound the Lorentz-violating non-SME terms of the most general Maxwell equations [31].

Concerning the UFF, we expect improvements by one order of magnitude within the next few years for laboratory experiments and up to five orders with dedicated satellite missions. These experiments, when carried through successfully, will definitely for the first time decide on the viability of certain quantum gravity predictions.

ACKNOWLEDGMENTS

We like to thank H. Dittus, F.W. Hehl, A. Kostelecký, A. Peters, and S. Schiller for fruitful discussions and ongoing cooperation. Financial support from the German Space Agency DLR, the German Academic Exchange Service DAAD, the Alexander von Humboldt-Stiftung, and CONACYT Grant 42191-F is acknowledged.

7. APPENDIX: STANDARD MODEL EXTENSION DEFINITIONS AND CONVENTIONS

Rich and extensive literature exists on the SME, so a full description is neither necessary nor possible here. However, we introduce briefly the definition and standard notation for Lorentz violating quantities in the SME.

Lagrangian. The SME starts from a Lagrangian formulation of the Standard Model, adding all possible observer Lorentz scalars that can be formed from the known particles and Lorentz tensors. Taken from the full SME that includes all known particles, the Lagrangian involving the Dirac field ψ of one fermion and the electromagnetic field $F^{\mu\nu}$ can be written as

$$\begin{aligned}\mathscr{L} &= \frac{i}{2}\bar{\psi}\Gamma_\nu D^\nu\psi - \frac{1}{2}\bar{\psi}M\psi + \text{h.c} \\ &\quad -\frac{1}{4}F^{\mu\nu}F_{\mu\nu} - \frac{1}{4}(k_F)_{\kappa\lambda\mu\nu}F^{\kappa\lambda}F^{\mu\nu} + \frac{1}{2}(k_{AF})^\kappa\varepsilon_{\kappa\lambda\mu\nu}A^\lambda F^{\mu\nu}, \end{aligned} \tag{46}$$

where h.c. denotes the Hermitian conjugate of the previous terms, and A^λ is the vector potential. The symbols Γ_ν and M are given by

$$\begin{aligned}\Gamma_\nu &= \gamma_\nu + c_{\mu\nu}\gamma^\mu + d_{\mu\nu}\gamma_5\gamma^\mu + e_\nu + if_\nu\gamma_5 + \frac{1}{2}g_{\lambda\mu\nu}\sigma^{\lambda\mu}, \\ M &= m + a_\mu\gamma^\mu + b_\mu\gamma_5\gamma^\mu + \frac{1}{2}H_{\mu\nu}\sigma^{\mu\nu}. \end{aligned} \tag{47}$$

The SME introduces such parameters into the Lagrangian of each type of fermion, as denoted by superscripts added to the field ψ as well as to the symbols a_μ, b_μ, $c_{\mu\nu}$, $d_{\mu\nu}$, e_μ, f_μ, $g_{\lambda\mu\nu}$, and $H_{\mu\nu}$. The γ_ν, γ_5 and $\sigma^{\mu\nu}$ are the conventional Dirac matrices, and D^ν is the usual covariant derivative. The tensors entering M have the dimension mass, the others are dimensionless. $H_{\mu\nu}$ is antisymmetric; $g_{\lambda\mu\nu}$ is antisymmetric in its first two indices. $c_{\mu\nu}$ and $d_{\mu\nu}$ are traceless. Gauge invariance and renormalizability excludes e_ν, f_ν, and $g_{\lambda\mu\nu}$, so these may be assumed to be either zero or suppressed relative to the other terms. Analogous terms are obtained in the framework of the generalized Dirac equation described in Sec.3.2.

Lorentz violation for the photons is encoded in the tensors $(k_{AF})^{\kappa}$ and $(k_F)_{\kappa\lambda\mu\nu}$. They correspond to the terms $\chi^{\kappa\lambda\mu\nu}$ and $\chi^{\kappa\lambda\mu}$ of the non-Lagrangian framework described in Sec.3.1.

Coordinate and field definitions. Some of the Lorentz violating parameters contained in one sector of the SME can be absorbed into the other sectors by coordinate and field redefinitions. For example, in a hypothetical world containing only photons and electrons, nine components of $(k_F)_{\kappa\lambda\mu\nu}$ could be moved into the nine symmetric components of $c_{\mu\nu}$. By definition, either the photon or the electron sector could be taken as conventional with respect to these parameters, with the Lorentz violation in the other sector. The presence of other particles, of course, changes this picture. We can still assume that one of the sectors is conventional, but then in general the other sectors are Lorentz-violating. Loosely speaking, in experiments where one compares the sectors against each other, only differential effects are meaningful.

REFERENCES

1. T. Quinn. Base units of the Systéme International d'unités, their accuracy, dissemination and international traceability. *Metrologia*, 31:515, 1995.
2. G. Amelino-Camelia. Are we at the dawn of quantum-gravity phenomenology? In J. Kowalski-Glikman, editor, *Towards Quantum Gravity, Lecture Notes in Physics 541*, page 1. Springer–Verlag, Berlin, 2000.
3. S. Sarkar. Possible astrophysical probes of quantum gravity. gr-qc/0204092.
4. T. Jacobson, S. Liberati, and D. Mattingly. Astrophysical bounds on Planck suppressed Lorentz violation. hep-ph/0407370.
5. G. Amelino-Camelia. Planck-scale Lorentz-symmetry test theories. astro-ph/0410076.
6. I. Antoniadis. Physics with large extra dimensions and non–Newtonian gravity at sub–mm distances. In D. Giulini, C. Kiefer, and D. Lämmerzahl, editors, *Quantum Gravity – From Theory to Experimental Search*, page 337. Springer–Verlag, Berlin, 2003.
7. T. Damour, F. Piazza, and G. Veneziano. Runaway dilaton and equivalence principle violations. *Phys. Rev. Lett.*, 89:081601, 2002.
8. T. Damour, F. Piazza, and G. Veneziano. Violations of the equivalence principle in a dilaton–runaway scenario. *Phys. Rev.*, D 66:046007, 2002.
9. V.A. Kostelecký. Signals for *CPT* and Lorentz violation in neutral–meson oscillations. *Phys. Rev.*, D 61:016002, 1999.
10. N.E. Mavromatos. Neutrinos and the phenomenology of *CPT* violation. hep-ph/0402005.
11. G. Amelino-Camelia. Gravity-wave interferometers as probes of a low-energy effective quantum gravity. *Phys. Rev.*, D 62:0240151, 2000.
12. S. Schiller, C. Lämmerzahl, H. Müller, C. Braxmaier, S. Herrmann, and A. Peters. Experimental limits for low-frequency space-time fluctuations from ultrastable optical resonators. *Phys. Rev.*, D 69:027504, 2004.
13. G. Amelino-Camelia and C. Lämmerzahl. Quantum-gravity-motivated Lorentz-symmetry tests with laser interferometers. *Class. Quantum Grav.*, 21:899, 2004.
14. M.J. Duff. A layman's guide to M-theory. hep-th/9805177.
15. C. Kiefer. *Quantum Gravity*. Oxford University Press, Oxford, 2004.
16. R. Gambini and J. Pullin. Nonstandard optics from quantum space–time. *Phys. Rev.*, D 59:124021, 1999.

17. J. Alfaro, H.A. Morales-Tecotl, and L.F. Urrutia. Quantum gravity corrections to neutrino propagation. *Phys. Rev. Lett.*, 84:2318, 2000.
18. J. Alfaro, H.A. Morales-Tecotl, and L.F. Urrutia. Loop quantum gravity and light propagation. *Phys. Rev.*, D 65:103509, 2002.
19. J. Alfaro, H.A. Morales-Tecotl, and L.F. Urrutia. Quantum gravity and spin-1/2 particle effective dynamics. *Phys. Rev.*, D 66:124006, 2002.
20. C.N. Kozameh and M.F. Parisi. Lorentz invariance and the semiclassical approximation of loop quantum gravity. *Class. Quantum Grav.*, 21:4063, 2004.
21. D. Colladay and V.A. Kostelecky. *CPT* violation and the standard model. *Phys. Rev.*, D 55:6760, 1997.
22. J. Ellis, N.E. Mavromatos, and D.V. Nanopoulos. Quantum–gravitational diffusion and stochastic fluctuations in the velocity of light. *Gen. Rel. Grav.*, 32:127, 2000.
23. J. Ellis, N.E. Mavromatos, D.V. Nanopoulos, and G. Volkov. Gravitational–recoil effects on fermion propagation in space-time foam. *Gen. Rel. Grav.*, 32:1777, 2000.
24. S. Coleman and S.L. Glashow. Cosmic ray and neutrino test of special relativity. *Phys. Lett.*, B 405:249, 1997.
25. D. Colladay and V.A. Kostelecky. Lorentz–violating extension of the standard model. *Phys. Rev.*, D 58:116002, 1998.
26. W.-T. Ni. Equivalence principles and electromagnetism. *Phys. Rev. Lett.*, 38:301, 1977.
27. M.P. Haugan and T.F. Kauffmann. A new test of the Einstein equivalence principle and the isotropy of space. *Phys. Rev.*, D 52:3168, 1995.
28. G. Amelino-Camelia, J. Ellis, N.E. Mavromatos, D.V. Nanopoulos, and S. Sarkar. Tests of quantum gravity from observations of gamma–ray bursts. *Nature*, 393:763, 1998.
29. R.C. Myers and M. Pospelov. Ultraviolet modifications of dispersion relations in effective field theory. *Phys. Rev. Lett.*, 90:211601, 2003.
30. C. Lämmerzahl. Quantum tests of foundations of general relativity. *Class. Quantum Grav.*, 14:13, 1998.
31. C. Lämmerzahl, A. Macias, and H. Müller. Lorentz invariance violation and charge (non–)conservation: A general theoretical frame for extensions of the Maxwell equations. *Phys. Rev.*, D:to appear, 2005.
32. C. Lämmerzahl. The geometry of matter fields. In deSabbata V. and Audretsch J., editors, *Quantum Mechanics in Curved Space–Time, NATO ASI series, series B: Physics*, volume 230, page 23. Plenum Press, New York, 1990.
33. J. Audretsch and C. Lämmerzahl. A new constructive axiomatic scheme for the geometry of space-time. In Majer U. and Schmidt H.-J., editors, *Semantical Aspects of Space-Time Geometry*, page 21. BI Verlag, Mannheim, 1993.
34. J. Audretsch, F.W. Hehl, and C. Lämmerzahl. Matter wave interferometry and why quantum objects are fundamental for establishing a gravitational theory. In J. Ehlers and G. Schäfer, editors, *Relativistic Gravity Research. With Emphasis on Experiments and Observations*, volume Lecture Notes in Physics **410**, page 368. Springer–Verlag, Berlin, 1992.
35. H.P. Robertson. Postulate versus observation in the Special Theory of Relativity. *Rev. Mod. Phys.*, 21:378, 1949.
36. R. Mansouri and R.U. Sexl. A test theory of special relativity: I. Simultaneity and clock synchronisation. *Gen. Rel. Grav.*, 8:497, 1977.
37. G. Cocconi and E.E. Salpeter. A search for anisotropy of inertia. *Nuovo Cim.*, X:646, 1958.
38. V.W. Hughes, H.G. Robinson, and V. Beltran-Lopez. Upper limit for the anisotropy of inertial mass from nuclear resonance experiments. *Phys. Rev. Lett.*, 4:342, 1960.
39. R.W.P. Drever. A search for the anisotropy of inertial mass using a free precession technique. *Phil. Mag.*, 6:683, 1961.
40. D.-E. Liebscher. The geometry of the Dirac equation. *Ann. Physik (Leipzig)*, 42:35, 1985.
41. U. Bleyer and D.-E. Liebscher. Mach's principle and causal structure. In Barbour J. and Pfister H., editors, *Mach's Principle, From Newton's Bucket to Quantum Gravity*, page 293. Birkhäuser, Boston, 1995.
42. V.A. Kostelecky and C.D. Lane. Constraints on Lorentz violation from clock–comparison experiments. *Phys. Rev.*, D 60:116010, 1999.

43. W.-T. Ni. A nonmetric theory of gravity. *preprint*, http://gravity5.phys.nthu.edu.tw:Montana State University, Bozeman, Montana, USA, 1973.
44. A. Kostelecky and M. Mewes. Signals for Lorentz violation in electrodynamics. *Phys. Rev.*, D 66:056005, 2002.
45. C. Lämmerzahl. Quantum tests of space–time structures. In P. Bergmann, V. deSabbata, G. Gillies, and P. Pronin, editors, *Proceedings of the International School of Cosmology and Gravitation, Course XV*, page 221. World Scientific, Singapore, 1998.
46. M. Gasperini. Testing the principle of equivalence with neutrino oscillations. *Phys. Rev.*, D 38:2635, 1988.
47. G.Z. Adunas, E. Rodriguez-Milla, and D.V. Ahluwalia. Probing quiantum violations of the equivalence principle. *Gen. Rel. Grav.*, 33:183, 2001.
48. T. Damour. Equivalence principle and clocks. In J. Trân Tanh Vân, J. Dumarchez, J. Reynaud, C. Salomon, S. Thorsett, and J.Y. Vinet, editors, *Gravitational Waves and Experimental Gravity*, page 357. World Publishers, Hanoi, 2000.
49. G. Dvali and M. Zaldarragia. Changing α with time: Implications for fifth–force–type experiments and quintessence. *Phys. Rev. Lett.*, 88:091303, 2002.
50. K. Nordtvedt. Space–time variation of physical constants and the equivalence principle. gr-qc/0212044.
51. A. Bauch and S. Weyers. New experimental limit on the validity of local position invariance. *Phys. Rev.*, D 65:081101(R), 2002.
52. B.E. Schaefer. Severe limits on variations of the speed of light with frequency. *Phys. Rev. Lett.*, 82:4964, 1999.
53. C. Percival. Quantum space-time fluctuations and primary state diffusion. *preprint*, quant-ph/9508021, 1995.
54. I.P. Percival and W.T. Strunz. Detection of space-time fluctuations by a model matter interferometer. *Proc. Roy. Soc. London*, A 453:431, 1996.
55. J. Ellis, S. Hagelin, D.V. Nanopoulos, and M. Srednicki. Search for violations of quantum mechanics. *Nucl. Phys.*, B 241:381, 1984.
56. H. Yu and L.H. Ford. Light-cone fluctuations in flat spacetimes with nontrivial topology. *Phys. Rev.*, D 60:084023, 1999.
57. J. Ellis, J.L. Lopes, N.E. Mavromatos, and D.V. Nanopoulos. Precision test of *CPT*–symmetry and quantum mechanics in the neutral kaon system. *Phys. Rev.*, D 53:3846, 1996.
58. O.W. Greenberg. *CPT* violation implies violation of Lorentz invariance. *Phys. Rev. Lett.*, 89:231602, 2002.
59. V.A. Kostelecky and S. Samuel. Spontaneous breaking of Lorentz symmetry in string theory. *Phys. Rev.*, D 39:1611, 1989.
60. T. Damour and A.M. Polyakov. The string dilaton and a least action principle. *Nucl. Physics*, B 423:532, 1994.
61. T. Damour and A.M. Polyakov. String theory and gravity. *Gen. Rel. Grav.*, 12:1171, 1996.
62. C. Wetterich. Crossover quintessence and cosmological history of fundamental constants. *Phys. Lett.*, B 561:10, 2003.
63. J. Ellis, N.E. Mavromatos, D.V. Nanopoulos, and A.S. Sakharov. Space-time foam may violate the principle of equivalence. *Int. J. Mod. Phys.*, A 19:4413, 2004.
64. T. Damour. Gravity, equivalence principle and clocks. to appear in the Proceedings of the Workshop on the Scientific Applications of Clocks in Space (JPL, Pasadena, November 7-8, 1996).
65. C. Wetterich. Cosmologies with variable Newton's "constant". *Nucl. Phys.*, B 302:645, 1988.
66. J. Ellis, N.E. Mavromatos, and D.V. Nanopoulos. String theory modifies quantum mechanics. *Phys. Lett.*, B 293:37, 1992.
67. E. Gravanis and N.E. Mavromatos. Putting Liouville string models of (quantum) gravity to test. hep-th/0108008.
68. J. Alfaro, M. Reyes, H.A. Morales-Tecotl, and L.F. Urrutia. On alternative approaches to Lorentz violation in loop quantum gravity inspired models. *Phys.Rev.*, D 70:084002, 2004.
69. G. Amelino-Camelia, L. Smolin, and A. Starodubtsev. Quantum symmetry, the cosmological constant and Planck scale. *Class. Quantum Grav.*, 21:3095, 2004.
70. R. Gambini, R. Porto, and J. Pullin. Loss of coherence from discrete quantum gravity. gr-qc/0305098.

71. N.R. Douglas and N.A. Nekrasov. Noncommutative field theory. *Rev. Mod. Phys.*, 70:977, 2001.
72. S. Doplicher, K. Fredenhagen, and J.E. Roberts. Space–time quantization induced by classical gravity. *Phys. Lett.*, B 331:39, 1994.
73. A. Matusis, L. Susskind, and N. Toumbas. The IR/UV connection in the non-commutative gauge theories. *JHEP*, 0012:002, 2000.
74. G. Amelino-Camelia, L. Doplicher, S. Nam, and Y.-S. Y.-S. Seo. Phenomenology of particle production and propagation in string motivated canonical noncommutative space–time. *Phys. Rev.*, D 67:085008, 2003.
75. J. Madore, S. Schraml, P. Schupp, and J. Wess. Gauge theory on noncommutative spaces. *Eur. Phys. J.*, C 16:161, 2000.
76. S. Majid and H. Ruegg. Bicrossproduct structure of kappa Poincaré group and noncommutative geometry. *Phys. Lett.*, B 334:348, 1994.
77. J. Lukierski, H. Ruegg, and W.J. Zakrzewski. Classical and quantum-mechanics of free κ–relativistic systems. *Ann. Phys.*, 243:90, 1995.
78. G. Amelino-Camelia and M. Arzano. Coproduct and star product in field theories on Lie algebra noncommutative space–times. *Phys. Rev.*, D 65:084044, 2002.
79. A. Agostini, G. Amelino-Camelia, and F. D'Andrea. Hopf algebra description on noncommutative space–time symmetries. hep-th/0306013.
80. J. Kowalski-Glikman and S. Nowak. Non-commutative space-time of doubly special relativity theories. hep-th/0204245.
81. S.M. Carroll, J.A. Harvey, V.A. Kostelecký, C.D. Lane, and T. Okamoto. Noncommutative field theory and Lorentz violation. *Phys. Rev. Lett.*, 87:141601, 2001.
82. C.E. Carlson, C.D. Carone, and R.F. Lebed. Coproduct supersymmetric noncommutative QED and Lorentz violation. *Phys. Lett.*, B 549:337, 2002.
83. T. Padmanabhan. Limitations on the operational definition of space–time events and quantum gravity. *Class. Quantum Grav.*, 4:L107, 1987.
84. D.V. Ahluwalia. Quantum measurement, gravitation, and locality. *Phys. Lett.*, B 339:301, 1994.
85. Y.J. Ng and H. Van Dam. Limit to space-time measurement. *Mod. Phys. Lett.*, A 9:335, 1994.
86. G. Amelino-Camelia. Limits on the measurability of space-time distances in the semi-classical approximation of quantum gravity. *Mod. Phys. Lett.*, A 9:3415, 1994.
87. G. Amelino-Camelia. On local observations in quantum gravity. *Mod. Phys. Lett.*, A 11:1411, 1996.
88. L.J. Garay. Quantum gravity and minimum length. *Int. J. Mod. Phys.*, A 10:145, 1995.
89. W.-T. Ni. . *Bull. Am. Phys. Soc.*, 19:655, 1974.
90. C. Lämmerzahl and F.W. Hehl. Riemannian light cone from vanishing birefringence in premetric vacuum electrodynamics. *Phys. Rev.*, D 70:105022, 2004.
91. H. Müller, S. Herrmann, C. Braxmaier, S. Schiller, and A. Peters. Modern Michelson–Morley experiment using cryogenic optical resonators. *Phys. Rev. Lett.*, 91:020401, 2003.
92. S.M. Caroll, G.B. Field, and R. Jackiw. Limits on a Lorentz– and parity–violating modifications of electrodynamics. *Phys. Rev.*, D 41:1231, 1990.
93. C. Lämmerzahl and Ch.J. Bordé. Testing the Dirac equation. In C. Lämmerzahl, C.W.F. Everitt, and F.W. Hehl, editors, *Gyros, Clocks, and Interferometers: Testing Relativistic Gravity in Space*, page 464. Springer–Verlag, Berlin, 2001.
94. C.M. Will. *Theory and Experiment in Gravitational Physics (Revised Edition).* Cambridge University Press, Cambridge, 1993.
95. C. Lämmerzahl. Constraints on space–time torsion from Hughes–Drever experiments. *Phys. Lett.*, A 228:223, 1997.
96. A. Camacho and A. Macías. New experimental proposals for testing Dirac equation. *Phys. Lett.*, B 582:229, 2004.
97. J. Ehlers, F.A.E. Pirani, and A. Schild. The geometry of free fall and light propagation. In L. O'Raifeartaigh, editor, *General Relativity, Papers in Honour of J.L. Synge*, page 63. Clarendon Press, Oxford, 1972.
98. V. Perlick. Characterisation of standard clocks by means of light rays and freely falling particles. *Gen. Rel. Grav.*, 19:1059, 1987.
99. B. Bertotti, L. Iess, and P. Tortora. A test of general relativity using radio links with the Cassini spacecraft. *Nature*, 425:374, 2003.

100. T.A. Jacobson, S. Liberati, D. Mattingly, and F.W. Stecker. New limits on planck scale Lorentz violation in QED. astro-ph/0309681.
101. G. Amelino-Camelia. Improved limit on quantum space–time modifications of Lorentz symmetry from observations of gamma–ray blazars. gr-qc/0212002.
102. G. Amelino-Camelia. Gravity-wave interferometers as quantum gravity detectors. *Nature*, 398:216, 1999.
103. Y.J. Ng and H. van Dam. Measuring the foaminess of space–time with gravity wave interferometers. *Found. Phys.*, 30:795, 2000.
104. G. Amelino-Camelia. A phenomenological description of quantum, gravity induced space–time noise. *Nature*, 410:1065, 2001.
105. S. Baeßler, B.R. Heckel, E.G. Adelberger, J.H. Gundlach, U. Schmidt, and H.E. Swanson. Improved tes of the equivalence principle for gravitational self–energy. *Phys. Rev. Lett.*, 83:3585, 1999.
106. K. Kuroda and N. Mio. Test of composition–dependent force by a free–fall interferometer. *Phys. Rev. Lett.*, 62:1941, 1989.
107. K. Kuroda and N. Mio. Limits on a possible composition–dependent force by a Galilean experiment. *Phys. Rev.*, 42:3903, 1990.
108. T.M. Niebauer, M.P. McHugh, and J.E. Faller. Galileian test for the fifth force. *Phys. Rev. Lett.*, 59:609, 1987.
109. P. Touboul. MICROSCOPE, testing the equivalence principle in space. *Comptes Rendus de l'Aced. Sci. Série IV: Physique Astrophysique*, 2:1271, 2001.
110. N. Lockerbie, J.C. Mester, R. Torii, S. Vitale, and P.W. Worden. STEP: A status report. In C. Lämmerzahl, C.W.F. Everitt, and F.W. Hehl, editors, *Gyros, Clocks, and Interferometers: Testing Relativistic Gravity in Space*, page 213. Springer–Verlag, Berlin, 2001.
111. J.P. Turneaure and S.R. Stein. . In J.D. Fairbank, B.S. Deaver, C.W.F. Everitt, and P.F. Michelson, editors, *Near Zero*, page 414. Freeman, New York, 1988.
112. C. Braxmaier, H. Müller, O. Pradl, J. Mlynek, A. Peters, and S. Schiller. Test of relativity using a cryogenic optical resolator. *Phys. Rev. Lett.*, 88:010401, 2002.
113. T. Kinoshita. The fine structure constant. *Rep. Prog. Phys.*, 59:1459-1492, 1996.
114. Thomas B. Bahder. Relativity of GPS measurement. *Phys. Rev.*, D 68:063005, 2003.
115. H. Meinke and F.W. Gundlach. *Hochfrequenztechnik*, volume 1-3. Springer-Verlag, Berlin, Heidelberg, New York, 4 edition, 1986.
116. Patrick G. O'Shea and Henry P. Freund. Free-electron lasers: Status and applications. *Science*, 292:1853, 2001.
117. C. Lämmerzahl and M.P. Haugan. On the interpretation of Michelson–Morley experiments. *Phys. Lett.*, A 282:223, 2001.
118. H. Müller, S. Herrmann, A. Saenz, A. Peters, and C. Lämmerzahl. Optical cavity tests of Lorentz invariance for the electron. *Phys. Rev.*, D 68:116006, 2003.
119. H. Müller, C. Braxmaier, S. Herrmann, A. Peters, and C. Lämmerzahl. Electromagnetic cavities and Lorentz invariance violation. *Phys. Rev.*, D 67:056006, 2003.
120. H. Müller and C. Lämmerzahl. Macroscopic matter in Lorentz tests. In V.A. Kostelecky, editor, *Proceedings of 3rd Conference on CPT*, page . World Scientific, Singapore, 2004.
121. H. Müller, S. Herrmann, A. Saenz, A. Peters, and C. Lämmerzahl. Test of Lorentz invariance using hydrogen molecules. *Phys. Rev.*, D 70:076004, 2004.
122. H. Müller. Tests of Lorentz invariance by use of vacuum and matter filled cavity resonators. hep-ph/0412385.
123. A.A. Michelson. The relative motion of the Earth and the luminiferous aether. *Am. J. Sci.*, 22:120–129, 1881.
124. A.A. Michelson and E.W. Morley. On the relative motion of the Earth and the luminiferous ether. *Am. J. Sci.*, 34:333, 1887.
125. R.J. Kennedy. A refinement of the Michelson–Moreley experiment. *Proc. Nat. Acad. Sci.*, 12:621, 1926.
126. R.J. Kennedy and E.M. Thorndike. Experimental establishment of the relativity of time. *Phys. Rev.*, 42:400, 1932.
127. A. Brillet and J.L. Hall. Improved laser test of the isotropy of space. *Phys. Rev. Lett.*, 42:549, 1979.
128. D. Hils and J.L. Hall. Improved Kennedy–Thorndike experiment to test special relativity. *Phys. Rev. Lett.*, 64:1697, 1990.

129. J.A. Lipa, J.A. Nissen, S. Whang, D.A. Stricker, and D. Avaloff. New limit on signals of Lorentz violation in electrodynamics. *Phys. Rev. Lett.*, 90:060403, 2003.
130. P. Wolf, S. Bize, A. Clairon, G. Santarelli, M.E. Tobar, and A.N. Luiten. Improved test of Lorentz invariance in electrodynamics. *Phys. Rev.*, D 70:051902, 2004.
131. H. Müller, S. Herrmann, A. Braxmaier, S. Schiller, and A. Peters. Precision tests of the isotropy of light propagation. *Appl. Phys.*, B 77:719, 2003.
132. S. Saathoff, S. Karpuk, U. Eisenbarth, G. Huber, S. Krohn, R. Muñoz-Horta, S. Reinhardt, D. Schwalm, A. Wolf, and G. Gwinner. Improved test of time dilation in special relativity. *Phys. Rev. Lett.*, 91:190403, 2003.
133. D.C. Champeney, G.R. Isaak, and A.M. Khan. A time dilation experiment based on the Mössbauer effect. *Proc. Phys. Soc.*, 85:583, 1965.
134. V.A. Kostelecký (ed.). Proceedings to the 3rd meeting on *CPT* and Lorentz invariance, Univ. of Indiana, Bloomington 2004. to appear.
135. D. Bear, R.E. Stoner, R.L. Walsworth, V.A. Kostelecky, and C.D. Lane. Limit on Lorentz and *CPT* violation of the neutron using a two–species Noble–gas maser. *Phys. Rev. Lett.*, 85:5038, 2000.
136. D. Bear, R.E. Stoner, R.L. Walsworth, V.A. Kostelecký, and C.D. Lane. Erratum: Limit on Lorentz and *CPT* violation of the neutron using a two-species noble-gas maser. *Phys. Rev. Lett.*, 89:209902(E), 2002.
137. R. Bluhm, V.A. Kostelecký, C.D. Lane, and N. Russell. Clock comparison tests of Lorentz and *CPT* symmetry in space. *Phys. Rev. Lett.*, 88:090801, 2002.
138. R. Bluhm, V.A. Kostelecky, C.D. Lane, and N. Russell. Probing Lorentz and *CPT* violation with space-based experiments. *Phys. Rev.*, D 68:125008, 2003.
139. J.D. Prestage, J.J. Bollinger, W.M. Itano, and D.J. Wineland. Limits for spatial anisotropy by use of nuclear–spin–polarized $^9Be^+$ ions. *Phys. Rev. Lett.*, 54:2387, 1985.
140. S.K. Lamoreaux, J.P. Jacobs, B.R. Heckel, F.J. Raab, and E.N. Fortson. New limits on spatial anisotropy from optically pumped ^{201}Hg and ^{199}Hg. *Phys. Rev. Lett.*, 57:3125, 1986.
141. T.E. Chupp, R.J. Hoara, R.A. Loveman, E.R. Oteiza, J.M. Richardson, and M.E. Wagshul. Results of a new test of local Lorentz invariance: A search for mass anisotropy in ^{21}Ne. *Phys. Rev. Lett.*, 63:1541, 1989.
142. C.J. Berglund, L.R. Hunter, D. Krause, E.O. Prigge, and M.S. Ronfeldt. New limits on local Lorentz invariance from hg and cs magnetometers. *Phys. Rev. Lett.*, 75:1879, 1995.
143. T. Kornack and M. Romalis. Operation of the K-^{3}He Co-magnetometer. In V.A. Kostelecký, editor, *Proceedings of the 3rd conference on CPT and Lorentz invariance*, page . World Scientific, Singapore, 2005.
144. R. Bluhm and V.A. Kostelecky. Lorentz and *CPT* tests with spin–polarized solids. *Phys. Rev. Lett.*, 84:1381, 2000.
145. E.G. Adelberger et al. . In P. Herszeg et al, editor, *Physics beyond the Standard Model*, page . World Scientific, Singapore, 1999.
146. L.-S. Hou, W.-T. Ni, and Y.-C. Li. Test of cosmic spatial isotropy for polarized electrons using rotatable torsion balance. *Phys. Rev. Lett.*, 90:201101, 2003.
147. R. Bluhm, V.A. Kostelecky, and N. Russell. *CPT* and Lorentz tests in Hydrogen and Antihydrogen. *Phys. Rev. Lett.*, 82:2254, 1999.
148. D.F. Phillips, M.A. Humphrey, E.M. Mattison, R.E. Stoner, R.F.C. Vessot, and R.L. Walsworth. Limit on Lorentz and *CPT* violation of the proton using a hydrogen maser. *Phys. Rev. D*, 63:111101, 2001.
149. Alan Kostelecký and Ralf Lehnert. Stability, causality, and Lorentz and *CPT* violation. *Phys. Rev. D*, 63:065008, 2001.
150. A. Kostelecky and M. Mewes. Cosmological constraints on Lorentz violation in electrodynamics. *Phys. Rev. Lett.*, 87:251304, 2001.
151. T. Jacobson, S. Liberati, and D. Mattingly. A strong astrophysical constraint on the violation of Special Relativity by quantum gravity. *Nature*, 424:1019–1021, 2003.
152. O. Gagnon and G.D. Moore. Limits on Lorentz violation from the highest energy cosmic rays. *Phys. Rev. D*, 70:065002, 2004.
153. G. Amelino-Camelia. Quantum-gravity phenomenology: status and prospects. *Mod. Phys. Lett.*, A 17:899, 2002.

154. J.R. Ellis, N.E. Mavromatos, D.V. Nanopoulos, and A.S. Sakharov. Quantum–gravity analysis of gamma–ray bursts using wavelets. *Astr. Astrophys.*, 402:409, 2003.
155. J.P. Norris, J.T. Bonnell, G.F. Marani, and J.D. Scargle. Glast, grbs and quantum gravity. astro-ph/9912136.
156. A. de Angelis. Glast, the Gamma–ray Large Area Space Telescope. astro-ph/0009271.
157. G. Amelino-Camelia. Improved limit on quantum-spacetime modifications of Lorentz symmetry from observations of gamma-ray blazars. gr-qc/0212002.
158. R.J. Gleiser and C.N. Kozameh. Astrophysical limits on quantum gravity motivated birefringence. *Phys. Rev.*, D 64:083007, 2001.
159. T. Kifune. Invariance violation extends the cosmic ray horizon? *Astrophys. J. Lett.*, 518:L21, 1999.
160. R. Aloisio, P. Blasi, P.L. Ghia, and A.F. Grillo. Probing the structure of space-time with cosmic rays. *Phys. Rev.*, D 62:053010, 2000.
161. G. Amelino-Camelia and T. Piran. Planck scale deformation of Lorentz symmetry as a solution to the ultraheigh energy cosmic ray and the TeV–photon paradoxes. *Phys. Rev.*, 64:036005, 2001.
162. O. Bertolami. Ultra high-energy cosmic rays and symmetries of space-time. *Gen. Rel. Grav.*, 34:707, 2002.
163. O. Bertolami. Threshold effects and Lorentz symmetry. hep-ph/0301191.
164. R. Lieu and L.W. Hillman. The phase coherence of light from extragalactic sources: Direct evidence against first-order Planck-scale fluctuations in time and space. *Astrophys. J.*, 585:L77, 2003.
165. Y.J. Ng, H. van Dam, and W.A. Christiansen. Probing Planck-scale physics with extragalactic sources?. *Astrophys. J.*, 591:L87, 2003.
166. D.H. Coule. Planck scale still safe from stellar images. *Class. Quantum Grav.*, 20:3107, 2003.
167. C. Lämmerzahl, C. Ciufolini, H. Dittus, L. Iorio, H. Müller, A. Peters, E. Samain, S. Scheithauer, and S. Schiller. An Einstein mission for improved tests of special and general relativity. *Gen. Rel. Grav.*, 36:2373, 2004.
168. S. Buchman, M. Dong, S. Wang, J.A. Lipa, and J.P. Turneaure. A space–based superconducting microwave oscillator clock. *Adv. Space Res.*, 25:1251, 2000.
169. Q.G. Bailey and V.A. Kostelecký. Lorentz–violating electrostatics and magnetostatics. *Phys. Rev.*, D 70:076006, 2004.

Radiation in Lorentz violating electrodynamics

R. Montemayor* and L.F. Urrutia†

*Instituto Balseiro and CAB, Universidad Nacional de Cuyo and CNEA, 8400 Bariloche, Argentina
†Departamento de Física de Altas Energías, Instituto de Ciencias Nucleares, Universidad Nacional Autónoma de México, Apartado Postal 70-543, 04510 México D.F.

Abstract. Synchrotron radiation is analyzed in the classical effective Lorentz invariance violating model of Myers-Pospelov. Within the full far-field approximation we compute the electric and magnetic fields, the angular distribution of the power spectrum and the total emitted power in the m-th harmonic, as well as the polarization. We find the appearance of rather unexpected and large amplifying factors, which go together with the otherwise negligible naive expansion parameter. This opens up the possibility of further exploring Lorentz invariance violations by synchrotron radiation measurements in astrophysical sources where these amplifying factors are important.

1. INTRODUCTION

Possible manifestations of quantum gravity could arise as modified energy-dependent dispersion relations violating the Lorentz symmetry. Such modifications were originally proposed in [1], opening up the door to quantum gravity phenomenology. The first heuristical derivation of such a type of dispersion relations from a theoretical framework was done in the context of a Loop Quantum Gravity (LQG) inspired model of corrections to flat space dynamics, leading to a consistent extension of Maxwell electrodynamics with linear correction terms in the quantum gravity scale [2]. An alternative approximation to LQG, inspired on Thiemann's regularization [3], extended such results to photons [4] and fermions[5]. For a review see for example [6]. Such estimations have to be understood as preliminary steps in the understanding of the semiclassical limit of LQG, which still remains an open problem. An alternative proposal for a quantum gravity scale modified Maxwell and fermionic dynamics, based on string theory, was developed in [7]. Recently, an effective field theory approach to the formulation of quantum gravity induced effects has been put forward [9]. The Gambini-Pullin (GP) [2] and Myers-Pospelov (MP) [9] models of modified electrodynamics lead to a frequency and helicity dependent speed of light. Even though both models differ in the details, the radiation regime behavior is similar, up to numerical factors of order one.

Using the synchrotron radiation from the Crab Nebulae, Jacobson et al. have found stringent bounds upon the parameters describing the modified dispersion relations [10]. They are based on the synchrotron radiation cut-off frequency ω_c, and use a heuristic kinematical estimation of ω_c, where photon and electron velocities are determined by the modified dispersion relations. These kinematical estimations do not embody birefringence effects such as the ones appearing in the GP and MP theories, which lead to new effects in the radiation spectrum.

CP758, *Gravitation and Cosmology: 2nd Mexican Meeting on Mathematical and Experimental Physics*
edited by A. Macías, C. Läemmerzahl and D. Nuñez

Here we examine this problem from the perspective of the MP effective theory, a dimension five modified electrodynamics which describes the dynamics arising from linear corrections at the quantum gravity scale. We develop a complete calculation of the synchrotron radiation. This allows the use of additional observational information, such as the polarization of the radiation, in order to set new bounds on the Lorentz violation parameters. Some of the results have been already presented in Ref.[8]. In this work we restrict ourselves to the classical regime, in such a way that the fine-tuning problems associated with the radiative corrections of the model are not considered [9, 11, 12].

2. MYERS-POSPELOV ELECTRODYNAMICS

The MP effective field theory includes Lorentz violations generated by dimension five operators, parametrized by a vector V^μ, and assumed to be suppressed by a factor M_P^{-1} ($M_P \approx 10^{19}$ GeV is the Planck mass). The electromagnetic field and the dynamics of the charge are affected by these perturbations and lead to distinctive signatures. Here we restrict ourselves to the case $V^\mu = (1,\mathbf{0})$, enough to put in evidence the main consequences of these violations.

The MP action for the electromagnetic field is

$$S_{MP} = \int d^4x \left[-\frac{1}{4} F_{\mu\nu}F^{\mu\nu} - 4\pi\, j^\mu A_\mu + \tilde{\xi}\,(V^\alpha F_{\alpha\delta})(V\cdot\partial)(V_\beta \tilde{F}^{\beta\delta}) \right], \tag{1}$$

where $\tilde{\xi} = \xi/M_P$, with ξ being a dimensionless parameter. We take the unperturbed velocity of light in vacuum, c, equal to 1. This is a gauge theory similar to Maxwellian electrodynamics, and implies the conservation of the electrical current $j^\mu = (\rho, \mathbf{j})$. The field tensor is $F_{\mu\nu} = \partial_\mu A_\nu - \partial_\nu A_\mu$, and the electric and magnetic fields are given by $E_i = F_{0i}$ and $B_k = -\frac{1}{2}\varepsilon_{kij}F_{ij}$, respectively. Hence $\mathbf{E}$ and $\mathbf{B}$ satisfy

$$\nabla\cdot\mathbf{B} = 0, \quad \nabla\times\mathbf{E} + \partial_t\mathbf{B} = 0. \tag{2}$$

In the rest frame $V^\alpha = (1,\mathbf{0})$, the equations of motion derived from the action (1) are

$$\nabla\cdot\mathbf{E} = 4\pi\rho, \tag{3}$$

$$-\partial_t\mathbf{E} + \nabla\times\mathbf{B} + \tilde{\xi}\partial_t\left(-\nabla\times\mathbf{E} + \partial_t\mathbf{B}\right) = 4\pi\mathbf{j}. \tag{4}$$

This dynamics gives place to an energy momentum tensor $T_{\mu\nu}$ with components

$$4\pi T^0_0 = \frac{1}{2}(\mathbf{E}^2 + \mathbf{B}^2) - \tilde{\xi}\mathbf{E}\cdot\partial_t\mathbf{B}, \tag{5}$$

$$4\pi\mathbf{S} = \mathbf{E}\times\mathbf{B} - \tilde{\xi}\mathbf{E}\times\partial_t\mathbf{E}, \tag{6}$$

which, outside the sources, satisfy $\partial_t T^{00} + \nabla\cdot\mathbf{S} = 0$. These expressions for the energy-momentum tensor and their conservation law are exact in $\tilde{\xi}$.

Alternatively, the equation of motion for $\mathbf{A}$ in the radiation gauge is

$$\partial_t^2\mathbf{A} - \nabla^2\mathbf{A} + 2\tilde{\xi}\nabla\times\partial_t^2\mathbf{A} = 4\pi\left(\mathbf{j} - (\nabla\nabla^{-2}\nabla)\mathbf{j}\right) \equiv 4\pi\mathbf{j}_T, \tag{7}$$

which in momentum space takes the form

$$\left(-\omega^2 + k^2 - 2i\tilde{\xi}\,\omega^2 \mathbf{k}\times\right)\mathbf{A}(\mathbf{k},\omega) = 4\pi\,\mathbf{j}_T(\mathbf{k},\omega). \tag{8}$$

In the circular polarization basis, where the components of a vector field $\mathbf{A}$ are $\mathbf{A}^{\pm} = \frac{1}{2}\left[\mathbf{A} - \left(\hat{\mathbf{k}}\cdot\mathbf{A}\right)\hat{\mathbf{k}} \pm i(\hat{\mathbf{k}}\times\mathbf{A})\right]$ the equations (8) decouple according to

$$\left(-\omega^2 + k^2 \pm 2\tilde{\xi}\,\omega^2 k\right)\mathbf{A}^{\pm} = 4\pi\mathbf{j}_T^{\pm}. \tag{9}$$

Thus, outside the sources the potential fields $\mathbf{A}^{\pm}$ propagate with well defined energy-dependent phase velocities $v_{\mp}^{-1} = \sqrt{1+\omega^2\tilde{\xi}^2} \pm \omega\tilde{\xi}$. In other words, the electromagnetic field in the MP theory propagates in vacuum as in a standard dispersive birefringent medium, where the modes with circular polarization have well defined refraction indices

$$n(\lambda z) = \sqrt{1+z^2} + \lambda z, \tag{10}$$

with $z = \tilde{\xi}\omega$ and $\lambda = \pm 1$.

The dynamics of a classical charged particle can be obtained from the geometrical optics limit of a scalar charged field. For this field the MP action is

$$S_{MP} = \int d^4x \left[\partial_\mu \varphi^* \partial^\mu \varphi - \mu^2 \varphi^* \varphi + i\tilde{\eta}\varphi^* \left(V\cdot\partial\right)^3 \varphi\right], \tag{11}$$

and leads to the dispersion relation $\left(p^0\right)^2 + \tilde{\eta}\left(p^0\right)^3 = \mathbf{p}^2 + \mu^2$, in the reference frame where $V^\alpha = (1,\mathbf{0})$. Here $\tilde{\eta} = -\eta/M_P$, where η is the dimensionless constant employed in the parameterization of Jacobson et al.. From the above dispersion relations, using the minimal interaction for the electromagnetic field and the geometric optics limit, we obtain the equation of motion for a massive charged particle interacting with a magnetic field, up to second order in $\tilde{\eta}$

$$\ddot{\mathbf{r}} = \frac{q}{E}\left(1 - \frac{3}{2}\tilde{\eta}E + \frac{9}{4}\tilde{\eta}^2 E^2\right)(\mathbf{v}\times\mathbf{B}), \tag{12}$$

with E being the energy of the particle. To simplify, we will consider the motion in a plane orthogonal to $\mathbf{B}$, which corresponds to the circular orbit

$$\mathbf{r}(t) = R\left(\cos\omega_0 t,\ \sin\omega_0 t,\ 0\right), \tag{13}$$

with radius $R = \beta/\omega_0$ and Larmor frequency

$$\omega_0 = \frac{|q|B}{E}\left(1 - \frac{3}{2}\tilde{\eta}E + \frac{9}{4}\tilde{\eta}^2 E^2\right). \tag{14}$$

The range of energies expected in astrophysical systems satisfies $E < (\mu/\tilde{\eta})^{1/2}$, with $\mu/E << 1$ and $\tilde{\eta}E << 1$ and hence the Lorentz factor γ can be approximated by

$$\gamma = \left(1-\beta^2\right)^{-1/2} \simeq \frac{E}{\mu}\left(1 + 2\frac{\tilde{\eta}E^3}{\mu^2}\right)^{-1/2}, \tag{15}$$

where we are using the standard definition $\beta = |\mathbf{v}|/c$.

3. RADIATION FIELD

The solution for the equation of motion (8) in the radiation approximation can be written as

$$\mathbf{A}(\omega,\mathbf{r}) = \mathbf{A}_+(\omega,\mathbf{r}) + \mathbf{A}_-(\omega,\mathbf{r}) = \frac{1}{r}\frac{2n(z)}{1+n^2(z)}\sum_{\lambda=\pm1} n(\lambda z)e^{in(\lambda z)\omega r}\mathbf{j}^\lambda(\omega,\mathbf{k}_\lambda), \qquad (16)$$

where $\mathbf{r} = r\hat{\mathbf{n}}$, $\mathbf{k}_\lambda = \omega n(\lambda z)\hat{\mathbf{n}}$. The fields $\mathbf{A}_+(\omega,\mathbf{r})$ and $\mathbf{A}_-(\omega,\mathbf{r})$ correspond to right and left circular polarization respectively. The electric and magnetic fields are $\mathbf{E}(\omega,\mathbf{r}) = i\omega\mathbf{A}(\omega,r)$ and $\mathbf{B}(\omega,\mathbf{r}) = \nabla\times\mathbf{A}(\omega,\mathbf{r})$. Let us remark that, contrary to the standard case, $\mathbf{E}$ and $\mathbf{B}$ are not orthogonal.

Although these fields provide a complete description of the radiation, the angular distribution of the power spectrum itself is very relevant from the phenomenological point of view. The energy emitted can be computed using the Poynting vector (6)

$$E = \int_{-\infty}^{\infty} dt\, \mathbf{n}\cdot\mathbf{S}(t,\mathbf{r})\, r^2 d\Omega, \qquad (17)$$

which is related to the power spectrum distribution by

$$E = \int_0^\infty d\omega \int d\Omega \frac{d^2E}{d\Omega d\omega} = \int_0^\infty d\omega \int d\Omega \int dT \frac{d^2P(T)}{dwd\Omega}. \qquad (18)$$

The final expression for the angular distribution of the energy spectrum in terms of the potentials results

$$\frac{d^2E}{d\Omega d\omega} = \frac{r^2\omega^2}{8\pi^2}\frac{1+n^2(z)}{n(z)}\left[\mathbf{A}_-(-\omega,\mathbf{r})\cdot\mathbf{A}_+(\omega,\mathbf{r}) + \mathbf{A}_-(\omega,\mathbf{r})\cdot\mathbf{A}_+(-\omega,\mathbf{r})\right]. \qquad (19)$$

To compute this distribution it is only necessary to express the products $\mathbf{A}_\mp(-\omega,\mathbf{r})\cdot\mathbf{A}_\pm(\omega,\mathbf{r})$ in terms of the current $\mathbf{j}(\omega,\, \mathbf{k})$ via the relation (16). This yields

$$\frac{d^2P(T)}{d\omega d\Omega} = \frac{\omega^2}{2\pi^2}\sum_{\lambda=\pm1}\frac{n^3(\lambda z)}{1+n^2(\lambda z)}\int_{-\infty}^{\infty} d\tau e^{-i\omega\tau} e^{i\omega n(\lambda z)\hat{\mathbf{n}}\cdot(\mathbf{r}(T+\tau/2)-\mathbf{r}(T-\tau/2))}\mathfrak{J}, \qquad (20)$$

for the angular distribution of the radiated power spectrum, with

$$\begin{aligned}\mathfrak{J} \;=\; & \frac{1}{2}\left[\mathbf{j}^*(T+\tau/2)\cdot\mathbf{j}(T-\tau/2) - n^{-2}(\lambda z)\rho^*(T+\tau/2)\rho(T-\tau/2)\right.\\ & \left.-i\lambda\hat{\mathbf{n}}\cdot\mathbf{j}^*(T+\tau/2)\times\mathbf{j}(T-\tau/2)\right]. \end{aligned} \qquad (21)$$

Another relevant characteristic of the radiation is its polarization. To describe the polarization, we will use the reduced Stokes parameters v, q, u, according to the definitions of Ref. [13], where they are written in terms of the circular polarization basis $\mathbf{e}_\pm$ and satisfy the constraint $1 = v^2 + q^2 + u^2$. Purely circular and linear polarizations are described by $v = \pm1$, $q = u = 0$ and $q = \pm1$, $v = u = 0$, respectively.

4. SYNCHROTRON RADIATION

In the case of the synchrotron radiation produced by a charge moving in a magnetic field we have $\rho(t,\mathbf{r}) = q\delta^3(\mathbf{r}-\mathbf{r}(t))$ and $\mathbf{j}(t,\mathbf{r}) = \rho(t,\mathbf{r})\mathbf{v}(t)$. To be explicit we introduce the direction of observation

$$\hat{\mathbf{n}} = (\sin\theta, 0, \cos\theta), \tag{22}$$

in the same coordinate system where $\mathbf{r}(t)$ was defined in Eq. (13). Recalling the corresponding expressions for $\mathbf{v}(t)$ we get

$$\begin{aligned} \mathbf{v}(T+\tau/2)\cdot\mathbf{v}(T-\tau/2) &= \beta^2\cos\omega_0\tau, \\ \mathbf{n}\cdot\mathbf{v}(T+\tau/2)\times\mathbf{v}(T-\tau/2) &= -\beta^2\cos\theta\sin\omega_0\tau \end{aligned} \tag{23}$$

and substituting in Eq. (21) we are left with

$$\mathfrak{J} = \frac{q^2}{2}\left[\beta^2\cos\omega_0\tau - n^{-2}(\lambda z) + i\lambda\beta^2\cos\theta\sin\omega_0\tau\right]. \tag{24}$$

The power spectrum can be computed from (20) by taking the time average over the macroscopic time T. We are interested in the synchrotron radiation of astrophysical objects where possible Lorentz violations could be tested, such as the emissions by supernova remnants (SNR) or gamma ray bursts (GRB). In such systems, frequencies of the order of $\omega = m\omega_0 \gtrsim 1\ GeV$ are detected, and there is also evidence of electrons of energies up to $10^4 TeV$. These range of energies correspond to a typical Larmor frequency $\omega_0 \simeq 10^{-30}\ GeV$, which means $m \gtrsim 10^{30}$, and to a Lorentz factor $\gamma \sim 10^9$. According to this, the interesting regime is characterized by $m \gg 1$ and $\gamma \gg 1$, but such that $m/\gamma \gg 1$. In this limit we get

$$\left\langle \frac{d^2P(T)}{d\omega d\Omega} \right\rangle = \sum_{m=0}^{\infty} \delta(\omega - m\omega_0)\frac{dP_m}{d\Omega}, \tag{25}$$

with $\omega_0 = \left(1+\frac{\tilde{\eta}\mu}{3}\gamma^3\right)|q\mathbf{B}|/(\mu\gamma)$ and the power angular distribution, in terms of the Bessel functions J_m (and their derivatives J'_m), is

$$\frac{dP_m}{d\Omega} = \frac{\omega^2 q^2}{2\pi}\left(1+2m\tilde{\xi}\omega\cos\theta\right)\left\{\left[J'_m(m\beta\sin\theta)\right]^2 + \left[\frac{J_m(m\beta\sin\theta)}{\tan\theta}\right]^2\right\}. \tag{26}$$

This expression shows an anisotropy to first order in $\tilde{\xi}$. The emission is suppressed (enhanced) when $m\tilde{\xi}\omega\cos\theta \simeq -1(+1)$. Note that both effects are amplified by the presence of the factor $m = \omega/\omega_0$ which can be very large in some regimes. When m is large the radiation becomes confined to a small angular range $\Delta\theta \simeq m^{-1/3}$ around $\theta = \pi/2$[14]. Thus, if we consider the radiation in the frontiers of the beam, i.e. $\theta \simeq \pi/2 \pm m^{-1/3}$, the anisotropy becomes significant when $\omega \simeq \left(\omega_0^2/\tilde{\xi}^3\right)^{1/5}$. The spatially integrated power radiated in the mth harmonic is approximately

$$P_m \simeq \frac{q^2 m\omega_0}{\sqrt{3}\pi R\gamma^2}\left[\frac{m_c}{m}\kappa\left(\frac{m}{m_c}\right) - \frac{2}{\gamma^2}K_{2/3}\left(\frac{m}{m_c}\right) + 2\left(m\omega_0\tilde{\xi}\beta\right)^2\frac{m^2}{\gamma^2}K_{2/3}\left(\frac{m}{m_c}\right)\right], \tag{27}$$

with m_c being the critical value $m_c = 3\gamma^3/2$. For large m and $1-\beta^2 > 0$ the behavior of the power radiated in the mth harmonic can be obtained using the asymptotic expressions for the MacDonald functions $K_{m/n}$. Within this limit there are two regimes of interest, according to the ratio between m and m_c. For the case $m/m_c >> 1$ we get

$$P_m \simeq \frac{1}{2R}\sqrt{\frac{m}{\pi\gamma}}\left(\frac{q^2 B}{E}\right)^2\left(1-3\tilde{\kappa}E+\frac{27}{4}\tilde{\kappa}^2E^2\right)\left[1+2\tilde{\xi}^2\left(\frac{m\omega}{\gamma}\right)^2\right]e^{-2m/3\gamma^3}, \quad (28)$$

while the complementary range $m/m_c << 1$ produces

$$P_m \simeq \frac{\sqrt[3]{9}\Gamma(2/3)}{\sqrt{3}\pi}\left(\frac{q^2 B}{E}\right)^2\left(1-3\tilde{\kappa}E+\frac{27}{4}\tilde{\kappa}^2E^2\right)\left[1+\left(\frac{m\omega_0\tilde{\xi}\beta}{\sqrt[3]{2}}\right)^2\frac{m^2}{\gamma^2}\right]m^{1/3}. \quad (29)$$

To characterize the polarization we can restrict our discussion to $\theta = \pi/2$, since the radiation is mostly concentrated around the plane of the orbit. In this case the effects of the Lorentz violation in the current are negligible, and only the explicit violations in the Maxwell equations affect the polarization of the radiation. The Stokes parameters take the form

$$\begin{aligned} v_m &= \frac{1-R_m^2}{1+R_m^2}, \\ q_m &= \frac{2R_m}{1+R_m^2}\cos([n(-z)-n(z)]\,\omega r), \\ u_m &= \frac{2R_m}{1+R_m^2}\sin([n(-z)-n(z)]\,\omega r), \end{aligned} \quad (30)$$

which give them in terms of the polarization index R_m

$$R_m(\theta=\pi/2) \simeq 1-2\frac{\tilde{\xi}m\omega}{\gamma^2}\frac{J_m(m\bar{\beta})}{J'_m(m\bar{\beta})}, \quad (31)$$

where $\bar{\beta} = \sqrt{1-\mu^2/E^2}$. In the large m limit and in the case $1-\beta^2 > 0$, one can write the Bessel function and its derivative in terms of the MacDonald functions. The latter can be further approximated according to $m/\hat{m}_c \lessgtr 1$ where $\hat{m}_c = 3\left(1-\bar{\beta}^2\right)^{-3/2}/2$. In such a way, if $m/\hat{m}_c >> 1$ the radiated power is exponentially damped and we get

$$R_m(\theta=\pi/2) \simeq 1-2\tilde{\xi}\omega m/\gamma. \quad (32)$$

Then we have, to first order in $\tilde{\xi}$,

$$v_m \simeq 2\omega m\tilde{\xi}/\gamma, \quad q_m \simeq \cos(2\tilde{\xi}\omega^2 r), \quad u_m \simeq -\sin(2\tilde{\xi}\omega^2 r). \quad (33)$$

In the opposite case, when $m/\hat{m}_c << 1$, we get

$$R_m(\theta=\pi/2) \simeq 1-\frac{2\tilde{\xi}\omega}{\gamma^2}\left(\frac{1}{4}\right)^{1/3}\frac{\Gamma(1/3)}{\Gamma(2/3)}m^{4/3}, \quad (34)$$

from which we have

$$v_m \simeq \left(\frac{1}{4}\right)^{1/3} \frac{\Gamma(1/3)}{\Gamma(2/3)} \frac{2\omega}{\gamma^2} m^{4/3} \tilde{\xi}\,, \qquad q_m \simeq \cos(2\tilde{\xi}\omega^2 r)\,, \qquad u_m \simeq -\sin(2\tilde{\xi}\omega^2 r). \quad (35)$$

In both cases powers of m appear as amplifiying factors. Note that q_m and u_m oscillate with a wave length $\Lambda = \pi/(\tilde{\xi}\omega^2)$. These oscillations come from the term $\pm\tilde{\xi}\omega^2 r$ of the phases in Eq. (16), i.e. they are a consequence of the birefringence of the vacuum.

5. FINAL REMARKS

The main results of a complete calculation of synchrotron radiation in the MP effective electrodynamics has been presented. This effective electrodynamics can be interpreted in terms of a standard parity violating birefringent and dispersive medium. A priori one can expect that the Lorentz violating effects would be of order $\tilde{\xi}\omega$. Instead we find amplifying factors proportional to m, which could render these modifications within or near the scope of present time observations.

When an astrophysical object is considered as a source of synchrotron radiation it is necessary to take into consideration the actual range of validity of the radiation approximation from where these results have been obtained. The propagation of the fields is characterized by the phase

$$n(\lambda z)\omega\left|\mathbf{r}-\mathbf{r}'\right| \simeq \omega r\left(1-\frac{\mathbf{n}\cdot\mathbf{r}'}{r}+\lambda\tilde{\xi}\omega-\lambda\tilde{\xi}\omega\frac{\mathbf{n}\cdot\mathbf{r}'}{r}+\frac{1}{2}\frac{r'^2}{r^2}\right), \quad (36)$$

where r' can be estimated by the radius of the orbit. In the radiation approximation the term proportional to $(r'/r)^2$ is neglected. If $|\xi\omega| > r'/r$ we can neglect only the term quadratic in $r'$and both $\tilde{\xi}$-dependent terms remain in the phase, which can be approximated, using $\mathbf{k}_\lambda = \omega n(\lambda z)\hat{\mathbf{n}}$, by

$$n(\lambda z)\omega\left|\mathbf{r}-\mathbf{r}'\right| \simeq n(\lambda z)\omega r\left(1-\hat{\mathbf{n}}\cdot\mathbf{r}'/r\right). \quad (37)$$

This is the full far-field approximation developed in the preceding sections, with the most general dependence on the index of refraction. An intermediate situation happens when $(r'/r)^2 < \left|\tilde{\xi}\omega\right| < r'/r$, in which case

$$n(\lambda z)\omega\left|\mathbf{r}-\mathbf{r}'\right| \simeq n(\lambda z)\omega r - \hat{\mathbf{n}}\cdot\mathbf{r}', \quad (38)$$

and hence the index of refraction still remains significant. Finally, if $\left|\tilde{\xi}\omega\right| < (r'/r)^2$ all the dependence on $\tilde{\xi}$ is negligible in the phase, which reduces to the usual expression

$$n(\lambda z)\omega\left|\mathbf{r}-\mathbf{r}'\right| \simeq \omega r - \hat{\mathbf{n}}\cdot\mathbf{r}'. \quad (39)$$

To be concrete we can consider the Crab Nebula, an SNR at 6000 ly. The synchrotron radiation spectrum shows a cutoff at $\omega \simeq 0.5\ GeV$, and there is evidence of electrons of energy $E_e \simeq 3 \times 10^6$ GeV producing this emission. We can estimate the magnetic field, the γ factor and the Larmor frequency using the unperturbed relations which give us $B \simeq 10^{-3}\ G$, $\gamma \simeq 6.0 \times 10^9$, and $\omega_0 \simeq 2 \times 10^{-30}$ GeV. Thus, in this case we have $r \sim 10^{36}$ GeV^{-1} and $r' \sim 10^{30}$ GeV^{-1}. Considering that $\tilde{\xi} \lesssim 10^{-19}$ GeV^{-1}, we get that not only $\left|\tilde{\xi}\omega\right| < r'/r$, but also $\left|\tilde{\xi}\omega\right| < (r'/r)^2$. Therefore all the information about the $\tilde{\xi}$ parameter is erased by the radiation approximation and the power spectrum only contains the parameter $\tilde{\eta}$, in agreement with Ref. [10]. When radiation from much more distant objects such as GRBs is considered, information about the parameter $\tilde{\xi}$ may become accessible. For example, in the case of Mkn 501 we have: $r \sim 10^8\ l.y. \sim 10^{40}$ GeV^{-1} and $r' \sim 10^{26}$ GeV^{-1}, which give $r'/r \sim 10^{-14}$. The photon energy is $\omega \sim 10^4$ GeV, and thus $\left|\tilde{\xi}\omega\right| \lesssim 10^{-15}$. In this way we have $(r'/r)^2 < \tilde{\xi}\omega < r'/r$, and hence the term $\tilde{\xi}\omega r$ in the phase becomes significant. Another example is GRB021206, at a distance of 10^{10} l.y., with a larger magnetic field and therefore a smaller radius for the electron orbits. Here $r'/r \sim 10^{-24}$, $\omega \sim 10^{-3}$ GeV, and $\left|\tilde{\xi}\omega\right| \lesssim 10^{-22}$, which yields $r'/r \ll \left|\tilde{\xi}\omega\right|$. In principle this case could allow access to the full far-field approximation. Summarizing, in the case of the CRAB nebula radiation the far-field approximation erases all the information about the $\tilde{\xi}$ parameter and only allows to test the parameter $\tilde{\eta}$ associated to the dynamics of the charged particle. The situation changes for extragalactic sources, such as in the case of GRB's. For these objects the far-field approximation could include $\tilde{\xi}$-LV terms and the m -amplifying factors become important, thus giving access to a new sector of the Lorentz violating parameters.

REFERENCES

1. G. Amelino-Camelia, J.R. Ellis, N.E. Mavromatos, D.V. Nanopoulos and S. Sarkar, Nature **393**, 763 (1998).
2. R. Gambini and J. Pullin, Phys. Rev. **D59**, 124021 (1999); R. Gambini and J. Pullin, in *Proceedings of the Second Meeting on CPT and Lorentz Symmetry*, ed. V.A. Kostelecky, World Scientific, 2002.
3. For a review see T. Thiemann, *Introduction to Modern Canonical Quantum General Relativity*, gr-qc/0110034.
4. J. Alfaro, H.A. Morales-Técotl and L.F. Urrutia, Phys. Rev. **D65**, 103509 (2002);
5. J. Alfaro, H.A. Morales-Técotl, L.F. Urrutia, Phys. Rev. Lett. **84**, 2318 (2000); J. Alfaro, H.A. Morales-Técotl and L.F. Urrutia, Phys. Rev. **D66**, 124006 (2002);
6. L.F. Urrutia, Mod. Phys. Lett. **A17**, 943 (2002); L.F.Urrutia, in *Particles and Fields: Proceedings of the Tenth Mexican School on Particles and Fields* eds. U. Cotti, M. Mondragón and G. Tavares-Velasco, AIP Conference Proceedings # 670, American Institute of Physics, 2003.
7. J.R. Ellis, N.E. Mavromatos and D.V. Nanopoulos, Phys. Rev. **D61**, 027503 (2000); J.R. Ellis, K. Farakos, N.E. Mavromatos, V.A. Mitsou and D.V. Nanopoulos, Astrophys. J. **535**, 139 (2000); J.R. Ellis, N.E. Mavromatos, D.V. Nanopoulos and G. Volkov, Gen. Rel. Grav. **32**, 1777 (2000).
8. R. Montemayor and L. Urrutia, Phys. Letts. **B 606**, 86 (2005)
9. R.C. Myers and M. Pospelov, Phys. Rev. Letts. **90**, 211601 (2003).
10. T. Jacobson, S. Liberati and D. Mattingly, Nature (London) **424**, 1019 (2003) ; ibid., Phys. Rev. **D67**, 124011 (2003) ; ibid., Phys. Rev. **D66**, 081302 (2002) ; T. Jacobson, S. Liberati, D. Mattingly and F. W. Stecker, Phys. Rev. Lett. **93**, 021101 (2004) .
11. A. Pérez and D. Sudarsky, Phys. Rev. Lett. **91**,179101 (2003)
12. J. Collins, A. Pérez and D. Sudarsky, L. Urrutia and H. Vucetich, Phys. Rev. Lett. **93**, 191301 (2004).

13. G.B. Rybicki and A.P. Lightman, *Radiative Processes in Astrophysics*, John Wiley & Sons, Inc., New York, 1979.
14. J. Schwinger, L.L. de DeRaad, K.A. Milton and W.-Y. Tsai, *Classical Electrodynamics*, Westview Press, Perseus Books Group, Boulder, 1998.

Deformed dispersion relations and Sagnac interferometr

A. Camacho and E. Castellanos

Departamento de Física, niversidad Autónoma Metropolitana–Iztapalapa, Apartado Postal 55–534, C.P. 09340, México, D.F., México

Abstract. The experimental requirements that could lead to the detection of the so–called deformed dispersion relations, in the context of Sagnac interferometry, are obtained.

1. INTRODUCTION

The possibility of a modification of the dispersion relation [1] appears in connection with the quest for a quantum theory of gravity. In the literature the ideas that contain this new dispersion relation teem. Indeed, we may find it in the context of non–commutative geometry [2, 3], or loop–quantum gravity models [4, 5], etc. One of the features that this fact entails is the presence of Lorentz symmetry only as an approximation for quantum space [6, 7].

There are several experimental proposals dedicated to the possible detection of this effect [1, 8], and a fleeting glimpse at them readily shows us that the aforementioned approaches fall within the so–called first–order coherence experiments [9]. Nevertheless, of this kind of experiments at least one of them is missing, namely, Sagnac's idea [9]. Here we analyze the use of Sagnac interferometry, in the realm of a deformed dispersion relation. The modifications upon the difference in the time of arrival, of the two involved light beams, will be deduced, as well as the corresponding phase difference. The experimental requirements that could lead to the detection of a deformed dispersion relation will be also calculated.

2. SAGNAC INTERFEROMETRY AND A DEFORMED DISPERSION RELATION

As previously mentioned, several quantum–gravity theories predict a modified dispersion relation [1, 4], which can be characterized, from a phenomenological point of view, by corrections hinging upon Planck's length, i.e., l_p

$$E^2 = p^2c^2\left[1-\alpha\left(E\sqrt{G/(c^5\hbar)}\right)^n\right]. \tag{1}$$

Recalling that

$$p = \hbar k, \tag{2}$$

CP758, *Gravitation and Cosmology: 2nd Mexican Meeting on Mathematical and Experimental Physics*
edited by A. Macías, C. Läemmerzahl and D. Nuñez

we conclude

$$k = \frac{E/(c\hbar)}{\left[1-\alpha\left(E\sqrt{G/(c^5\hbar)}\right)^n\right]^{1/2}}. \tag{3}$$

The modifications are quite small, then the following expansion is justified

$$k = \frac{E}{c\hbar}\left[1+\frac{\alpha}{2}\left(E\sqrt{G/(c^5\hbar)}\right)^n+\frac{3}{8}\alpha^2\left(E\sqrt{G/(c^5\hbar)}\right)^{2n}+\ldots\right]. \tag{4}$$

In the realm of deformed dispersion relations the speed of light is not constant [7]. Forsooth, in this case

$$v = c\left[1-\alpha\left(E\sqrt{G/(c^5\hbar)}\right)^n\right]^{1/2}\left[1\frac{\alpha n\left(E\sqrt{G/(c^5\hbar)}\right)^n}{1-\alpha\left(E\sqrt{G/(c^5\hbar)}\right)^n}\right]^{-1}. \tag{5}$$

Setting $\alpha = 0$ in this last equation renders the usual speed of light. From now on we will write $E_p = \sqrt{c^5\hbar/G}$.

Now we introduce a Sagnac interferometer [9], whose radius and angular velocity are b and Ω, respectively. Additionally, we will assume that this device is immersed in a region where a gravitational potential exists, namely

$$U(z) = gz. \tag{6}$$

The zero of this potential will be chosen at the lowest point of the interferometer. The plane where this apparatus is located has its normal vector perpendicular to the z–axis. Assume now that at the highest point of the interferometer a light beam, with frequency $\tilde{\nu}$, enters the device, and it is split up into two parts, one rotating in the same direction as the interferometer, whereas the second beam travels in the opposite direction.

Each beam is subject to a gravitational shift

$$\nu = \frac{\tilde{\nu}}{1+\Delta U/v^2}. \tag{7}$$

Hence

$$v = c\left[1-\alpha\left(\frac{n}{2}+1\right)\left(\frac{\hbar\tilde{\nu}}{E_p}\right)^n\left(1+\frac{\Delta U}{v^2}\right)^{-n}\right]. \tag{8}$$

This last expression leads us to a third–order polynomial, which renders, approximately, the speed of the light beam

$$v^3 + c\left[\alpha\left(\frac{n}{2}+1\right)\left(\frac{\hbar\tilde{\nu}}{E_p}\right)^n - 1\right]v^2 + n\Delta U v - cn\Delta U = 0. \tag{9}$$

A fleeting glimpse at (9) shows that if we set $\alpha = 0$ and $\Delta U = 0$, then $v = c$ is indeed a solution. In other words, we recover the usual situation.

A rough approximation renders the solution

$$v = c\left[1 - n\Delta U/c^2 - \frac{\alpha}{3}\left(\frac{n}{2}+1\right)\left(\frac{\hbar\tilde{\nu}}{E_p}\right)^n\right]. \tag{10}$$

This last expression will be employed in order to find the time that each one of our beams requires to meet the interferometer. Let us now proceed to deduce the interference pattern, and in order to do so let us introduce the angle θ, the one measures the rotation of the interferometer. Here $\theta = 0$ coincides with the point at which the beam enters into the interferometer. From our previous definitions we have for the z–coordinate of the interferometer

$$z = b[1 + \cos\beta]. \tag{11}$$

Since the beam enters the interferometer at the point

$$z_0 = 2b, \tag{12}$$

then

$$\Delta U = gb[\cos\beta - 1]. \tag{13}$$

β is the angle described by the interferometer, whereas, θ is related to the rotation of the beams. Additionally, we have

$$v = b\frac{d\theta}{dt}. \tag{14}$$

The travelled time, required by the beams, as a function of the angle is deduced joining expressions (10) and (14)

$$t - t_0 = \frac{b}{c}\int_0^{\infty}\left[1 - \frac{\alpha}{3}\left(\frac{n}{2}+1\right)\left(\frac{\hbar\tilde{\nu}}{E_p}\right)^n - \frac{2gb}{c^2}[\cos(\theta) - 1]\right]^{-1} d\theta. \tag{15}$$

For the sake of clarity let us set $t_0 = 0$, and then we may integrate (15) quite easily [10]

$$\begin{aligned} t &= \frac{2b}{c}\left\{\left[1 - \frac{\alpha}{3}\left(\frac{n}{2}+1\right)\left(\frac{\hbar\tilde{\nu}}{E_p}\right)^n + \frac{2gb}{c^2}\right]^2 - 4\left(\frac{2gb}{c^2}\right)^2\right\}^{-1/2} \\ &\times \arctan\left\{\left(\left[1 - \frac{\alpha}{3}\left(\frac{n}{2}+1\right)\left(\frac{\hbar\tilde{\nu}}{E_p}\right)^n + \frac{2gb}{c^2}\right]^2 - 4\left(\frac{2gb}{c^2}\right)^2\right)^{1/2}\right. \\ &\left.\times\left[1 - \frac{\alpha}{3}\left(\frac{n}{2}+1\right)\left(\frac{\hbar\tilde{\nu}}{E_p}\right)^n\right]^{-1}\tan\frac{\theta}{2}\right\}. \end{aligned} \tag{16}$$

Hence, approximately

$$\frac{ct}{2b}\left\{\left[1 - \frac{\alpha}{3}\left(\frac{n}{2}+1\right)\left(\frac{\hbar\tilde{\nu}}{E_p}\right)^n + \frac{2gb}{c^2}\right]^2 - \left(\frac{2gb}{c^2}\right)^2\right\}^{1/2} = \frac{\theta}{2}. \tag{17}$$

The beam moving in the opposite direction, with respect to the displacement of the interferometer, meets the detection device after it has moved an angle equal to

$$\beta_1 = t_1\Omega. \tag{18}$$

The angle described by the beam reads then

$$\theta_1 = 2\pi - \beta_1. \tag{19}$$

As of the remaining beam, it meets the interferometer after this has rotated an angle

$$\beta_2 = t_2\Omega. \tag{20}$$

The angle described by the second beam is given by

$$\theta_2 = 2\pi + \beta_2. \tag{21}$$

Introducing these two times, t_1 and t_2, into (17) renders

$$t_1 = \frac{2\pi b}{c}\left[\gamma + \frac{b\Omega}{c}\right], \tag{22}$$

$$t_2 = \frac{2\pi b}{c}\left[\gamma - \frac{b\Omega}{c}\right], \tag{23}$$

$$\gamma = \left\{\left[1 - \frac{\alpha}{3}\left(\frac{n}{2}+1\right)\left(\frac{\hbar\tilde{\nu}}{E_p}\right)^n + \frac{2gb}{c^2}\right]^2 - \left(\frac{2gb}{c^2}\right)^2\right\}^{1/2}. \tag{24}$$

These two last expressions entail ($\Delta t = t_2 - t_1$)

$$\Delta t = \frac{4\pi b^2\Omega}{c^2}\left\{\left[1 - \frac{\alpha}{3}\left(\frac{n}{2}+1\right)\left(\frac{\hbar\tilde{\nu}}{E_p}\right)^n + \frac{2gb}{c^2}\right]^2 - \left(\frac{2gb}{c^2}\right)^2 - \frac{b^2\Omega^2}{c^2}\right\}^{-1}. \tag{25}$$

3. CONCLUSIONS

For the sake of clarity let us consider the case in which $n = 1$. Under this condition (25) turns out to be

$$\Delta t = \frac{4\pi b^2\Omega}{c^2}\left\{1 + \alpha\frac{\hbar\tilde{\nu}}{E_p} - 4\frac{gb}{c^2} - \frac{\alpha^2}{4}\left(\frac{\hbar\tilde{\nu}}{E_p}\right)^2 + 2\alpha\frac{gb}{c^2}\frac{\hbar\tilde{\nu}}{E_p} + \left(\frac{b\Omega}{c}\right)^2\right\}. \tag{26}$$

The optical path and phase differences, associated to (26), read

$$\Delta L = c\Delta t, \tag{27}$$

$$\Delta\theta = \frac{\Delta L}{\tilde{\lambda}}. \tag{28}$$

In consequence

$$\Delta\theta = \frac{4\pi b^2\Omega}{c\tilde{\lambda}}\left\{1 + \alpha\frac{\hbar\tilde{\nu}}{E_p}\left(1 - \frac{\alpha}{4}\frac{\hbar\tilde{\nu}}{E_p}\right) - 4\frac{gb}{c^2}\left(1 - \frac{\alpha}{2}\frac{\hbar\tilde{\nu}}{E_p}\right) + \left(\frac{b\Omega}{c}\right)^2\right\}. \tag{29}$$

Setting $\alpha = 0$ we recover the usual value, $\Delta\theta^{(0)}$, for the phase shift [9]. In the present model the feasibility of this experimental proposal hinges upon the difference $\Delta\theta -$

$\Delta\theta^{(0)}$. Forsooth, if $\delta\theta$ denotes the corresponding experimental resolution, then the requirement, to be fulfilled, reads

$$\delta\theta < \left|\Delta\theta - \Delta\theta^{(0)}\right|. \tag{30}$$

This may be cast in the following approximate form

$$\delta\theta < \frac{4\pi b^2\Omega}{c\tilde{\lambda}}\alpha\frac{\hbar\tilde{\nu}}{E_p}. \tag{31}$$

Considering the case of an optical transition, i.e., $\hbar\tilde{\nu} \sim 10^{-12}$ erg. (31) (and also assuming that $\delta\theta \sim 10^{-6}$ [9]) entails

$$b^2\Omega > 10^{27}\ \mathrm{cm}^2/\mathrm{s}. \tag{32}$$

A very stringent restriction, since it requires, either, very high angular velocities, or very large radii. If we now impose a smoother experimental condition

$$b^2\Omega \sim 10^5\ \mathrm{cm}^2/\mathrm{s}, \tag{33}$$

then we obtain

$$\hbar\tilde{\nu} > 1\ \mathrm{erg}. \tag{34}$$

ACKNOWLEDGMENTS

This research was supported by CONACYT Grant 42191–F. A. Camacho would like to thank A.A. Cuevas–Sosa and A. Macías for useful discussions and literature hints.

REFERENCES

1. G. Amelino–Camelia, Int. J. Mod. Phys. **D9**, 1633 (2003).
2. G. Amelino–Camelia and T. Piran, Phys. Rev. **D64**, 036005 (2001).
3. G. Amelino–Camelia, Nature **418**, 661 (2000).
4. R. Gambini and J. Pullin, Phys. Rev. **D59**, 124021 (1999).
5. L. Smolin, LANL hep–th/0303185.
6. G. Amelino–Camelia, Nature **398**, 216 (1999).
7. G. Amelino–Camelia, J. Ellis, N. E. Mavromatos, D. V. Nanopoulos, and S. Sarkar, Nature **393** (London), 763 (1998); J. Alfaro, H. A. Morales–Tecotl, and L. F. Urrutia, Phys. Rev. Lett. **84**, 2318 (2000).
8. G. Amelino–Camelia and C. Lämmerzahl, Class. Quantum Grav.**21**, 899–915 (2004).
9. L. Mandel and E. Wolf, *Optical Coherence and Quantum Optics*, Cambridge University Press, Cambridge (1995).
10. I. S. Gradshteyn and I. M. Ryzhik, *Table of Integrals, Series and Products*, Academic Press, New York (1980).

Experimental Tests of the Equivalence Principle and Newton's Law in Space

H. Dittus and C. Lämmerzahl

ZARM, University of Bremen, Am Fallturm, 28359 Bremen, Germany

Abstract. Tests of the Weak Equivalence Principle (WEP) by means of torsion balances have confirmed the WEP nearly to the 10^{-13} level. At this level, the coupling to changing gravitational environment on earth as well as seismic and thermal noise seem to establish a severe threshold for the improvement of accuracy. In order to overcome these difficulties, experiments in space offer new ranges of accuracy: Space platforms and satellites in a nearly free fall along geodesics offer unique possibilities to improve the accuracy of WEP-tests by several orders of magnitude which is important as some quantum gravity inspired theories predict a WEP-violation. At present, two satellite projects are planned and scheduled to test the WEP for neutral bulk matter: the French mission MICROSCOPE approved to be launched early 2008 is designed to measure on the 10^{-16} level and the ESA/NASA mission STEP which is still under study, will be precise on the 10^{-18} level. Modern space technology is not only interesting for WEP tests of neutral bulk matter, but can be used to carry out comprehensive EP-tests (e.g. Lunar Laser Ranging). Beside tests with bulk matter, complementary tests with comparable results for other kind of matter or couplings (charged particles, atoms, neutrons, anti-matter, or spin) are required. New ideas for space experiments to test the WEP for anti-matter or charged matter have been discussed recently. Space also enables to test Newton's Law on very small and huge distances. In the range below 0.1 mm, one can make use of the advantage of the weightlessness environment. For long range measurements above 1 astronomical unit, deep space satellites are and will be used to probe the gravitational field with improving accuracy.

1. INTRODUCTION

In 1648, *Galileo Galilei* wrote in his famous "Discorsi et dimonstrazioni matematiche intorno a due nuove scienze": The experiment made with two moveables, as different as possible in weight, made to fall from height in order to observe whether they are of equal speed, labors under certain difficulties. If the height is very tall, the medium that must be opened and driven aside by the impetus of the falling body will be of greater prejudice to the small momentum of a very light moveable than to the force of a very heavy one, and over a long distance the light one will remain behind. But in a small height it may be doubtful whether there is really no difference, at least it is unobservable. It was *Isaac Newton* who, maybe as the first, recognized the importance of the Equivalence Principle. In his "Philosopiae Naturalis Principia Mathematica" (1687), he discusses in detail the fact well known to him, that two bodies in the gravitational field of a third one experience the same acceleration independently on their mass and composition. The 6th Theorem related to his Gravitational Law states: "That all bodies gravitate towards every planet; and that the weights of bodies toward any other planet, at equal distances from the centre of the planet, are proportional to the quantities of matter which they

CP758, *Gravitation and Cosmology: 2nd Mexican Meeting on Mathematical and Experimental Physics*
edited by A. Macías, C. Läemmerzahl and D. Nuñez

severally contain." Although *Newton* had carried out his own experiments with pendula, the decisive proof of the Theorem for him had been the astronomical observations of the planetary trajectories which demonstrates that the gravitational force is balanced by the centripetal (inertial) force. His theory implicates the validity of inertial and gravitational force as and the Universality of Free Fall results from *Newton*'s Gravitational Law.

In contrast to this, *Albert Einstein* revalorized the Principle of Equivalence in his theory of General Relativity as fundamental hypotheses (Hypothesis of complete equivalence). Therefore, experimental tests of the Equivalence Principle are fundamental and highly important with respect to the experimental validation of General Relativity. Usually, different formulations of the Equivalence Principle are distinguished. The Weak Principle of Equivalence (WEP) states that all point-like particles follow the same trajectory if the particles have the same initial conditions and only gravity is acting. The important fact is that the trajectory, independent of the particle properties like mass and composition characterizes the gravitational interaction. This independence from particle properties allows to identifiy the gravitational interaction with geometry [1]. The strong formulation of the Equivalence Principle (gravity can locally be transformed away by choosing a particular frame of reference) allows to fix the structure of the geometry. It states, that in a freeling falling laboratory (e.g. a satellite on an ideal geodesic) all freely falling point-like particles with some mass (locally) move along straight lines with constant velocity which is the usual definition of weightlessness. Therefore, the Strong Equivalence Principle implies that locally the action of any gravitational field can be compensated by choosing an adequate coordinate system.

Following *Einstein*'s formulation of 1907, one can define the Einstein Equivalence Principle (EEP) as an extension of the strong formulation. The EEP states that the outcome of any local, non-gravitational experiment is (1) independent of the orientation and velocity of the laboratory relative to the gravitational field and (2) is independent of where and when in the gravitational field the experiment is performed [2]. The EEP provides that not only the WEP but also Local Lorentz Invariance (LLI) referring to the first requirement and Local Position Invariance (LPI) which is described by the second statement are valid. A local experiment is performed within a space-time region small enough that tidal forces are not effective. A direct consequence of the LPI is the Universality of the Gravitational Redshift which states that all clocks despite of the working mechanism or physical principles they are based on behave in the same way and experience the same redshift (change of frequency) in a gravitational field. Therefore, tests of the validity of the Equivalence Principle do not only concentrate on the perfomance of classical Galilean type free fall experiments, but cover also tests to prove LLI and LPI, in particular UGR.

General Relativity requires another extension, because the validity of the EEP must be valid also for bodies with non-negligible self-gravitation. Self-gravitation of a an extended body like a planet or star must contribute to inertial ans gravitational mass in the same way wherefore an extended body in the gravitational field of a second extended massive object behaves like a point-like particle and, related to its centre of mass, moves like a test particle. This fundamental hypothesis is usually named the Strong Equivalence Principle with gravitation. It does not fix the geometry but the structure of the field equations and includes the WEP, SEP, and EEP. Important experimental tests of the Strong Equivalence Principle with gravitation have been performed by means of Lunar

TABLE 1. Experimental tests of the Weak Equivalence Principle for different kind of matters and materials.

neutral bulk matter	$\eta_E \leq (-0,2 \pm 2,8) \cdot 10^{-12}$	Su et al.[3] (torsion balance experiments with Al and Be)*
atoms	$\eta_E \leq 10^{-9}$	
neutrons	$\eta_E \leq 10^{-3}$	Schmiedmayer [4] (neutron scattering)
charged matter (electrons)	$\eta_E \leq 10^{-1}$	Witteborn, Fairbank [5]
magnetic matter (spin coupling)	$A < 10^{-11}$	hyperfine splittings of hydrogen
anti-matter		

* A more precise test is possible with Lunar Laser Ranging (LLR) measuring the distance between moon and earth freely falling in the Sun's gravitational field, see below. Nevertheless, a WEP test provides the validity of the Equivalence Principle for bodies with non-negligible self gravitation [6].

Laser Ranging.

In the following, we like to summarize the experimental tests performed up to now. We start with a review of the WEP tests in Sec.2, report briefly on tests of LLI and LPI in Sec.3, and in Sec.4 give an overview on planned and approved space missions to test the Equivalence Pirnciple.

2. EXPERIMENTAL PROOFS OF THE WEAK EQUIVALENCE PRINCIPLE

The Equivalence Principle is valid for all kinds of structureless matter and and states of matter. In General Relativity, elementary particles, atoms, macroscopic bulk matter, charged and neutral matter as well as magnetic and non-magnetic masses experience the Universality of Free Fall (UFF). Therefore, experiments to test the Equivalence Principle have to be carried out with all kind of matter. Practically, the most precise, and in particular the historical tests, are or have been performed with neutral bulk matter only. The reason for that is that experiments in which non-gravitational interactions become dominant need appropriate shielding (e.g. electrostatic fields for charged matter, magnetostatic forces for spin coupling tests, surface tension for fluids, and van der Waals forces for microscopic test masses). Table 1 shows the present experimental limits for WEP tests for different kind of matter.

η_E is the Eötvös parameter which is used as a measure for an eventual violation of the WEP. For two test masses of different composition a and b with inertial mass m_{ia} and m_{ib} as well as gravitational mass m_{ga} and m_{gb}, η_E is defined by

$$\eta_E \leq 2 \left| \frac{\left((m_i/m_g)_a - (m_i/m_g)_b\right)}{\left((m_i/m_g)_a + (m_i/m_g)_b\right)} \right| . \tag{1}$$

Proofs of the WEP are of great importance because it is a precondition of SEP and EEP and experiments can be performed on laboratory scale easily.

2.1. Experiments with neutral bulk matter

Experiments to prove the WEP have been carried out seriously by *Isaac Newton* the first time. He used masses of different composition (a and b) on pendula and compared their pendulum periods $T_{a,b} = 2\pi\sqrt{(m_i/m_g)_{a,b}/(l/g)}$. If the pendulum length l and the gravitational acceleration g at the laboratory are the same for all experiments, any difference of the periods would indicate a WEP-violation, because the simple relation

$$2\frac{T_a^2 - T_b^2}{T_a^2 + T_b^2} = 2\left|\frac{((m_i/m_g)_a - (m_i/m_g)_b)}{((m_i/m_g)_a + (m_i/m_g)_b)}\right| = \eta_E\,, \tag{2}$$

which can be derived from eq.1. It was *Friedrich Wilhelm Bessel* in the 19th century who improved the pendulum experiments to an accuracy of $\eta_E \leq 10^{-5}$. The accuracy of pendulum experiments is limited. In order to measure the pendulum period, one needs the approximation $\sin x = x$ in the non-linear differential equation. For precision test, this approximation leads to systematic errors because the pendulum period is not longer independent of the amplitude which decreases during integration time.

Around the year 1900, *Roland Baron v. Eötvös* carried out experiments with a torsion balance invented by him to test the WEP. In such a torsion balance, two test masses of different composition are placed at both ends of a small bar which itself is hanging on a thin torsion fibre which is about 1 m long. Any changing gravitational force acting on the test masses causes a torque, if the ratio of inertial and gravitational mass of the test masses is different. The feed-back force can be measured. Torsion blances can be operated in two different modes: (i) In the solar mode one uses the gravitational field of the Sun which gravitational force acting on the test masses is balanced by the centrifugal force of the rotating earth. A resulting torque would be caused by an eventual imbalance of the gravitational mass attracted by the sun and the inertial mass accelerated by the earth rotation of a test mass. (ii) In the terrestrial mode one turns the torsion balances within 24 h around the fibre axis synchronously to the Earth rotation. Because of the Earth's rotation, the gravity vector on the Earth's surface is inclined by a small angle with respect to the vertical. Therefore, an imbalance of inertial and gravitational force would result in a periodic variation of this angle[1]. *Eötvös* et al. were able to confirm $\eta_E \leq 10^{-8}$ for many different materials [7]. Much better results have been attained by *Dicke* et al. [8] and *Braginsky* [9]. In solar mode experiments, they determined $\eta_E = (1.3 \pm 1.0) \cdot 10^{-11}$ for gold und aluminum and $\eta_E = (-0.3 \pm 0.9) \cdot 10^{-12}$ for platinum and aluminum, respectively. Todays best results to test the WEP with terrestrial mode experiments have been attained by *Adelberger* et al. at the University of Washington in Seattle with improving results since 1987 (eg.[12, 13]). The results vary for differnt materials between $\eta_E = (-1.9 \pm 2.5) \cdot 10^{-12}$ for the comparison of beryllium and copper and $\eta_E = (-0.2 \pm 2.8) \cdot 10^{-12}$ for the comparison of beryllium and aluminum [3].

Although all experimental WEP tests failed so far, recent theoretical developments suggest a violation at some level. Theoretical approaches to a quantum theory of gravity

[1] A detailed description of modern torsion balances can be found in [10] and [11].

and to unifying the four fundamental interactions like string theory are predicting WEP violations. In string theory motivated dilaton scenarios it might be possible that the WEP is violated at the 10^{-13} level [14, 15]. Also from quintessence models a violation of the WEP at the level of 10^{-14} is predicted. The interesting feature of the latter prediction is that the parameters of the underlying theory are fixed by cosmological observations. This are driving motivations to carry out much more precise experiments on satellites.

2.2. Lunar Laser Ranging

Equivalence Principle tests have also been done with Lunar Laser Ranging (LLR) using retro-reflectors which had been placed on the Moon's surface during the Apollo and Luna missions at 5 different locations between 1969 and 1973. Aligned perfectly to Earth, the retro-reflectors are reflecting pulsed laser light sent from earth. Since only 1 single photon of 10^{19} photons sent from Earth to Moon returns, a special data analysis had to be developed. It needed enormous efforts to improve the LLR resolution by a factor of 1,000 during the last 30 years. A single photon can only be detected by comparison of its arrival time with its pre-calculated one. The resolution for Earth-Moon distance measurements attainable today is in the range of 1 cm. If one considers the iron-dominated Earth and the silica-dominated Moon as two test masses freely falling in the solar gravitational field, the distance measurements can be used to determine the Eötvös factor to about $\eta \leq 10^{-13}$ [16].

2.3. Experiments with quantum matter

There are two possibilities to carry through experiments with quantum matter: First one can observe the evolution of a single "free" quantum system in a gravitational field and, second, one can make spectroscopy of bound systems. The first class of experiments are the quantum analoque of free fall experiments. The free fall of atoms can be observed with very high precision usinf atomic interferometry. With this technique, the acceleration of freely falling atoms has been compared with freely falling bulk matter what led to a verifaction of the WEP kn the quantum level azt the level of 10^{-9} [17].

The other type of exyperiments examines the property of quantum matter using spectroscopy. Spectroscopy or, equivalently, atomic clock experiments, can search directly for shifts in the energy spacings of atomic states that may depend on the velocity of the atoms or on the position in its gravitational environment. Such deviations would indicate preferred frame or preferred location effects. Typical experiments to search for relative shifts between the frequencies of ground-state hyperfine transitions depending on the orientation of an atom in the gravitational field are Hughes-Drever-like experiments [18, 19, 20, 21, 22]. This can be used to contrain, e.g., a hypothetical of space–time torsion [23].

2.4. Spin-coupling experiments

Particles with spin have an internal structure and, assuming the conventional minimal coupling procedure, couple to the space–time curvature, see e.g. [24]. Thus, the WEP does not apply to these particles. However, these terms are many orders of magnitude too small to be of importance in any experiment, so that it makes sense to search for a violation of the WEP up to the order these competetive terms may play a role. This is tantamount to a search for anomalouls spin couplings. Such anomalous spin coupling have been proposed by, e.g., Moody and Wilczek [25] in order to solve the strong *PC* puzzle, see [26] for a review. These couplings consist in spin–mass and spin–spin couplings which lead to different forces experienced by a polarized body in a gravitational field for a different orientation of the polarization, and to extra forces due to the polarization content of he gravitating body. All searches for such extra forces yeild no result [27, 28, 29].

2.5. Experiments with charged matter

Tests of the Universality of Free Fall for charged matter are motivated by the fact that within certain frameworks a violation of the UFF is connected with charge non-conservation [30]. Considering a connection of UFF and UGR (e.g. [31]), the most favourable model for a violation of the UGR is a time dependent fine structure constant caused by a time-varying electron charge. Therefore, tests of the UFF for charged matter can be interpreted as UGR tests, too.

Testing the validity of the EP is tantamount to testing the minimal coupling procedure. In other words, by doing EP tests one searches for an anomalous coupling of the gravitational field. As far as charges are concerned, an anomalous couling means a coupling of the charge of particles with the gravitational field which is beyond the standard coupling given by the minimal coupling procedure. On the non–relativistic Newtonian level the most simplest anomalous coupling of charge with the gravitational field is an interaction term of the structure $\sim qU$ where q is the charge of the particle and U the Newtonian potential. The the Hamiltonian of a charged particle in a gravitationa field reads

$$H = -\frac{\hbar^2}{2m}\left(\nabla - \frac{iq}{\hbar c}A\right)^2 + mU + \kappa qU + \lambda q^2 U \tag{3}$$

where the parameters κ and λ possess the dimension mass/charge and mass/charge^2, respectively. The last two terms $mU + \kappa qU = (m + \kappa q)U$ can be interpreted in the way that the gravitational mass depends on the charge. Additions of that type have also been introduced by Moody and Wilczek [25] in order describe anomalous couplings between the spin and the gravitational field. The relation between spin and space–time symmetries is obvious since spin and mass characterizes representations of the Lorentz group. However, charge is also related to space–time symmetries through the *CPT* theorem. Therefore, it may make sense also in this respect to search for anomalous charge couplings.

Furthermore, it has been shown by Rohrlich [32] that for the stability of an extended electron it is required to have an effective dependence on the square of the electron charge. In addition, it has been shown by Ni [33] that generalized Maxqwell equations in general violate the Universality of Free Fall. The additional non–geometric force acting the charged particle depends on the square of the electric field and, thus, again on the square of the charge.

Motivated by all these considerations, we can now take as our general mocdel to have charge–dependent inertial and gravitational masses [34]. It is important to note thst it is possible to choose param,eters in such a way that neutral systems made up of bound charged particles exactly fulfill the UFF while isolated charged particles may violate it. As a result, we can introduce a charge dependent Eötvös coefficient $\eta = \eta^0 + \kappa_1 \frac{q_1}{m_1} - \kappa_2 \frac{q_2}{m_2}$ (we restrict to linear charge dependence only), where η^0 is the ordinary Eötvös parameter for the masses. The most important case is, of course, a comparison between the free fall of a charged and a neutal particle described by $\eta = \eta^0 + \kappa \frac{q}{m}$.

If all elctromagnetic fields are shielded, charged particles without internal structure must fall along the same path as neutral particles[2]. So far, results of only one experiment dedicated to test the UFF of charged particles have been reported. Within the experiment, carried out by *Witteborn* and *Fairbank* with electrons, an accuracy of $\eta \leq 0.1$ had been attained [5]. The reson for the relatively low accuracy was the appearance of stray electric fields and of gravity-induced electric fields. It can be shown that the disturbances induced by the stray fields can reduced by orders of magnitude by doing appropriate experiments in space [34]

2.6. Experiments with anti-matter

Another type of matter is antimatter for which the validity of the WEP has been questioned in, e.g., supergravity theories [35] which led to a wide discussion on that question [36, 37]. The capability to produce neutral anti-matter at CERN's Antiproton Decelerator gives an unique possibility to carry out the first WEP tests with anti-matter. Although anti-protons or positrons have been available so far, effective shielding of electromagnetic fields and the reduction of patch effects as it has been discussed for tests with charged particles had been a serious obstacle for WEP-tests with anti-matter. Antihydrogen is produced by assembling it from anti-protons and positrons in recombination processes. Anti-protons slowed down to energies in the kV-range and positrons cooled down to 4 K by sympathetic cooling with cold electrons via Coulomb interaction are made to interact in a double-well potential. Within the process, about 10^5 anti-hydrogen atoms per hour can be produced [38]. The most important parameter for gravity experiments with anti-hydrogen is the temperature. Within the K-range, the velocity of the

---

[2] The statement is negelecting the fact that accelerated charged particles are subject to a back action of the field of the field of the electric field of the accelerated charge to its motion resulting in an acceleration $\sim \frac{2}{3}(e^2/c^3)(d/dt^3)x$. The question, controversally discussed, is whether a charged particle at a fixed position at Earth's surface or a freely falling particle radiates

atoms is still higher than 100 m/s. Free fall experiments need cooling below the 1 mK range related to velocities below 1 m/s at least. *Walz* and *Hänsch* [38] propose to use a pulsed Lyman-α source for laser cooling and additional sympathetic cooling. Positive anti-hydrogen ions (a bound system of an anti-proton and two positrons) can be stored in a conventional ion trap and cooled down by sypathetic cooling with laser cooled ordinary Be^{+}-ions. The additional positroncan be removed by a photodetachment laser pulse. The measurement of the time between a photodetachment and the signal from an annihilation detector can be used to determine the gravitational acceleration of the anti-hydrogen atoms. For a drop height of $l = 10$ cm corresponding to a time-of-flight $\tau = 0.143$ s, the error of the measured gravitational acceleration is given by

$$\delta g = \sqrt{\left(\frac{2}{\tau}\delta l\right)^2 + \left(\frac{2}{\tau}\delta z\right) + \left(\frac{2}{\tau}\delta v\right)}. \tag{4}$$

With $\delta r = \sigma/\sqrt{N} = [1\ \text{mm}]/\sqrt{N}$ is the initial vertical position uncertainty averaged over N atoms in the cloud and $\delta v = \sqrt{kT/m}/\sqrt{N} = [0.41\text{m/s}]/\sqrt{N}$ the uncertainty of the initial velocity of a single atom, $\delta g/g \approx 10^{-5}$ results. It is obvious, that an increase of the time-of flight τ in a weightlessness environment will increase the accuracy by orders of magnitude.

3. TESTS OF LOCAL LORENTZ INVARIANCE AND LOCAL POSITION INVARIANCE

For a more general discussion of the Equivalence Principle, one has to explore the structure of the gravitational field. Within Einstein's General Relativity, the gravitational field is described by a Riemannian space–time metric. This is encoded in the Einstein Eqivalence Principle (EEP) which states that beside the validity of the WEP, also Special Relativity is locally valid, i.e. in regions of space-time which are small enough, and physics does not depend on the position in the gravitational field. Therefore, proofs of the EEP must also focus on the isotropy and constancy of light speed as well as on preferred-location effects.

Isotropy of light speed has been tested very precisely with Michelson-Morley and Kennedy-Thorndike experiments. Within the framework of the Mansouri-Sexl test theory, in the low velocity regime one can describe the speed of light as function of velocity and orientation with respect to the cosmic background [39, 40]:

$$c(v, \vartheta) = 1 + \left(\beta - \alpha + (\beta - \delta)\cos^2\vartheta\right) v^2. \tag{5}$$

Best results for the isotropy of light speed with a Michelson-Morley experiment have been reported to $|\beta - \alpha| \leq 1.5 \cdot 10^{-9}$ [41]. The constancy of light has been tested with Kennedy-Thorndike experiments at a level of $|\beta - \delta| \leq 7 \cdot 10^{-7}$ [42].

A direct consequence of the EEP is the gravitational redshift and the relativistic Doppler shift which can be observed best by monitoring the relative rates of different types of atomic clocks for dependence on the clock velocities through or clock location in a gravitational field. In a low velocity and weak field regime the frequency ratio shift can be measured with respect to preferred frame and preferred location effects by

$$\delta\left(f_1/f_2\right) = \alpha_{ij} v^i v^j + \beta_{ij} U^{ij}, \tag{6}$$

where U^{ij} denotes the Newtonian potential. α_{ij} characterize perferred frame effects and would be measured in a Doppler shift experiment, β_{ij} is related to an eventual preferred frame effect related to deviation of the Gravitational Redshift [2]. In an experiment employing trapped lithium atoms moving at 6.4% of light speed special relativistic time dilation has been tested to $|\alpha_{ij}| \leq 2 \cdot 10^{-7}$ [43]. The Universality of the Gravitational Redshift has been confirmed best by a comparison of the rate of two clocks, one an atomic fountain cloch and the other a H-maser, during their motion in the varying gravitational potential of the Sun at the position of the Earth during a one year orbit. As a result, LPI has been verified with a $2 \cdot 10^{-5}$ precision [44]. An *absolute* measurement of the gravitational redshift has been performed by comparing two identical hydrogen maser clocks to a precision of $|\beta_{ij}| \leq 7 \cdot 10^{-5}$ [45]. This test was the Gravity Probe A (GP-A) experiment, which has been performed on a ballistic rocket flight. The maser on the rocket has been compared to a second on ground.

4. TEST OF NEWTON'S LAW

Experimental tests of the WEP are linked to the question what a WEP violation could cause and how these tests can reject non-Newtoian theories. Usually these theories are characterized by the introduction of an additional scalar field (e.g. the dilaton field) resulting in a hypothetical "5th force" which would lead to a WEP violation. A 5th force with range λ_5 and coupling strength α_5 would modify the Newtonian potential [46]

$$U_5(r) = G\frac{m_i m_j}{r}\left(1 + \alpha_5 e^{-r/\lambda_5}\right). \tag{7}$$

The additive term in eq.7relates to a Yukawa-like potential. $\lambda_5 = \hbar/(m_5 c)$ is the Compton wavelength of the hypothetical exchange quanta with mass m_5. Although a huge number of experiments to measure 5th force effects have been carried out during the last 15 years, there is no evidence of its existence [46] (see Fig.1). WEP tests and tests of Newton's $1/r^2$-law Eq.(7) are strongly related. WEP tests must be carried out with view to composition as well as distance dependence. Nevertheless, these tests leave open wheter deviations are related to time and position dependence of gravitation or to an additional interaction. The Newtonian inverse square law has been tested with an accuracy of 10^{-8} at distances up to 10^7 km and to an accuracy of about 10^{-3} in the sub-millimeter range. Recent publications on Doppler shift measurements for the Pioneer 10 and 11 deep space spacecrafts reported an anomalous acceleration directed towards the sun at distances between 10 and 20 astronomical units [47]. There are also many theoretical motivations for examining the inverse square law at short ranges. Speculations about the possibility that gravitational interaction could extend to more compactified dimensions than the other forces or about new particles and phenomena beyond the Standard Model could explain the weakness of gravity. As a consequence of that, Newton's Law should

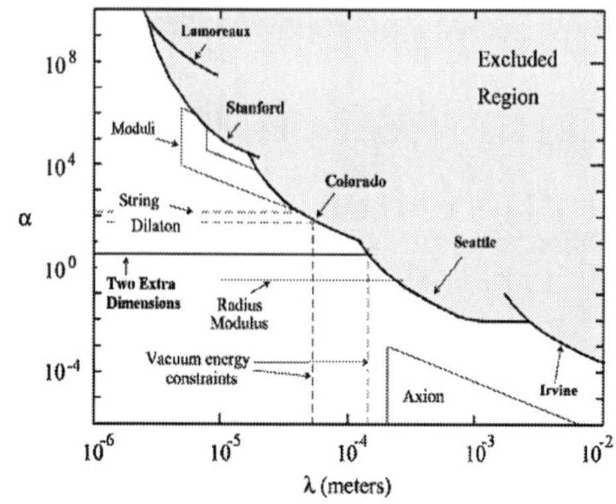

FIGURE 1. Present limits on Non–Newtonian forces for small distances.

be modified at very small distances. As one need deep space satellite missions to explore the large distance range, precise laboratory tests for investigations in the μm-range seem to be possible and have been proposed recently. Recent experiments have employed a rigid membrane to separate a moving source mass from the detector in order to avoid disturbances by the Casimir force and variations in the electrical potential caused by patch effects. Recently, *Speake* et al. reported on the possibility to search for violations of the inverse square law at separations of only 1μm [48]. There analysis of a new experimental set-up suggests that a cryogenic, weightlessness environment is necessary to make the test feasible, as thermal and vibrational noise could be reduced to a certain limit.

5. SPACE TESTS WITH SATELLITES

Satellite-based experiments to test the WEP yield to measure $\eta \leq 10^{-15}$ on MICROSCOPE (Micro-satellite à trainee Componsée pour l'Observation du Principe d'Équivalence) and $\eta \leq 10^{-19}$ on STEP (Satellite Test of Equivalence). MICROSCOPE [49] is an approved mission to be launched 2008[3]. STEP [50, 51] is a project under study at NASA and ESA. Although Galilean-type free fall experiments to test the WEP seem to be the most efficient and simple, because they use the full gravitational acceleration of the Earth's gravitational field, they did not play any role as precision tests. Sensing and machining requirements are extremely high, because on Earth the free fall time is limited to only some seconds, the conditions change from between free fall tests (e.g. influenced by tidal forces), and periodic meassurement as they are possible wih pendulum and torsion balance experiments are impossible. In order to measure the differential acceleration of two test masses during free fall of 100 m related to an experimental time of $t \approx 4.5$ s with an accuracy of $\delta a = 10^{-13}$ m / s^2, one has

[3] MICROSCOPE is a mission within the CNES Space Programme in France

to determine the position with a maximum error of $\delta x = \delta a/2 \cdot t^2 \approx 10^{-12}$ m which is about 1% of the atomic radius only. Even the position of the centres of mass (CoM) of both test masses have to be known very precisely. Through to inhomogeneities of the earth gravitational field, the gravitational acceleration decreases of about 0,3 mGal / m = $3 \cdot 10^{-6}$ m / ($s^2 \cdot$ m), related to about 10^{-12} m / s^2 per 1μm. For a measurement to determine $\eta \leq 10^{-12}$, the difference of the CoMs must be fixed to better than 1 μm, and the test masses have to be machined with an appropriate accuracy along all axis. All Galilean-type free fall experiments performed so far, the attained accuracies had not been better than $\eta \leq 10^{-11}$ (e.g. [52, 53]. Because of principal reasons, it can be shown that free fall experiments on earth cannot be expected to be better than $\eta_E < 10^{-14}$ [?].

Even todays experiments with torsion balances might be limited with respect to accuracy. One inportant error source is the stiffness of the torsion fibre $k_{tor} = \pi \mu d^4/(32l)$ [10], where l is the length, d the diameter and μ the shear modulusl of the fibre. Typical tungsten fibres are 30 μm thick and about 1 m long. With a shear modulus of $\mu = 10^{-10}$ Nm / rad, a minimum torsion angle of 10^{-7}rad, related to a differntial acceleration of 10^{-14}m / s^2, can be measured.

Satellites experience a periodic free fall along their orbital trajectory. For a circular low-earth orbit, two test masses would have a difference of their orbital height of only $\delta r = h_1 - h_2 = \eta_E r_1/3$, where $r_1$is the radius of the orbit of one test mass. With $\eta_E = 10^{-13}$ it results $\delta r = 0,2$ μm. The measurement of such tiny displacements of orbital heights is not possible due to many disturbing effects, wherefore one has to couple the test masses mechanically in a way that they can move only along an axis perpendicular to the orbital plane. As one test mass moves on a circular orbit, the other one moves on a slightly elliptical one, and the differential position with time is a sinusoidal signal. Two test masses arranged in a way that their centres of mass coincide form a differential accelerometer. For coinciding CoMs, the test masses would separate during the first quarter of their orbit and would coincide after half orbital time. During second half of the orbital time, the process would run in the opposite direction (see Fig.2). As at STEP the relative motion of the test masses will be observed and measured by a SQUID (Superconducting Quantum Interference Device)-based sensing system, the WEP test on MICROSCOPE is performed by controlling the relative motion of both test masses at null with capacitors sensing and electrostatic actuators so that any WEP-violation appears through the measured forces necessary to nullify this relative motion. Attitude and orbit of the satellites STEP and MICROSCOPE have to be controlled precisely. The satellite follows the test mass and shields them from disturbung forces (drag-free control). Drag-free control guarantees that the satellite moves on a free fall trajectory unless disturbances like air drag, solar radiation pressure etc. are acting. The satellite spin (ω_{spin}) modulates the WEP-violation (ω_{WEP}) signal and avoid couplings from disturbances appearing with orbital frequency (ω_{orbit}): $\omega_{WEP} = \omega_{spin} + \omega_{orbit}$. The acceleration acting on the two test masses in each accelerometer can be split in two parts: the differential mode signal (i.e. the WEP-violation signal) and the common mode signal which is used to control an ensemble of μN-thrusters of the satellite [55].

As it has been pointed out, tests of the Einstein Equivalence Principle also need additional experiments to prove Local Lorentz Invariance as well as Lorentz Position Invariance. For tests it might be of advantage to use the large variations of the gravitational potential and the satellite velocity with respect to the cosmic background along the orbital

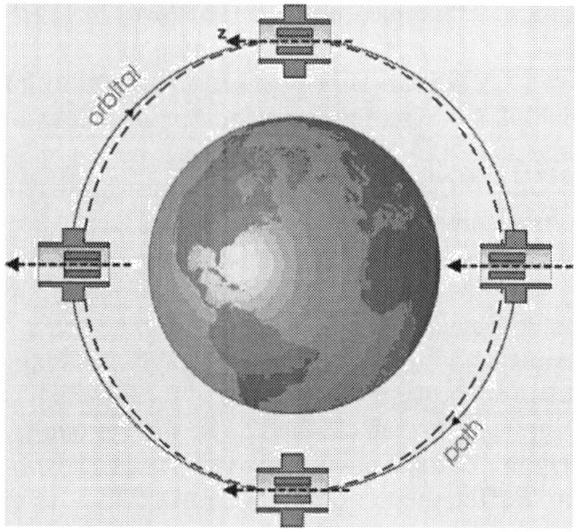

FIGURE 2. Schematics of the STEP and MICROSCOPE experiment. The dashed line indicates the path of the second test mass.

trajectory. A satellite with very precise attitude and orbit control on a highly elliptical orbit around the earth could provide an ideal environment in order to perform interferometer experiments on a much higher precision level than on earth. The satellite OPTIS [56] has been designed for carrying out Michelson-Morley and Kenndy-Thorndike type experiments as well as experiments to test the Universality of Redshift.

5.1. STEP

STEP (Satellite Test of Equivalence Principle) will fly in a circular sun-synchronous orbit at an orbital height of abut 500 km. The test masses are distributed in pairs to 4 differential accelerometers which can be operated simultanously or alone. The differential acceleration of each test mass pair can be measured iductively with a SQUID-read-out system which requires cooling down to 2 K to attain superconductivity. The test masses have cylinder-symmetric shape and can move freely only along the cylinder axes (see Fig.2). Each test mass pair is constrained from moving radially by superconducting magnetic bearings. Motion along the cylindrical axis can be measured precisely by the SQUID read-out system, as motion in all other degrees of freedom is measured by a capacitive system with lower resolution. Inductive sensing is done with pick-up coils facing the top ends of the test masses. The magnetic reaction force from these coils provides a very weak retaining spring in the axial direction. Due to this reaction, a displacement is a measure of the difference in acceleration experienced by each test mass pair. Movement of a superconducting-coated test mass towards one pick-up coil and away from the opposite, changes the inductance. Becasue of flux conservation in the superconducting loop, this forces a current trough a third coil (SQUIDinput-coil) in parallel with the first two. This net-current is roughly proportional to the displacement. The read-out system is able to sense differntial as well as common mode movements of the test masses. The common mode system is used as the input signal for the drag-free attitude and orbit control of the satellite. The test mass positions are adjusted electrostatically with a set of electrodes surrounding each test mass. The test masses can be centred to within 10^{-10}m to reduce the efffects of tidal forces. The capacitive system also enables to measure the charges of the test masses as response signal to voltages applied to the electrodes. The SQUID-based sensing system enables the measurement of differential accelerations with a resolution of about 10^{-18} m / s^2 during an integration time of about 20 orbital time periods [51].

In order to avoid systematic errors, three different materials have been choosen. The test mass pairs Pt-Ir/Be, Nb/Be, and Pt-Ir/Nb guarantees a cyclic closure condition and the fourth test mass pair Pt-Ir/Be is duplicated. The test mass shapes have been designed as quasi-spherical monopoles in order to minimize the coupling of higher order gravitational moments [54]. The four differential accelerometers are arranged in pairs in the two directions orthogonal to the orbital plane and mounted in a quartz block within a high vacuum chamber inside a helium Dewar (see Fig.4). The whole assembly is covered by a superconducitng shield. The temperature stability is less than 0.5 mK per orbit. STEP is a highly optimized experiment. Most error sources (systematic and stochastic) were evaluated analytically. The analysis includes etimates of specific disturbances to the masses due to thermal noise, gas pressure, electrical forces, magnetic forces, tidal effects, radiation pressure to the spacecraft, and vibrations. It also includes effects from disturbances to the SQUID-read-out system, like noise, changes to the supeconductors, thermal stability, and mechnical displacements. Further effects to be considered are: electrical potentials in the qaurtz housing (patch effects), losses from eddy current damping, the radiometer effect, tidal effects from helium sloshing in the Dewar, cross-coupling to the differntial mode of the accelerometers, SQUID-noise, changes in the penetration depth in the superconducters due to insufficient coating of the test masses, and momentum transfer from penetrating particle radiation. Satellite and experiment design is based on this multi-parameter error model to keep the science requirement to measure $\eta \leq 10^{-19}$.

5.2. MICROSCOPE

Free fall experiments on MICROSCOPE will test the WEP to an accuracy of $\eta \leq 10^{-16}$. In contrast to STEP, the MICROSCOPE experiment is performed by controlling the relative motion of both test masses at null (closed loop control) so that any WEP-violation appears through the measured forces necessary to nullify this relative motion. Therefore, the differential accelerometers can be operated with a much larger bandwidth, but a lower resolution than with the open loop design is possible. The experimental baseline of MICROSCOPE is to carry out the experiments on-board a drag-free satellite with two servo-controlled electrostatic accelerometers, one implented inside the other one. Thus, the two test masses are centred to avoid influences of gravity gradient fluctuations. The configuration senses differential accelerations acting on both test masses. Due to the closed loop control of the test mass movements, influences of tidal effects (gravity gradient) and other environmental effects are eliminated. The capacitive sensing does not need a low temperature environment, wherefore tmeperature stability in the satellite becomes an important task. The differntial accelerometers are constructed around a high density proof mass. The housing structure is made glas ceramics. A set of electrodes for capacitive sensing and is surrounding each test mass. Another set of electrodes is needed to control all 6 degrees of freedom within the closed loop circle. The experiment consists of two differential accelerometers to be operated at room temperature. Test masses of the first differential accelerometer have cylindrical shape and are made from platin and titanium, respectively. The second differential accelerometer contains two cylindical test masses both made from platin to enable the measurement of systematic errors. The cylinders' symmetry axis is directed inside the orbital frame (x-axis), resulting in a signal variation periodic with orbit frequency modulated by the satellite spin in the case of WEP-violations. The satellite itself spins about the axis perpendicular to the orbital plane (y-axis) varying the orientation of the Earth gravity field in the instrument reference frame and enabling to discriminate systematic disturbances. With signal integration over 20 orbits (100,000 s), differential accelerations will be measured with a resolution of 10^{-15} m/s^2· Hz$^{-1/2}$ at the sum of orbital and spinning frequency which will be about 10^{-3}Hz. A 10^{-15} rejection of the accelerations applied in common mode of the test masses leads to a 10^{-8} m/s^2· Hz$^{-1/2}$ requirement for the darg-free compensation. The MICROSCOPE satellite will fly on a sun-synchronous circular orbit at an inclination of about 97°. Orbital height is about 600 km and its excentricity must be less than 10^{-3}. The thrust system of the satellite for drag and torque compensation consists of 4 clusters of 2 micro-propulsion ion thrusters (FEEPs) giving a thrust to controlled from the 0.1 to 20 μN range.

5.3. OPTIS

The mission idea OPTIS have been developed for a variety of improved tests of Special and General Relativity. In particular, tests of the EEP have been porposed. Table 2 shows the scientific tests planned on OPTIS [56]. The OPTIS satellite is planned to launched into a highly elliptical orbit in order to take advantage of a large periodic variation of the earth gravitational poential between apogee and perigee and a substantial variation of the satellites's velocity with respect to the cosmic background along its orbital trajectory. The satellite carries three lasers highlystabilized by optical cavities. Additional atomic clocks and a frequency comb allow the direct comparison of optical and atomic clock frequency. The drag-free controlled satellite allows a Michelson-Morley interferometric experiment to test the isotropy of light speed by use of the cavities. Within a Kennedy-Thorndike test the optical frequency has to be compared with an atomic clock. The velocity change along the elliptical trajectory around the earth varies between +4 and -6.5 km/s. The relative gravitational potential difference between apogee and perigee varies by a factor 2.7 which allows precise tests of the Universality of Gravitational Redshift. A Laser link to the satellite tracks its trajectory very precisely

TABLE 2. The scientific objectives of the OPTIS mission: present status and planned accuracy.
* Accuracies are given in terms of Mansouri–Sexl parameters (see below),
** The authors of [61] claims a different, more pessimistic error budget. See also [62, 63],
*** Using a newly developed method for analyzing LAGEOS satellites data, the present accuracy for testing the Newtonian potential will increase to 10^{-11} [64]. However, the influence of certain tiny non–gravitational perturbations of thermal origin has not been considered in that study. See also [65].

	test	**method**	**present accuracy**	**OPTIS accuracy**
1	isotropy of speed of light *	cavity–cavity comparison	$1.5 \cdot 10^{-9}$ [41]	10^{-12}
2	constancy of speed of light *	cavity–clock comparison	$7 \cdot 10^{-7}$ [42]	10^{-8}
3	time dilation – Doppler effect	laser link	$2 \cdot 10^{-7}$ [43]	10^{-9}
4	universality of gravit. redshift I	cavity–clock comparison	$1.7 \cdot 10^{-2}$ [57]	10^{-4}
5	universality of gravit. redshift II	clock–clock comparison	$2.5 \cdot 10^{-5}$ [44]	10^{-7}
6	absolute gravitational redshift	time transfer	$1.4 \cdot 10^{-4}$ [45]	10^{-8}
7	Lense-Thirring effect**	laser tracking	0.3 [58]	10^{-3} [59]
8	Einstein perigee advance	laser tracking	$3 \cdot 10^{-3}$ [60]	$6 \cdot 10^{-4}$ [59]
9	test of Newton potential***	laser tracking	10^{-5} [46]	10^{-12} [59]

in order to carry out a variety of additional tests for General Relativity (see Table 2).

6. CONCLUSION AND OUTLOOK

Today, our basic theories of physics are questioned. They cannot be correct, because it is not possible to unify them along the lines given by conventional quantum theory. As all approaches to attack this problem predict violations of the underlying assumptions and hypotheses of the basic theories, it is of great interest to improve the experimental search for New Physics according to deviations from standard theoretical models. It requires an increasing experimental accuracy which can be achieved only under well defined conditions. Space, however, is an unique environment and will be used more and more to carry out high precision experiments for Fundamental Physics. In particular, modern space technology and recent developments of experimental techniques seem to enable more precise experimental tests of the Equivalence Principle than om earth.

Space is a unique laboratory for gravitational physics. The missions and experiments approved and planned up to now are based on state-of-the-art technology, usually. Experimental accuracy is limited by space technology mainly. As the laboratory techniques have been developed for decades, their adaptation to space conditions is a challenging and time-consuming task, now. Along many studies, various proposals for dedicated missions and co-experiments on platforms, like the ISS, have been developed. In any case, technology development for sensors and actuators is the most important issue. Experimental gravitational physics is based on the measurement of distance, acceleration, time, and frequency. Challenging experiments set the goals, and the development of completely new technology for these experiments and missions is useful and promising, anyway for our future

REFERENCES

1. C. Lämmerzahl. On the equivalence principle in quantum mechanics. *Gen. Rel. Grav.*, 28:1043, 1996.
2. M.P. Haugan and Lämmerzahl. Principles of equivalence: Their role in gravitation physics and experiments that test them. In C. Lämmerzahl, C.W.F. Everitt, and F.W. Hehl, editors, *Gyros, Clocks, and Interferometers: Testing Relativistic Gravity in Space*, page 195. Springer–Verlag, Berlin, 2001.
3. Y. Su, B.R. Heckel, E.G. Adelberger, J.H. Gundlach, M. Harris, G.L. Smith, and H.E. Swanson. New test of the universality of free fall. *Phys. Rev.*, D 50:3614, 1994.
4. J. Schmiedmayer. *Nucl.Instr.Methods Phys.Res.* A 284:59, 1989.
5. F.C. Witteborn and W.M. Fairbank. Experimental comparison of the graviational force on freely falling electrons and metallic electrons. *Phys. Rev. Lett.*, 19:1049, 1967.
6. K. Nordtvedt. Equivalence Principle for Massive Bodies, II: Theory. *Phys.Rev.*, 169:1017, 1968.
7. R. v.Eötvös, V. Pekár, and E. Fekete. Beiträge zum Gesetz der Proportionalität von Trägheit und Gravität. *Ann.Phys.*, 68:11, 1922.
8. P.G. Roll, R. Krotkov, R.H. Dicke. *Ann. Phys. (N.Y.)*, 26:442, 1967.
9. V. Braginsky, V. Panov. *Sov. Phys. JETP*, 34:463, 1972.
10. Y.T. Chen and A. Cook. *Gravitational Experiments in the Laboratory.* Cambridge University Press, Cambridge, 1993.
11. E. Adelberger. Searches for New Long-range Forces: Equivalence Principle Violation and Planck-scale Physics. Report, WEIN98 Conference, 1998 [www.npl.washington.edu/eotwash/].
12. G.L. Stubbs, E.A. Adelberger, F.J. Raab, J.H. Gundlach, B.R. Heckel, K.D. McMurry, H.E. Swanson, and R. Wtanabe. Search for an intermediate-range interaction. *Phys. Rev. Lett.*, 58:1070, 1987.
13. G.L. Smith, C.D. Hoyle, J.H. Gundlach, E.G. Adelberger, B.R. Heckel, and H.E. Swanson. Short range test of the equivalence principle. *Phys. Rev.*, D 61:022001, 2000.
14. T. Damour, F. Piazza, and G. Veneziano. Runaway dilaton and equivalence principle violations. *Phys. Rev. Lett.*, 89:081601, 2002.
15. T. Damour, F. Piazza, and G. Veneziano. Violations of the equivalence principle in a dilaton–runaway scenario. *Phys. Rev.*, D 66:046007, 2002.
16. K. Nordtvedt. Lunar laser ranging – a comprehensive probe of the post–newtonian long range interaction. In C. Lämmerzahl, C.W.F. Everitt, and F.W. Hehl, editors, *Gyroscopes, Clock, Interferometers, ...: Testing Relativistic Gravity in Space*, volume LNP 562, page 317. Springer, Berlin, 2001.
17. A. Peters, K.Y. Chung, and S. Chu. Measurement of gravitational acceleration by dropping atoms. *Nature*, 400:849, 1999.
18. V.W. Hughes, H.G. Robinson, and V. Beltran-Lopez. Upper limit for the anisotropy of inertial mass from nuclear resonance experiments. *Phys. Rev. Lett.*, 4:342, 1960.
19. R.W.P. Drever. A search for the anisotropy of inertial mass using a free precession technique. *Phil. Mag.*, 6:683, 1961.
20. J.D. Prestage, J.J. Bollinger, W.M. Itano, and D.J. Wineland. Limits for spatial anisotropy by use of nuclear–spin–polarized $^9Be^+$ ions. *Phys. Rev. Lett.*, 54:2387, 1985.
21. S.K. Lamoreaux, J.P. Jacobs, B.R. Heckel, F.J. Raab, and E.N. Fortson. New limits on spatial anisotropy from optically pumped ^{201}Hg and ^{199}Hg. *Phys. Rev. Lett.*, 57:3125, 1986.
22. T.E. Chupp, R.J. Hoara, R.A. Loveman, E.R. Oteiza, J.M. Richardson, and M.E. Wagshul. Results of a new test of local Lorentz invariance: A search for mass anisotropy in ^{21}Ne. *Phys. Rev. Lett.*, 63:1541, 1989.
23. C. Lämmerzahl. Constraints on space–time torsion from Hughes–Drever experiments. *Phys. Lett.*, A 228:223, 1997.
24. C. Lämmerzahl and G. Neugebauer. The Lense–Thrring effect: From the basic notions to the observed effects. In C. Lämmerzahl, C.W.F. Everitt, and F.W. Hehl, editors, *Gyroscopes, Clock, Interferometers, ...: Testing Relativistic Gravity in Space*, volume LNP 562, page 31. Springer, Berlin, 2001.
25. J.E. Moody and F. Wilczek. New macroscopic forces? *Phys. Rev.*, D 30:130, 1984.
26. J.E. Kim. Light pseudoscalar, particle physics and cosmology. *Phys. Rep.*, 150:1, 1987.
27. R.C. Ritter, C.E. Goldblum, W.-T. Ni, G.T. Gillies, and C.C. Speake. Experimental test of equivalence principle with polarized masses. *Phys. Rev.*, D 42:977, 1990.

28. R.C. Ritter, L.I. Winkler, and G.T. Gillies. Search for anomalous spin-dependent forces with a polarized-mass torsion pendulum. *Phys. Rev. Lett.*, 70:701, 1993.
29. W.-T. Ni, T.C.P. Chui, S.-S. Pan, and B.-Y. Cheng. Search for anomalous spin–spin interaction between electrons using a DC SQUID. *Physica*, B 194:153, 1994.
30. S. Landau, P.D. Sisterna, and H. Vucetich. Charge conservation and the equivalence principle. gr-qc/0105025.
31. G. Dvali and M. Zaldarragia. Changing α with time: Implications for fifth–force–type experiments and quintessence. *Phys. Rev. Lett.*, 88:091303, 2002.
32. F. Rohrlich. The equivalence principle revisited. *Found. Phys.*, 30:621, 2000.
33. W.-T. Ni. Equivalence principles and electromagnetism. *Phys. Rev. Lett.*, 38:301, 1977.
34. H. Dittus, C. Lämmerzahl, and H. Selig. Testing the universality of free fall for charged particles. *Gen. Rel. Grav.*, 36:571, 2004.
35. J. Scherk. Anti–gravity: a crazy idea? *Phys. Lett.*, B 88:265, 1979.
36. M.M. Nieto and T. Goldman. The arguments against "antigravity" and the gravitational acceleration of antimatter. *Phys. Rep.*, 205:222, 1991.
37. E.G. Adelberger, B.R. Heckel, C.W. Stubbs, and Y. Su. Does antimatter fall with the same acceleration as ordinary matter? *Phys. Rev. Lett.*, 66:850, 1991.
38. J. Walz and T.W. Haensch. A proposal to measure antimatter gravity using ultracold antihydrogen atoms. *Gen. Rel. Grav.*, 36:561, 2004.
39. H.P. Robertson. Postulate versus observation in the Special Theory of Relativity. *Rev. Mod. Phys.*, 21:378, 1949.
40. R. Mansouri and R.U. Sexl. A test theory of special relativity: I. Simultaneity and clock synchronisation. *Gen. Rel. Grav.*, 8:497, 1977.
41. H. Müller, S. Herrmann, C. Braxmaier, S. Schiller, and A. Peters. Modern Michelson–Morley experiment using cryogenic optical resonators. *Phys. Rev. Lett.*, 91:020401, 2003.
42. P. Wolf, S. Bize, A. Clairon, A.L. Luiten, G. Santarelli, and M.E. Tobar. Tests of relativity using a microwave resonator. *Phys. Rev. Lett.*, 90:060402, 2003.
43. S. Saathoff, S. Karpuk, U. Eisenbarth, G. Huber, S. Krohn, R. Muñoz-Horta, S. Reinhardt, D. Schwalm, A. Wolf, and G. Gwinner. Improved test of time dilation in special relativity. *Phys. Rev. Lett.*, 91:190403, 2003.
44. A. Bauch and S. Weyers. New experimental limit on the validity of local position invariance. *Phys. Rev.*, D 65:081101(R), 2002.
45. R.F.C. Vessot, M.W. Levine, E.M. Mattison, E.L. Blomberg, T.E. Hoffmann, G.U. Nystrom, B.F. Farrel, R. Decher, P.B. Eby, C.R. Baughter, J.W. Watts, D.L. Teuber, and F.D. Wills. Test of relativistic gravitation with a space–borne hydrogen maser. *Phys. Rev. Lett.*, 45:2081, 1980.
46. E. Fischbach and C.L. Talmadge. *The Search for Non–Newtonian Gravity*. Springer–Verlag, New York, 1999.
47. J.D. Anderson, P.A. Laing, E.L. Lau, A.S. Liu, M.M. Nieto, and S.G. Turyshev. Study of the anomalous acceleration of Pioneer 10 and 11. *Phys. Rev.*, D 65:082004, 2002.
48. C.C. Speake, G.D. Hammond, and C. Trenkel. The feasibility of testing the inverse square law of gravitation at Newtonian and at mass separations of 1 μm. *Gen. Rel. Grav.*, 36:503, 2004.
49. P. Touboul. Space accelerometers: Present status. In C. Lämmerzahl, C.W.F. Everitt, and F.W. Hehl, editors, *Gyroscopes, Clock, Interferometers, ...: Testing Relativistic Gravity in Space*, volume LNP 562, page 274. Springer, Berlin, 2001.
50. N. Lockerbie, J.C. Mester, R. Torii, S. Vitale, and P.W. Worden. STEP: A status report. In C. Lämmerzahl, C.W.F. Everitt, and F.W. Hehl, editors, *Gyros, Clocks, and Interferometers: Testing Relativistic Gravity in Space*, page 213. Springer–Verlag, Berlin, 2001.
51. T. Sumner. Equivalence principle measurements. *Gen. Rel. Grav.*, 36:2331, 2004.
52. K. Kuroda and N. Mio. Test of composition–dependent force by a free–fall interferometer. *Phys. Rev. Lett.*, 62:1941, 1989.
53. T.M. Niebauer, M.P. McHugh, and J.E. Faller. Galileian test for the fifth force. *Phys. Rev. Lett.*, 59:609, 1987.
54. N. A. Lockerbie, A.V. Veryashkin, X. Xu: Class.Quantum Grav. **10**, 2419 (1994)
55. S. Theil: Satellite ans Test Mass Dynamics Modelling and Observation for Drag-free Satellite Control of the STEP Mission, Doct. Thesis (Diss.) University of Bremen (2002)
56. C. Lämmerzahl, C. Ciufolini, H. Dittus, L. Iorio, H. Müller, A. Peters, E. Samain, S. Scheithauer,

and S. Schiller. An Einstein mission for improved tests of special and general relativity. *Gen. Rel. Grav.*, 36:2373, 2004.

57. J.P. Turneaure, C.M. Will, B.F. Farrel, E.M. Mattison, and R.F.C. Vessot. Test of the principle of equivalence by a null gravitational red–shift experiment. *Phys. Rev.*, 27:1705, 1983.
58. I. Ciufolini. The 1995-99 measurements of the Lense–Thirring effect with using laser-ranged satellites. *Class. Quantum Grav*, 17:2369, 2000.
59. L. Iorio, I. Ciufolini, S. Schiller, H. Dittus, and C. Lämmerzahl. On the possibility of measuring the Lense–Thirring effect with a LAGEOS–LAGEOS II–OPTIS–mission, 2003. preprint gr-qc/0211013.
60. I.I. Shapiro. Solar system tests of General Relativity: recent results and present plans. In N. Ashby, D.F. Bartlett, and W. Wyss, editors, *General Relativity and Gravitation*, page 313. Cambridge University Press, Cambridge, 1990.
61. J. Ries, R.J. Eanes, and B.D. Tapley. Lense–Thirring precession determination from laser ranging to artificial satellites. In R.J.. Ruffini and C. Sigismondi, editors, *Nonlinear Gravitodynamics*, page 201. World Scientific, Singapore, 2003.
62. L. Iorio. The impact of the static part of the Earth's gravity field on some tests of general relativity with satellite laser ranging. *Celest. Mech & Dyn. Astron.*, 86:277, 2003.
63. L. Iorio and A. Morea. The impact of the new Earth gravity models on the measurement of the Lense–Thirring effect. *Gen. Rel. Grav.*, 36:to appear, 2004, [gr-qc/0304011].
64. L. Iorio. Constraints on a Yukawa gravitational potential from laser data of LAGEOS satellites. *Phys. Lett.*, 298:315, 2002.
65. D.M. Lucchesi. LAGEOS II perigee shift and Schwarzschild gravitoelectric field. *Phys. Lett.*, A 318:234, 2003.

The Pioneer Anomaly: The Data, its Meaning, and a Future Test

[1]Michael Martin Nieto*, Slava G. Turyshev† and John D. Anderson†

*Theoretical Division (MS-B285), Los Alamos National Laboratory, University of California, Los Alamos, New Mexico 87545
†Jet Propulsion Laboratory, California Institute of Technology, Pasadena, CA 91109, U.S.A.

Abstract. The radio-metric Doppler tracking data from the Pioneer 10/11 spacecraft, from between 20-70 AU, yields an unambiguous and independently confirmed anomalous blue shift drift of $a_t = (2.92 \pm 0.44) \times 10^{-18}$ s/s^2. It can be interpreted as being due to a constant acceleration of $a_P = (8.74 \pm 1.33) \times 10^{-8}$ cm/s^2 directed towards the Sun. No systematic effect has been able to explain the anomaly, even though such an origin is an obvious candidate. We discuss what has been learned (and what might still be learned) from the data about the anomaly, its origin, and the mission design characteristics that would be needed to test it. Future mission options are proposed.

1. THE DATA

1.1. The Pioneer missions and the anomaly

The Pioneer 10/11 missions, launched on 2 March 1972 (Pioneer 10) and 4 December 1973 (Pioneer 11), were the first to explore the outer solar system. After Jupiter and (for Pioneer 11) Saturn encounters, the two spacecraft followed escape hyperbolic orbits near the plane of the ecliptic to opposite sides of the solar system. (See Figure 1.) Pioneer 10 eventually became the first man-made object to leave the solar system.

The Pioneers were excellent crafts with which to perform precise celestial mechanics experiments. This was due to a combination of many factors, including (1) their attitude control – spin-stabilized, with a minimum number of commanded attitude correction maneuvers using thrusters, (2) power design – Radioisotope Thermoelectric Generators (RTGs) on extended booms aiding the stability of the craft and also reducing the heat systematics, and (3) precise Doppler tracking – with sensitivity to resolve small frequency drifts at the level of mHz/s). The result was the most precise navigation in deep space to date.

By 1980 Pioneer 10 had passed a distance of $\sim$20 AU (Astronomical Units) from the Sun and the acceleration contribution from solar-radiation pressure on the craft (directed away from the Sun) had decreased to less than 4×10^{-8} cm/s^2. At that time an anomaly

[1] Speaker: M.M.N. Email addresses: mmn@lanl.gov, turyshev@jpl.nasa.gov, john.d.anderson@jpl.nasa.gov

CP758, *Gravitation and Cosmology: 2nd Mexican Meeting on Mathematical and Experimental Physics* edited by A. Macías, C. Läemmerzahl and D. Nuñez

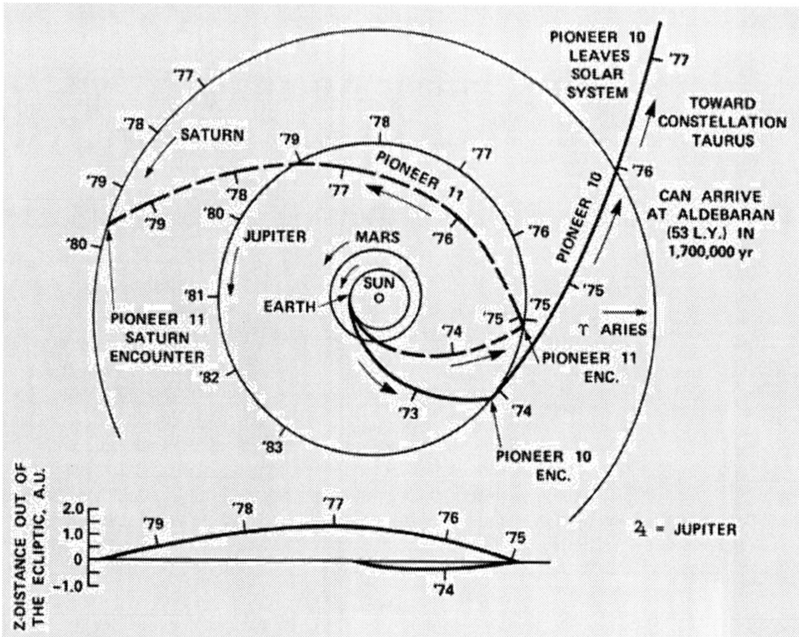

FIGURE 1. The early trajectories of the Pioneer 10 and 11 spacecraft, out to the orbit of Saturn.

in the Doppler signal became evident. Therefore, from time to time various pieces of the early data were monitored for residuals by different analysts who used different criteria for judgments. But there never was a detailed look at the systematics nor a comprehensive look at the whole.

However, as shown in Figure 2, these analyses strongly indicated an anomaly past 10 AU. (All the data averages were of order 8×10^{-8} cm/s^2, with statistical errors that decreased with distance and were less than 2×10^{-8} cm/s^2.)

Subsequently, the Pioneer Collaboration was formed among the present authors and Philip Laing, Eunice Lau, and Tony Liu to perform a NASA-sponsored analysis of the radio-metric tracking data from the Pioneer 10/11 spacecraft. To make sure that the radiation pressure was not an overwhelming systematic, the data from 1987.0 was considered. This covered distances between 20 - 70 AU from the Sun. (The inner part from Pioneer 11, the outer part from Pioneer 10.)

This data consistently indicated the presence of an anomalous, small, constant Doppler frequency drift. In figure 3 we show Pioneer 10 Doppler residuals using The Aerospace Corporation's CHASMP navigation code.

Although the most obvious explanation would be that there is a systematic origin to the effect, perhaps generated by the spacecraft themselves from excessive heat or propulsion gas leaks, none has been found; that is, no unambiguous, onboard systematic has been discovered [2, 4]. The largest systematics were indeed from on-board the craft but, as recounted in Table 1, they did not come near to explaining the anomaly [2, 4].

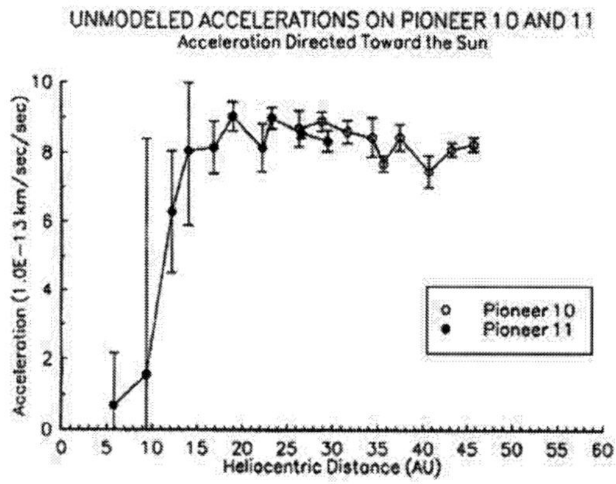

FIGURE 2. A plot of the early unmodeled accelerations of Pioneer 10 and Pioneer 11, from about 1981 to 1989 and 1977 to 1989, respectively [2].

After taking into account all systematics, a blue shift in $(\Delta v/v)$ was determined of $a_t = (2.92 \pm 0.44) \times 10^{-18}$ s/s^2. It can be interpreted as being due to a constant acceleration of $a_P = (8.74 \pm 1.33) \times 10^{-8}$ cm/s^2 directed towards the Sun [1]-[3].

Attempts to find a convincing systematic explanation have not succeeded [2, 4]. This inability to explain the anomalous acceleration of the Pioneer spacecraft with conventional physics has contributed to the growing discussion about its origin.

Attempts to verify the anomaly using other spacecraft have proven disappointing [1, 2]. This is because the Voyager, Galileo, Ulysses, and Cassini spacecraft navigation data all have their own individual difficulties for use in an independent test of the anomaly (see below). In addition, many of the deep space missions that are currently being planned either will not provide the needed navigational accuracy and trajectory stability of under 10^{-8} cm/s^2 (i.e., Pluto Express, InterStellar Probe) or else they will have significant on-board systematics that mask the anomaly (i.e., JIMO – Jupiter Icy Moons Orbiter).

The acceleration regime in which the anomaly was observed diminishes the value of

TABLE 1. Large (on-board) Systematics (Biases and Uncertainties) [2].

Description of error budget constituents	Bias 10^{-8} cm/s^2	Uncertainty 10^{-8} cm/s^2
a) Radio beam reaction force	$+1.10$	± 0.11
b) RTG heat reflected off the craft	-0.55	± 0.55
c) Differential emissivity of the RTGs		± 0.85
d) Non-isotropic radiative cooling of the spacecraft		± 0.48
e) Expelled Helium produced within the RTGs	$+0.15$	± 0.16
f) Gas leakage		± 0.56

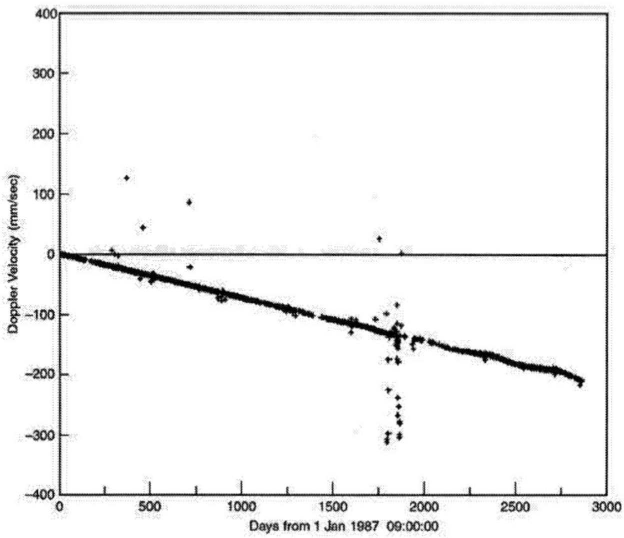

FIGURE 3. CHASMP best fit for the Pioneer 10 Doppler residuals with the anomalous acceleration taken out [1, 2]. After adding one more parameter to the model (a constant radial acceleration) the residuals are distributed about zero Doppler velocity with a systematic variation ~ 3.0 mm/s on a time scale of $\sim$ 3 months. The outliers on the plot were rejected from the fit. [The quality of the fit may be determined by the ratio of residuals to the downlink carrier frequency, $\nu_0 \approx 2.29$ GHz.]

using modern disturbance compensation systems for a test. For example, the systems that are currently being developed for the LISA and LISA Pathfinder missions, are designed to operate in the presence of a very low frequency acceleration noise (at the mHz level), while the Pioneer anomalous acceleration is a strong constant bias in the Doppler frequency data. Further, currently available DC (constant) accelerometers are a few orders of magnitude less sensitive than is needed for a test. Also, should the anomaly not really be a force but rather an effect that universally affects frequency standards [2], the use of accelerometers will shed no light on what is the true nature of the observed anomaly.

2. THE DATA'S MEANING

2.1. What the data tells us

The major study [2] demonstrated that:

For Pioneer 10. Between about 40 and 70.5 AU (data taken between 1987.0 and 1998.5), there was an experimental signal of

$$a_{P(expt)}^{Pio10} = (7.84 \pm 0.01) \times 10^{-8} \text{ cm/s}^2. \tag{1}$$

For Pioneer 11. Between about 22.4 and 71.7 AU (data taken between 1987.0 and 1990.8), there was an experimental signal of

$$a_{P(expt)}^{Pio10} = (8.55 \pm 0.02) \times 10^{-8}\ \mathrm{cm/s^2}. \tag{2}$$

As stated above, the analysis for both Pioneers with all systematics included yielded

$$a_P = (8.74 \pm 1.33) \times 10^{-8}\ \mathrm{cm/s^2}. \tag{3}$$

The effect was seen only on these small ($\sim$ 250 kg) craft on hyperbolic orbits. There were indicative signals from the Galileo and Ulysses craft, but the data was not conclusive.[2] The Voyager data is too imprecise to test for the anomaly.[3] On the other hand, the data is NOT SEEN on large, bound, astronomical bodies.

Finally, remember that, although the anomaly is commonly *interpreted* as an acceleration, the fundamental data is a Doppler shift. The anomaly could be something else, like a time acceleration. That being said, What do we only "suspect" or do not know?

To begin with, we have no real idea how far out the anomaly goes. We only know that it is roughly a constant between 20 and 70 AU, that *before* Saturn encounter (at around 10 AU) and the transition to hyperbolic orbit, Pioneer 11 did not show the anomaly, and after 10 AU Pioneer 10 roughly showed the anomaly.

An important goal is to perform a good detailed analysis of this early data.

2.2. Lessons learned from studying the Pioneer data

The lessons learned from the Pioneers are a guide on how to build a spacecraft that can accurately investigate the anomaly. Among the most important features of the Pioneer 10/11 spacecraft were their attitude control system, navigation and communication, on-board power, thermal design, and mission design [5, 6, 7]. (See Figure 4.)

Precision attitude control. The attitude control system should enable precise acceleration reconstruction along each spacecraft axis. This can be done with spin-stabilized attitude control, as was implemented on the Pioneers. This is the most preferable option. It allows for a minimum number of attitude correction maneuvers which are, because of the maneuver-associated propulsive gas leaks, notoriously difficult to model. Leakage from thrusters of the propulsion system is the major navigation problem for 3-axis stabilized vehicles, but its impact is minimal for spin-stabilized spacecraft.[4] With spin-

---

[2] Galileo was only spin-stabilized during Earth-Jupiter cruise. Although this data set was useful to verify the Deep Space Network hardware, it came from so close in to the Sun that it was too highly correlated with the solar radiation pressure to yield a positive result. Ulysses had to have an excessive number of maneuvers due to a failed nutation damper. Although the analysis was indicative, individual errors were as much as an order of magnitude larger than the effect [2].

[3] The Voyagers are three-axis stabilized. The resulting often-used gas jets yield a navigation error of $\sim 10^{-6}$ cm/s^2, which is an order of magnitude larger than the Pioneer anomaly [2].

[4] If one were to chose 3-axis stabilization, the uses of accelerometers and/or reaction wheels would be critical points of concern.

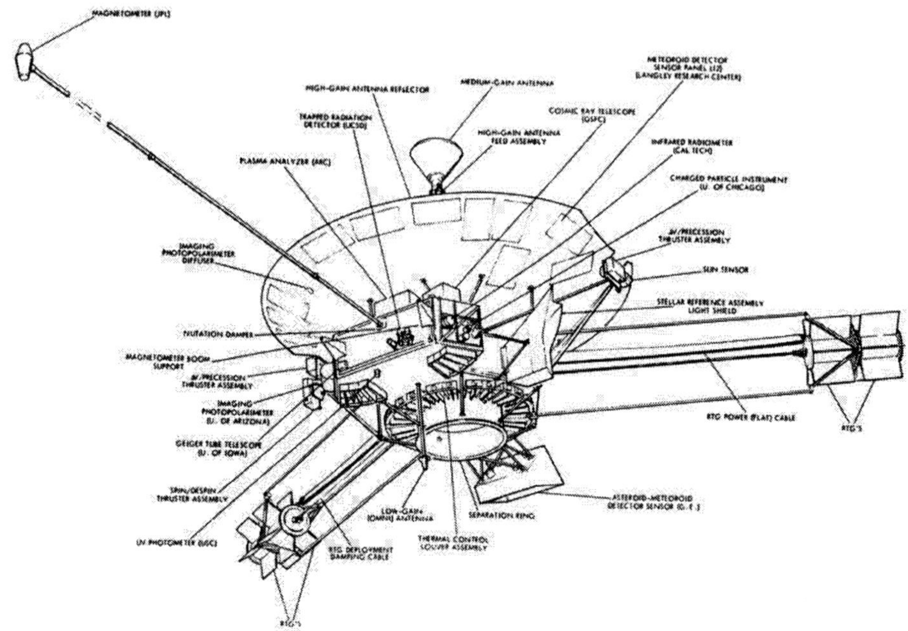

FIGURE 4. A drawing of the Pioneer spacecraft.

stabilization the spacecraft spin behavior can also be precisely monitored. The understanding of the spin history, coupled with knowledge of all possible sources of torque, will provide auxiliary information on the anomaly.

On-board propulsion system. For the reasons discussed in the attitude control requirements above, one would need precisely calibrated thrusters, propellant lines, and fuel gauges and knowledge of the propellant mass usage history. However, currently available sensors may not be accurate enough to yield the required long-term 3-dimensional acceleration sensitivity. Autonomous real-time monitoring and control of their performances would also be helpful. This would come, however, at the cost of an increased telecommunication data rate.

Navigation and communication. As with the attitude control system, the navigation and communication system should allow a 3-dimensional acceleration reconstruction at the level of $\sim 0.01 \times 10^{-8}$ cm/s^2 for each vector component. Having both Doppler and range tracking, and possibly VLBI and/or ΔDOR, would allow the precise measurements of plane-of-the-sky angles that are needed for 3-dimensional acceleration reconstruction. mrad pointing should be sufficient to enable precise attitude reconstruction. The preferred communications frequencies are X- and Ka-band with significant dual-band tracking for possible dispersive media corrections. Alternatively, optical tracking could be employed. This will be further evaluated in the future.

On-board power for deep space. No deep-space mission can accomplish its goals without a reliable source of on-board power. For now, this must be provided by RTGs. The location of the RTGs is a very critical choice, as it must satisfy the requirements for inertial balance, stability, and thermal isolation. For a spin-stabilized option, one would want to position the RTGs as far away as practical from the bus. Having the RTGs on extended booms aids the stability of the craft and also reduces the heat systematics.

Thermal design requirements. Thermal design is one of the most critical design issues for a mission to explore the Pioneer anomalous signal. The entire spacecraft and/or probe should, as much as possible, be heat-balanced and heat-symmetric fore/aft. In particular, the emitted radiant heat from the RTGs must be symmetrical in the fore and aft directions and the thermal louvers should be placed on the sides to eliminate fore/aft thermal recoil force due to the release of excess radiant heat.[5] One should have a precise knowledge of all heat sources - RTGs, electronics, thrusters, etc. Also, an active control of all heat dissipation channels would be an additional critical aid.

Finally, it is important to have a precise knowledge of how the spectral properties of the materials, from which the spacecraft surface is composed, will degrade. This is a challenging issue to discuss quantitatively. The difficulty lies firstly in the precise folding of the reflective insulation blankets and in the precision painting of all the external surfaces. But it is still hard to predict the exact behavior of all the surfaces on the spacecraft, especially after long exposure to the space environment (i.e., solar radiation, dust, planetary fly-byes, etc.). Knowing this all would result in a precise knowledge of the future history of the 3-dimensional vector of any residual thermal recoil force.

Modeling external non-gravitational forces. The analysis presented in [2] also emphasized the importance of a precise modeling of the solar radiation pressure. For distances below 10 AU, the capability of measuring, monitoring, and compensating for the effects of solar radiation pressure may be very important in achieving the precision radio-science objectives. In addition, similar information on the electrical charge accumulated on the spacecraft would be very important information for the purposes of precise orbit determination and attitude control.

Hyperbolic, solar system escape trajectories. The Pioneer anomaly was found on spacecraft following hyperbolic escape trajectories at distances between 20 and 70 AU out from the Sun [2]. Although it might have been present closer in, this has only been imprecisely studied [1, 2, 5, 6]. For this reason, and also to reduce the effect of external systematics, the experiment should reach distances greater than 20 AU from the Sun. Achieving small orbit transfer times to regions further than 20 AU would yield a concept validation and technology infusion to other missions having a demand for rapid access to distant regions of the solar system. Thus, arriving to this distance from the Sun in less than 10 years would be most desirable.

To yield a direct test for any velocity-dependence in the signal, one also wants the craft to have a significantly different velocity than the Pioneers. All this means that when the

[5] For a 3-axis stabilized craft it would be harder to balance recoil forces and torques.

craft reaches deep space it should be in a high-velocity, hyperbolic, escape orbit. As a baseline, the desirable mission duration should be shorter than 10 years to 20 AU.

To summarize. We observe that the Pioneers were "accidentally" built in a way that yielded very precise orbits; newer craft will need special designs to surpass the accuracies of the Pioneers. Effects that have normally been considered to be relatively unimportant, such as rejected thermal radiation, gas leaks, and radio beam reaction, now turn out to be critical for the precise navigation of science craft in the 21 century. It is hard not to emphasize the most successful feature and main Pioneer lesson for a potential spacecraft and mission design to test the anomalous acceleration - make it simple!

3. A FUTURE TEST

3.1. Mission objectives

The physics objectives of a mission to test the Pioneer anomaly would be two-fold:[6]

1. to fly a deep-space experiment that is capable of achieving a 3D accuracy of $\sim 0.01 \times 10^{-8}$ cm/s^2 for small *unmodeled* DC accelerations. This quantity is equivalent to a level of $\sim 3 \times 10^{-21}$ s/s^2 in clock accelerations.
2. to determine the direction and physical origin of any anomaly discovered.

The experiment could be performed on a dedicated mission or else considered for a probe that is jettisoned in deep space from a large mission (Sec. 3.6).

3.2. Fore/aft symmetric deep-space mission

We begin by discussing a fore/aft symmetric spacecraft that can be considered a baseline concept for any successful mission [6]. It is designed to perform a general test of the anomaly as it was observed. It is also designed to eliminate systematics and allow precise tracking from distances in the inner solar system to 40 AU and beyond. Its design is specifically motivated by the lessons learned from the Pioneer missions described in Section 2.2.

For a nuclear powered spacecraft, the major navigation systematic in deep space is thermal emission generated by the spacecraft's power system. This is because, with either space-craft centered RTGs or else a space-craft centered nuclear reactor, there are many to hundreds of kW of heat power generated. This produces up to hundreds or more W of electrical power in the bus. The heat dissipation can produce a non-isotropic force on the craft which can dominate a force the size of the Pioneer anomaly, especially if the craft is light. For example, only ~ 63 W of directed power could have explained

[6] With the growing interest in the anomaly, there have recently been a number of proposals for missions to test the anomaly [5]-[10].

the anomaly the 241 kg Pioneer craft with half its fuel depleted. Therefore, if those RTGs had been placed "forward" they obviously would have yielded a huge systematic

The heat systematic will be eliminated by making the heat dissipation fore/aft symmetric. In a stroke of serendipitous luck, the Pioneer RTGs, with $\sim$ 2,500 W of heat, were placed at the end of booms.[7] This meant they had little thermal effect on the craft. Further, the rotation of the Pioneer craft and their RTG fin structure design meant the radiation was extremely symmetrical fore-aft, with very little heat radiated in the direction towards the craft. The same concept will be used for this mission, with perhaps shielding of the RTGs to further prevent anisotropic heat reflection.

The electrical power in the equipment and instrument compartments must also be radiated out so as not to cause an undetected systematic. For the Pioneers the central compartment was surrounded by insulation. There were louvers forward to be open and let out excess heat early in the mission and to be closed and retain heat later on when the electrical power was less. The electrical power degrades faster than the radioactive decay because *any* degradation of the thermoelectric components means the electric power degrades from this *on top of* the degradation of the input heat due to the 87.74 half-life of the ^{238}Pu.[8] For this mission, the louvers will be on the side of the compartment so that they radiate in an axially symmetric manner as the spacecraft rotates.

The most unique design feature of this mission is the symmetric radio-beam: The dual for/aft symmetric spacecraft design uses two identical and simultaneously transmitting radio-antennae aligned along the spin-axis and facing in the opposite direction. This is the yo-yo concept described in Ref. [6] and shown in Fig. 5. By implementing such a design, one can essentially eliminate the radio-beam communication bias. A spacecraft design such as this has never been proposed before.

This symmetric antenna design also allows us to further minimize the heat systematic. Any heat reaching the back of the two antennas. despite insulation placed in, around, and in the support of the bus, will be reflected symmetrically fore/aft. This choice would also eliminate any remaining fore/aft asymmetry in the acceleration estimation since, by rotating the craft 180°, one could cancel out any difference in the two signals.

Thus, this fore/aft symmetric design greatly reduces the size of any possible heat systematic by its simplicity.[9] Preliminary analysis (see Ref. [6]) suggests that even with existing technology one could balance the fore/aft geometry of the spacecraft to minimize the possible differential heat rejection systematic to the level $\leq 0.03 \times 10^{-8}$ cm/s^2. (With improved technology one could reach our desired goal.)

A final factor in the spacecraft heat transfer mechanism is the optical properties with time of the spacecraft surfaces. However, such changes did not seem to affect the Pioneer results despite their rugged voyages [2, 4]. Further, this mission's use of rotating the antennas (described above) will obviate any residual effect by cancellation.

---

[7] Because they were the first deep space craft, the Pioneer engineers were worried about the effects of nuclear radiation on the main bus electronics. Placing the RTGs at the end of booms was the solution.

[8] For the Pioneers, the time from launch to when the Pioneer 10 electrical power had been reduced to 50% was about 20 years.[4]

[9] Recall that for the Pioneers, contributions to the detected anomaly of order 10^{-8} cm/s^2 came individually from the RTGs and power dissipation [2, 4].

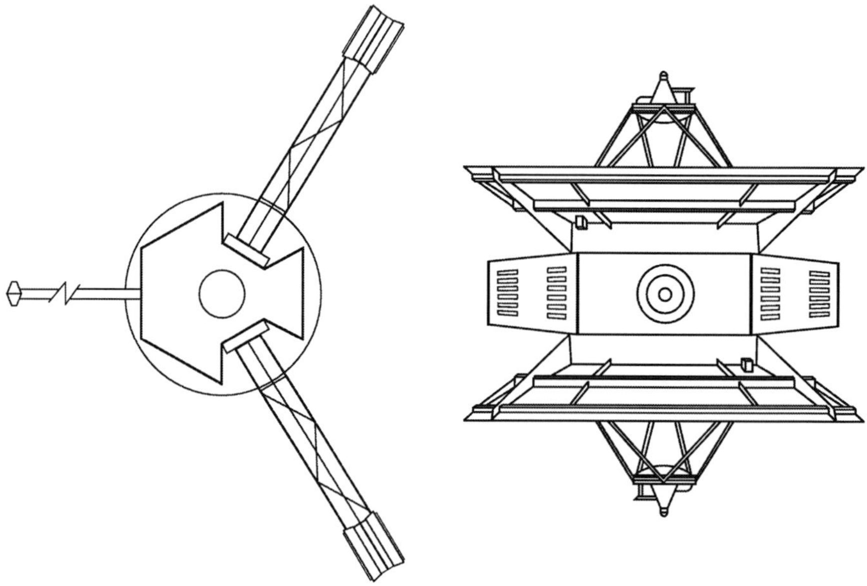

FIGURE 5. The top (left) and side (right) views (different scales) of "yo-yo" craft concept [6]. The scale of the circular antenna is on the order of 2 to 2.5 m. The RTGs are deployed on the left. There also is an indication of a third long boom where an instrument package to detect interstellar matter could be placed. Depending on the final mission objectives this instrument package could be replaced by a third RTG. The side view shows the louvers radiating to the side and the antennas, taken and modeled from the Cassini Cassegrain antenna [11].

3.3. The Direction of the Anomaly

For the Pioneers at the 20-70 AU studied [2], the directions (1) towards the Sun, (2) towards the Earth, (3) along the direction of motion of the craft, or (4) along the spin axis could not easily be distinguished. To obtain a determination of the actual direction would be very important, since these directions would tend to indicate a physical origin that is (1) new dynamical physics originating from the Sun, (2) a time signal anomaly, (3) a drag or inertial effect, or (4) an on-board systematic.

At 20 AU these directions are of order 3 degrees apart (the maximum angle subtended by the Sun and the Earth (even more depending on the hyperbolic escape velocity vector). In Figure 6 we show the angles at which these forces would act for a hyperbolic trajectory in the ecliptic, between 20 and 40 AU. The eccentricity is 5 and the craft travels at approximately a terminal velocity of 5 AU/yr. (a, the minimum distance from the hyperbola to its intersecting asymptotes, is 1.56 AU.) The reference curve (1) at zero degrees is the constant direction towards the Sun. Other angles are in reference to this. Starting to the right in the plane for definiteness, the angle towards the Earth (2) is a cosine curve which is modified by an $1/r$ envelope as the craft moves further out. The angle from the Sun to the trajectory line is shown in (3). Finally, the direction along the spin axis (4) is a series of decreasing step functions. This indicates two maneuvers per

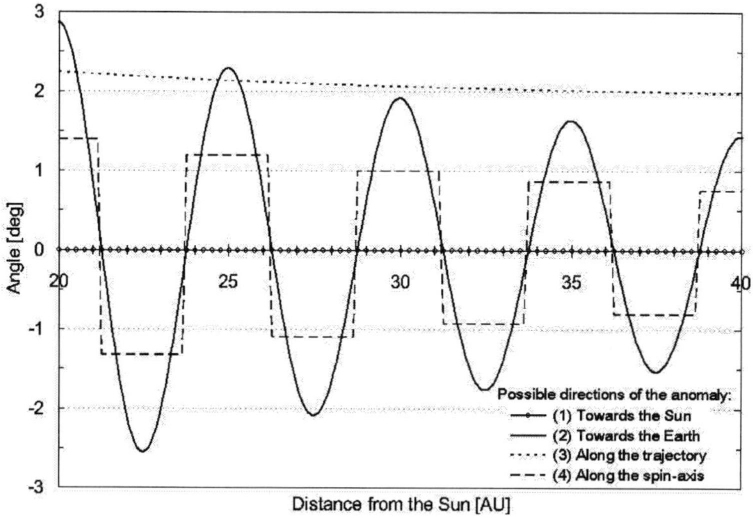

FIGURE 6. The signatures for four possible directions of the anomalous acceleration acting on the proposed spacecraft. The signatures are distinctively different and are easily detectable with the proposed mission. (See the text.)

year to place the antenna direction between the maximum Earth direction and the null Sun direction, performed as the Earth passes from one side of the Sun to the other.

The use of 3-D navigation discussed above would result in a precise spacecraft positioning with respect to the solar system barycentric reference frame. As with the Pioneers, the accuracy of the determination will depend on the properties of the antenna radiation pattern. Highly pointed radiation patterns are available for higher communication frequencies. In order to be on the safe side, one can use a standard X-band antenna with a 0.5° angular resolution. Therefore, if the anomaly is directed towards the Sun (1), the radiometric tracking described above will be able to establish such a direction with sufficiently high accuracy.

If the anomaly is directed towards the Earth (2), the current accuracy of the Earth's ephemerides will be a key to determining this fact. Furthermore, in this case one would clearly see a dumped sinusoidal signal that is characteristic to this situation (see Figure 6). The use of standard hardware would enable one to accurately establish this direction with a high signal-to-noise ratio.

An almost-linear angular change approaching the direction towards the Sun (also highly correlated with the hyperbolic trajectory) would indicate a trajectory-related source for the anomaly (3). This situation will be even more pronounced if the spacecraft were to perform a planetary fly-by. In the case of a fly-by, a sudden change in the anomaly's direction will strongly suggest a trajectory-related source for the anomaly.

Finally, a step-function-like behavior of the anomaly, strongly correlated with the maneuver history, would clearly support any anomaly directed along the spin-axis (4). As a result, a combination of the standard navigation methods addressed above in combination of the symmetric spacecraft design would again enable one to discriminate between these four different directions of the anomaly with a sufficiently high accuracy.

It is clear that these four possible anomaly directions all have entirely different characters. The symmetric fore/aft mission concept was designed with this issue in mind. In Ref. [6] it was determined that the main features of the signatures of Fig. 6 could be distinguished over a year.[10]

Thus, this mission will also provide evidence on the origin of the anomaly, by helping to determine its direction. Determining which of the possible physical causes for the anomaly is the correct one will be important in the more general frameworks of the solar system ephemerides as well as spacecraft design and navigation.

3.4. Formation flying mission concept

As these developments have been going on, interest in the anomaly has spread. In particular a collaboration has been formed among members of the European science community, which we have joined. It has presented a proposal [12] to the European Space Agency (ESA) in answer to its call, Cosmic Vision Themes for 2015-2025.. This collaboration is interested in producing a mission that will also develop new technologies, so as to involve the space technology community.

Formation flying technologies have become one of the top priorities for both ESA and NASA. In particular, a recent NASA Research Announcement calls for development of precision formation flying as an enabling technology to meet future high-priority science objectives [13]. Due to limited launch vehicle fairing sizes and the need to phase optical elements over long distances on flexible structures, separated spacecraft formation flying is the only viable means to satisfy many demands of modern science missions.

Therefore, as a reference design, the collaboration choose to emphasize a concept that relies on formation flying [9]. The design has a primary craft that is robust and able to nurture itself for 7-12 years in the environment of deep space. The craft would communicate high-precision Doppler and range data to Earth, with X- or Ka-band, or perhaps even via optical communication.

The idea is to avoid the inherent problems of self disturbance of an inertial sensor on board the primary spacecraft by placing the inertial reference mass(es) (i.e. subsatellite(s)) outside the craft at a sufficient, but not too large distance. A laser ranging sensor that employs a mW laser monitors the 3-dimensional vector of mutual separation between the spacecraft and subsatellite. Any subsatellite is covered with corner-cube retro-reflectors that enable precise laser ranging similar to that currently used for satellite and lunar laser ranging. Figure 7 shows a generic concept [9].[11]

To enable this concept three requirements must be fulfilled: i) the subsatellite must be at a sufficient distance from the main craft so that any radiant heat from the primary craft will not affect the motion of the subsatellite; ii) the reflected laser light must also

[10] The use of antennas with highly pointed radiation patterns and of star pointing sensors would create even better conditions for resolving the true direction of the anomaly.

[11] A related, but more complicated, idea has also been proposed [14].

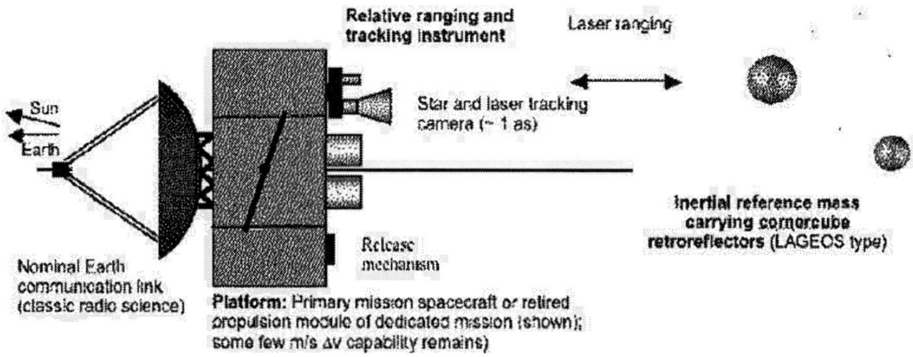

FIGURE 7. Formation flight scenario [9]: an active spacecraft (or retired propulsion module) tracking a locally detached, formation flying reference inertial masses via ranging and angular sensors. The spheres have been previously gently released from the spacecraft. Additional on-board sensors may include magnetometer, solar radiation intensity monitors, and a dust spectrometer including both charge and inertial sensors.

not cause a significant force on the subsatellite; iii) the primary craft must be able to laser range to the subsatellite to determine its relative position history; i.e., the primary craft must be able to "follow" the subsatellite. The objective of this architecture is to precisely measure the motion of the inertial reference mass and to eliminate the effect of any disturbance of the primary spacecraft on the subsatellite.

Conceptually a combination of two observables related to the precise distance determination between the Earth and the spacecraft and with that between the spacecraft and the subsatellite would yield a more accurate distance between the Earth and the subsatellite. In principle the spacecraft does not have to be dynamically quiet as long as the relative distance and lateral position remain within the laser tracking capability.[12]

The relative range, range rate and lateral tracking is a 3-dimensional two-step process and needs to be sufficiently accurate in relative angle determination between main link to Earth and local link to reference. Despite the advantages, the mission operation has some added complexity. Also, the error in the two-step measurement process has not been studied for a noisy main craft. This needs to be done to be sure the concept works.

The spacecraft has to provide an accurate release mechanism for the reference masses. Depending on its stabilization mechanism, its other properties could be similar to those of the fore/aft symmetric mission; it would not need to perform many large maneuvers, the dedicated payload, including reference masses, could be in the range of 150-250 kg, it would use the only source of autonomous power in deep space, RTGs. It would need at least about 50-75 W of electrical power to run the communications and heat the fuel (for maneuvering). Since the electrical thermocouples degrade with time, the power at launch should be of order 140 W.

[12] Current estimates suggest that tolerable distances are up to few km with a lateral imbalance of up to 5 degrees.

3.5. Propulsion Options

The launch vehicle is a major consideration for any deep-space mission. Propulsion systems are quite literally the driving force behind any effort to get a payload into space, especially on the interplanetary orbit.

To test the Pioneer anomalous acceleration in the most suitable environment, one wants to reach a distance greater than 15 AU from the Sun. In this region one can clearly distinguish any effect from solar radiation pressure, interplanetary magnetic fields, as well as solar and interplanetary plasmas. A fast transfer orbit is very desirable, to allow reaching the target region in a reasonable time. Therefore, a large solar system escape velocity is desired (say, more than 5-10 AU/yr). In contrast, the Pioneers and Voyagers are cruising at velocities of about 2 and 3 AU/yr. One needs something faster than that.

The obvious first idea is a very energetic chemical propulsion rocket. An escape terminal velocity of $\sim$ 5 AU/yr is achievable with current mission design technologies and existing heavy launch vehicles (Ariane 5, Proton, Delta IV, or Atlas V) [6, 15].

However, it must be noted that purely chemical propulsion, even using present day powerful and expensive (!) launchers, cannot achieve the desired short mission times without a powerful chemical kick stage and/or suitable planetary gravity-assist maneuvers. That immediately sets severe constraints on any mission profile. On the other hand, if a chemical launch vehicle were to be used for a dedicated mission, with gravity-assist flybys, a mission could also address the question of if the Pioneer 11 anomaly started near its Saturn flyby, when it reached escape velocity.

Further, the use of chemical propulsion is at the limit of its capabilities to satisfy the needs for deep-space exploration. For this reason, both ESA and NASA have initiated programs to study alternative propulsion methods for their deep space exploration missions. For this reason, the new collaboration also decided to consider various new technologies that might be employed. They include 1) nuclear electric, 2) solar electric, and 3) solar thermal propulsion.

But the choice with most widespread interest in Europe is solar sails. The Europeans have had a long interest in this idea, as exemplified by the ESA/German Odissee concept [16]. With ESA's [16, 17] and NASA's [18, 19] separate interest in this solar sail technology, it may prove useful and also could yield science [20].

The ultimate hope is to obtain sail materials with a weight of 1 g per m^2 of sail area and light weight structures to support a sail of size 200 m in length. For now this goal is in the future, but the advent of multi-cm long nanotubes may be pointing the way. With such a sail, one can envision accelerating to velocities over 14 AU/yr by the time the orbit of Jupiter is reached, jettisoning the sail, and having the main mission coast.

3.6. Testing the anomaly as part of a larger mission

Given that any mission to very deep space will need a large terminal velocity and the concomitant commitment of effort and funds, the question naturally arises of if a test of the Pioneer anomaly can be done, not as a dedicated mission but rather, as a probe on a very large mission that would be jettisoned to fly on separately after the requisite distance and speed had been obtained [5, 6, 10]. In principle this could happen with a mission that used nuclear reactor power.

As an example, it might be considered for an add-on to a mission that would use a reactor-ion engine, such as Prometheus [21]. Indeed, this idea has already been discussed in the context of the InterStellar Probe mission to the heliopause [22]. Such a marriage may prove to be an interesting possibility for a test of the Pioneer Anomaly.

ACKNOWLEDGMENTS

We all are thankful for the many contributions of the original Pioneer collaboration (Sec. 1.1) as well as those from the new collaboration [12]. M.M.N. thanks Alfredo Macías and Claus Lämmerzahl for their kind hospitality and acknowledges support by the U.S. DOE. The work of S.G.T and J.D.A. was performed at the Jet Propulsion Laboratory, California Institute of Technology, under contract with the National Aeronautics and Space Administration.

REFERENCES

1. Anderson, J. D., Laing, P. A., Lau, E. L., Liu, A. S., Nieto, M. M., and Turyshev, S. G., *Phys. Rev. Lett.* **81**, 2858-2861 (1988), e-print gr-qc/9808081.
2. Anderson, J. D., Laing, P. A., Lau, E. L., Liu, A. S., Nieto, M. M., and Turyshev, S. G., *Phys. Rev. D* **65**, 082004/1-50 (2002), e-print gr-qc/0104064.
3. Markwardt, C., eprint gr-qc/0208046.
4. Anderson, J. D., Lau, E. L., Turyshev, S. G., Laing, P. A., and Nieto, M. M., *Mod. Phys. Lett. A* **17**, 875-885 (2002), e-print gr-qc/0107022.
5. J. D. Anderson, S. G. Turyshev, and M. M. Nieto, Int. J. Mod. Phys. D. **11**, 1545-1551 (2002), eprint gr-qc/0205059.
6. M. M. Nieto and S. G. Turyshev, Class. Quant. Grav. **21**, 4005-4023 (2004), eprint gr-qc/0308017.
7. S. G. Turyshev, M. M. Nieto, and J. D. Anderson, Proceedings of the 35th COSPAR Scientific Assembly (to be published), eprint gr-qc/0409117.
8. Bertolami and M. Tajmar, "Gravity Control and Possible Influence on Space Propulsion: A Scientific Study," ESA Report CR (P) 4365 (2002).
9. U. Johann and R. Förstner, "Enigma," EADS Astrium GmbH Pioneer proposal to ESA(2004).
10. A. Rathke, in: *Frontier Science 2004: Physics and Astrophysics in Space* (to be published), eprint astro-ph/0409373.
11. http://saturn.jpl.nasa.gov/spacecraft/instruments-cassini-radar.cfm
12. H. Dittus, C. Lämmerzahl, S. Theil, B. Dachwald, W. Seboldt, W. Ertmer, E. Rasel, R. Foerstner, U. Johann, F. W. Hehl, C. Kiefer, H.-J. Blome, R. Bingham, B. Kent, T. J. Sumner, O. Bertolami, J. L. Rosales, B. Christophe, B. Foulon, P. Touboul, P. Bouyer, S. Reynaud, C. J. de Matos, C. Erd, J. C. Grenouilleau, D. Izzo, A. Rathke, J. D. Anderson, S. W. Asmar, S. G. Turyshev, M. M. Nieto, and B. Mashhoon, "A Consolidated Cosmic Vision Theme Proposal to Explore the Pioneer Anomaly," submitted to ESA (Oct. 2004).
13. "New Millennium Program: Space Technology - 9 (ST9)," NASA Research Announcement # NNH04ZSS002N, 25 August 2004. http://research.hq.nasa.gov/code_s/nra/current/NNH04ZSS002N/index.html
14. K. Penanen and T. Chui, in: *Second International Conference on Particle and Fundamental Physics in Space (SpacePart'03)*, Nucl. Phys. B (to be published), eprint gr-qc/0406013.
15. http://www.aero.org/publications/crosslink/winter2004/07.html
16. See http://www.kp.dlr.de/solarsail/Welcome.html

17. B. Dachwald, AIAA J. Guidance, Control, and Dynamics **27**, 66-72 (2004).
18. C. E. Garner, W. Layman, S. A. Gavit, and T. Knowles, in: *Space Technology and Applications International Forum (STAIF - 2000)*, ed. M. S. El-Genk, AIP Conf. Proc. 504, ed. M. S. El-Genk, (Am. Inst. Phys., Melville, NY., 2002) p 947-961.
19. P. C. Liewer, R. A. Mewaldt, J. A. Ayon, and R. A. Wallace, in: *Space Technology and Applications International Forum (STAIF - 2000)* AIP Conf. Proc. 504, ed. M. S. El-Genk (Am. Inst. Phys., Melville, NY, 2002) p. 911-916.
20. M. M. Nieto and S. G. Turyshev, Int. J. Mod. Phys. D **13**, 899-906 (2004), eprint astro-ph/0308108.
21. http://spacescience.nasa.gov/missions/prometheus.htm
22. H. O. Funsten, presentation to the Natl. Acad. Sci. panel on the use of nuclear power in space (2004).

Pioneer's Anomaly and the Solar Quadrupole Moment

Hernando Quevedo

Instituto de Ciencias Nucleares
Universidad Nacional Autónoma de México
A.P. 70-543, México D.F. 04510, México

Abstract. The trajectories of test particles moving in the gravitational field of a non-spherically symmetric mass distribution become affected by the presence of multipole moments. In the case of hyperbolic trajectories, the quadrupole moment of an oblate mass induces a displacement of the trajectory towards the mass source, an effect that can be interpreted as an additional acceleration directed towards the source. Although this additional acceleration is not constant, we perform a general relativistic analysis in order to evaluate the possibility of explaining Pioneer's anomalous acceleration by means of the observed Solar quadrupole moment, within the range of accuracy of the observed anomalous acceleration. We conclude that the Solar quadrupole moment generates an acceleration which is of the same order of magnitude of Pioneer's constant acceleration only at distances of a few astronomical units.

INTRODUCTION

Pioneer 10/11 are spacecraft which were launched more than thirty years ago to explore the outer region of the Solar system. They were also the first missions to enter the edge of interstellar space. Due to their especial structure, they also represent an ideal system to carry out precision celestial mechanics experiments which permit acceleration estimations to the level of $10^{-8} cm/s^2$. After encountering the outer planets Jupiter and Saturn, the two spacecraft followed hyperbolic orbits near the plane of the ecliptic to opposite sides of the Solar system. A careful analysis of orbital data from Pioneer 10/11 has been reported in [1, 2] which indicates the existence of a very weak, long-range acceleration, $a_P = (8.74 \pm 1.33) \times 10^{-8} cm/s^2$, directed towards the Sun. The most intriguing result about this acceleration is that it is constant, with a single measurement accuracy averaged over 5 days. Moreover, it has been measured at distances contained within the approximate range of 20 to 60 astronomical units (AU) from the Sun.

The most conservative hypothesis about the origin of this anomalous acceleration is that this kind of effects on Pioneer's tracking data must be due to some technological reasons related to the configuration and equipment of the spacecraft. However, the analysis reported in [1, 2] seems to have ruled out all possible non-gravitational forces generated inside and outside the spacecraft. Moreover, the authors have taken into account the accepted values of the errors in the determination of the planetary ephemeris, Earth's

CP758, *Gravitation and Cosmology: 2nd Mexican Meeting on Mathematical and Experimental Physics*
edited by A. Macías, C. Läemmerzahl and D. Nuñez

orientation, precession, and nutation. More adventurous explanations of the anomalous acceleration suggest the existence of new physical phenomena, especially inspired by the fact that a_P is of the order of cH, where c is the velocity of light in empty space and H is the Hubble constant. For instance, in [3, 4, 5] it is suggested that the expansion of the Universe induces local tidal forces which generate the anomalous acceleration. In [6] a Kaluza-Klein model is proposed in which the fifth dimension becomes dynamical with an expansion rate that generates a_P. A scale dependent cosmological term has been used in [7] which affects the motion of test particles. A different explanation is suggested in [8] by means of a time-dependent Newtonian gravitational constant that produces a long-range acceleration. More sophisticated explanations have been proposed in the literature (see, for instance [2], for a recent review) which include dark matter and stringy inspired scalar fields.

In this work, we investigate the possibility of explaining the anomalous acceleration by means of the Solar quadrupole moment. Clearly, the additional acceleration due to the quadrupole is not constant, but decreases with distance. In this sense the, quadrupole moment could immediately be ruled out as an explanation for a_P. However, our approach is different. Since a_P has been observed only at distances less than 60 AU, with an accuracy of around 15%, our goal is to determine whether the acceleration generated by the quadrupole is of the same order of a_P and can be fitted within its range of accuracy. This idea is based upon the fact that the quadrupole moment of the Sun has been determined in an exact manner only recently [9], and has been given the value of $J_2 = -2.28\ 10^{-7} \pm 15\%$, where J_2 is a dimensionless parameter to be introduced below. Before this value was known, different methods generated completely different values of J_2, sometimes with differences of several orders of magnitude. (In fact, such a great ambiguity in the value of J_2 could easily be used to explain a_P in the complete range of observation!) With this new value, we will show that the quadrupole moment of the Sun produces an acceleration which is of the same order of a_P at distances of a few AU's, but is several orders of magnitude less than a_P at distances between 20 and 60 AU's.

We will analyze the relativistic hyperbolic motion of test particles in the gravitational field of a mass with quadrupole moment. For the sake of simplicity, we will assume that the quadrupole is due only to the asphericities of the mass distribution, preserving the axial symmetry, and will neglect the rotation of the source. This allows us to use a static, axisymmetric solution of Einstein's vacuum field equations as the metric which describes the spacetime. In particular, we will use the Erez–Rosen metric for our analysis. Moreover, we will assume that the motion is confined to the equatorial plane of the source. This is in good agreement with the actual path of Pioneer 10/11 which move near the plane of the ecliptic.

GRAVITATIONAL FIELD OF A MASS WITH QUADRUPOLE MOMENT

In this section we present a static axisymmetric solution of Einstein's vacuum field equations which was first discovered by Erez and Rosen [10]. This solution generalizes Schwarzschild solution to include an arbitrary mass quadrupole moment. Originally, this solution was obtained in prolate spheroidal coordinates, but for the purposes of this work spherical coordinates (t, r, θ, φ) are more appropriate. In these coordinates, it can be shown that it is symmetric with respect to the equatorial plane, $\theta = \pi/2$, so that it becomes a geodesic plane. For the sake of simplicity we will limit ourselves to the investigation of free motion on the equatorial plane. Accordingly, the Erez-Rosen solution can be written as [11]

$$ds^2 = \left(1 - \frac{2m}{r}\right) e^{2q\psi} dt^2 - \left(1 - \frac{2m}{r}\right)^{-1} e^{2q(\gamma-\psi)} dr^2 - r^2 e^{-2q\psi} d\varphi^2 , \qquad (1)$$

where the constant m represents the Schwarzschild mass, q is a constant parameter that determines the quadrupole moment, and ψ and γ are, in general, functions of r and θ. On the equatorial plane, these functions take the form

$$\psi = -\frac{1}{4}\left[\frac{1}{2}\left(3\frac{r^2}{m^2} - 6\frac{r}{m} + 2\right)\ln\left(1 - \frac{2m}{r}\right) + 3\left(\frac{r}{m} - 1\right)\right] , \qquad (2)$$

$$\gamma \approx \frac{1}{10}\frac{m^4}{r^4} . \qquad (3)$$

The general form of the function γ is rather cumbersome. Here we quote only the leading term in the limit $r \to \infty$ which is sufficient for the analysis we will perform in the next section.

In general, the Erez-Rosen solution is asymptotically flat and this allows us to determine its multipole moments in a covariant manner. For instance, using the Geroch-Hansen [12] definition of multipole moments one can show that the monopole is m and the quadrupole moment is $(2/15)qm^3$. Higher multipoles are given in terms of the monopole and quadrupole moments. So the parameters m and q acquire a clear physical meaning. As expected, in the limiting case $q = 0$, the Erez-Rosen metric (1) coincides with the Schwarzschild solution on the equatorial plane. The hypersurface $r = 2m$ turns out to be singular, in accordance with the black hole uniqueness theorems. An additional singularity is situated at $r \to 0$. Outside the hypersurface $r = 2m$, the spacetime is completely regular.

At large distances from the source, the expansion of the metric component g_{tt} leads to

$$g_{tt} \approx 1 - \frac{2m}{r}\left[1 - \frac{2}{15}q\left(\frac{m}{r}\right)^2\right] . \qquad (4)$$

Later on we will use this expression to determine the acceleration generated by the quadrupole moment.

MOTION OF TEST PARTICLES

Let us consider the motion of test particles on the equatorial plane of the Erez-Rosen metric (1). The geodesic equation can be reduced to a set of first order differential equations due to the existence of two constants of motion p_t and p_φ associated with the time-translation and the axial symmetry, respectively. For the case of a massive test particle, the geodesic equations can be written as:

$$\left(1-\frac{2m}{r}\right)e^{2q\psi}\dot{t} = p_t \ , \tag{5}$$

$$r^2 e^{-2q\psi}\dot{\varphi} = p_\varphi \ , \tag{6}$$

$$\dot{r}^2 = e^{-2q\gamma}\left[p_t^2 - \left(1-\frac{2m}{r}\right)e^{2q\psi} - \left(1-\frac{2m}{r}\right)\frac{p_\varphi^2}{r^2}e^{4q\psi}\right] \ , \tag{7}$$

where a dot represents the derivative with respect to the affine parameter along the geodesic. It is possible to perform a detailed analysis of these set of geodesic equations [11] and find out all the effects of the quadrupole moment on the motion of test particles. Here, however, we are interested only on hyperbolic motion. Hence, let us consider the special case $r = r(\varphi)$ so that $\dot{r} = (dr/d\varphi)\dot{\varphi}$, and introduce the following notations:

$$u = \frac{2m}{u} \ , \quad \alpha = \frac{2m}{p_\varphi} \ . \tag{8}$$

Then, the geodesic equation (7) reduces to

$$(u')^2 = F(u) := e^{-2q\gamma}\left[u^3 - u^2 + \alpha^2 u e^{-2q\psi} + \alpha^2 e^{-2q\psi}\left(p_t^2 e^{-2q\psi} - 1\right)\right] \ , \tag{9}$$

where a prime denotes the derivative with respect to the angle coordinate φ. Since the expression $(u')^2$ is positive for a timelike geodesic, we must demand that the condition $F(u) \geq 0$ be satisfied. Accordingly, from Eq.(9) we obtain

$$\alpha^2 p_t^2 \geq (1-u)e^{2q\psi}\left(\alpha^2 + u^2 e^{2q\psi}\right) \ . \tag{10}$$

Because the left-hand side of this inequality is a positive definite constant, this implies that the roots of the equation $F(u) = 0$ must be contained in the interval $(1-u) > 0$ or, according to Eq.(8), in the interval $r \in (2m, \infty)$. As mentioned in the last section, the hypersurface $r = 2m$ contains in general a curvature singularity. Thus, hyperbolic motion is confined to the region where the Erez-Rosen metric is regular.

On the other hand, simple algebraic manipulations show that the equation $F(u) = 0$ can be expressed as a quadratic algebraic equation for the function $\exp(2q\psi)$, i.e.,

$$e^{4q\psi} + \frac{\alpha^2}{u^2}e^{2q\psi} - \frac{\alpha^2 p_t^2}{(1-u)u^2} = 0 \ . \tag{11}$$

Since $\exp(2q\psi) > 0$, we need to consider only the positive root of this equation which can be expressed as

$$q = \frac{1}{2\psi}\ln\left[\frac{\alpha^2}{2u^2}\left(\sqrt{1+4\frac{p_t^2}{\alpha^2}\frac{u^2}{1-u}} - 1\right)\right] \ . \tag{12}$$

Now, the function ψ is positive in the interval $r \in (2m, \infty)$. Then, from Eq.(12) we obtain that

$$\frac{\alpha^2}{2u^2}\left(\sqrt{1+4\frac{p_t^2}{\alpha^2}\frac{u^2}{1-u}}-1\right) > 1 \quad \text{for} \quad q > 0\,, \tag{13}$$

$$\frac{\alpha^2}{2u^2}\left(\sqrt{1+4\frac{p_t^2}{\alpha^2}\frac{u^2}{1-u}}-1\right) < 1 \quad \text{for} \quad q < 0\,. \tag{14}$$

These conditions can be represented in an equivalent manner as

$$f(u) := u^2(u-1) + \alpha^2(u-1) + \alpha^2 p_t^2 > 0 \quad \text{or} \quad < 0 \quad \text{for} \quad q > 0 \quad \text{or} \quad q < 0\,, \tag{15}$$

respectively. On the other hand, as can be seen from Eq.(9), the function $f(u)$ coincides with the function $F(u)$ for vanishing quadrupole moment ($q = 0$), i.e., $f(u)$ determines the motion of test particles in the Schwarzschild spacetime. This is a nontrivial result. In fact, suppose that the equation $F(u) = 0$ has a root at $u = u_0$. Then at u_0 the value of q is related to the value of u_0 by means of (12) with u replaced by u_0. If, for instance, $q < 0$, then according to Eq.(15) the Schwarzschild function $f(u_0)$ is negative. In the case of hyperbolic motion, the value $u = u_0$ corresponds to a vanishing "velocity", i.e., $u'(u = u_0) = 0$ or, equivalently, $\dot{r} = 0$, a condition which can be satisfied only at the perihelion of the hyperbolic trajectory. At this point, the motion of test particles in the Schwarzschild spacetime is not allowed, because there $f(u_0) < 0$. From the form of the functions $F(u)$ and $f(u)$ one can show that in general

$$f(u) < F(u) \quad \text{for} \quad q < 0 \qquad \text{and} \quad f(u) > F(u) \quad \text{for} \quad q > 0\,. \tag{16}$$

Let us consider the case of an oblate configuration ($q < 0$). Hyperbolic motion corresponds to the case where $F(u = 0) > 0$ and possesses (at least) one positive root, say u_2. The motion is confined to the region $0 < u < u_2$ with $r_2 = 2m/u_2$ being the perihelion distance. At $u = u_2$, according to (16), we have that $f(u_2) < 0$ and so no motion is allowed in the Schwarzschild spacetime. If we "move" from u_2 towards $u = 0$, we will find a point, say u_1, where $f(u_1) = 0$. Starting from this point and up to the point $u = 0$, where $f(u = 0) > 0$, the motion is allowed in the Schwarzschild as well as in the Erez–Rosen spacetime. For the trajectory in the Schwarzschild spacetime the perihelion distance is given by $r_1 = 2m/u_1$ and at this point, according to (16), $F(u = u_1) > 0$, that is, the trajectory in the Erez–Rosen spacetime would correspond to a non-zero radial velocity. So we see that the perihelion distance of a hyperbolic path becomes affected by the presence of an oblate quadrupole moment in such a way that

$$r_2 < r_1\,, \tag{17}$$

a result that trivially follows from the fact that $u_1 < u_2$. Thus an oblate quadrupole moment reduces the perihelion distance.

Let us denote by $r_{ER}(\varphi)$ the hyperbolic path of a test particle in the Erez–Rosen spacetime with mass m and a negative quadrupole q, and by $r_S(\varphi)$ the corresponding path in the Schwarzschild spacetime which is obtained from $r_{ER}(\varphi)$ by setting $q = 0$.

Notice that the mass m must be the same in order to be able to compare both trajectories. Clearly, $r_S(\varphi)$ must be a solution of the equation $(u')^2 = f(u)$, whereas $r_{ER}(\varphi)$ is a solution of $(u')^2 = F(u)$ with the same mass. As we have shown above, at the perihelion we have that $r_{ER}(\varphi_{per}) < r_S(\varphi_{per})$. Moreover, since in this case $f(u) < F(u)$ for all values of u, we obtain that in general

$$r_{ER}(\varphi) < r_S(\varphi) \tag{18}$$

at all points of the trajectory. The difference between these two trajectories will become smaller as r increases. Moreover, r_{ER} will approach r_S asymptotically at $r \to \infty$, where the contribution of the mass quadrupole moment becomes negligible. If we were analyzing the motion of a test particle in the gravitational field of a mass with negative quadrupole moment, ignoring the contribution of the quadrupole, we would expect a path $r_S(\varphi)$, whereas the measured path would be $r_{ER}(\varphi)$, which is always less than $r_S(\varphi)$. Then, one could interpret this result as due to an additional acceleration directed towards the source of the gravitational field. In fact, the presence of a quadrupole moment generates an "effective" acceleration which leads to deformations of the trajectories of test particles. In the next section we will estimate this effective acceleration.

To finish this section we mention that in the case of a prolate quadrupole moment ($q > 0$) the effect is opposite, that is, the perihelion distance increases and the effective acceleration would be directed outwards the gravitational source. This case, however, is not expected to occur in astrophysical bodies because any deviations from spherical symmetry are usually generated by rotation which leads to oblate configurations.

THE EFFECTIVE ACCELERATION

To find the exact hyperbolic orbit of a test particle in the gravitational field described by the Erez–Rosen metric one must solve the differential equation (9) in the range where this kind of motion is allowed. This can be done by applying standard numerical methods. Another possibility is to consider the case of motion at large distances from the source, where we can use a Taylor expansion of the functions ψ and γ that enter the function $F(u)$ in Eq.(9), and preserve the leading terms only. The resulting function $F(u)$ turns out to be a cubic polynomial of u. In this case, the differential equation (9) can be solved analytically as $\varphi = \varphi(u)$ (see [13] for an analysis of the resulting solutions), an expression from which, in principle, we could derive the acceleration due to the quadrupole moment q. Nevertheless, the resulting expression is rather cumbersome and an analysis seems to be possible only by applying numerical procedures.

To evaluate the acceleration a_q induced by the quadrupole moment, we will use the analogy with the method applied to measure the Solar quadrupole. Indeed, for an axially symmetric distribution of matter, for instance the Sun, the contribution of the quadrupole moment to the exterior gravitational field can be expressed in the metric component g_{tt} as

$$g_{tt} = 1 - 2\frac{m}{r}\left[1 + J_2\left(\frac{R_0}{r}\right)^2 P_2(\cos\theta)\right] , \tag{19}$$

where J_2 is a dimensionless constant, $P_2(\cos\theta)$ is the Legendre polynomial, and R_0 is the radius of the Sun. Evaluating this expression on the equatorial plane ($\theta = \pi/2$) and comparing it with Eq.(4), we can see that both expressions coincide if

$$q = 60 J_2 \left(\frac{R_0}{r}\right)^2 . \tag{20}$$

Then, from Eq.(19) we can read the expression for the gravitational potential whose derivative with respect to the radial coordinate produces the leading term of the acceleration. Thus, we obtain

$$a_q = -\frac{3 J_2 R_0 m}{2 r^4} . \tag{21}$$

For the evaluation of this quantity we take, according to [9], the value of $J_2 = -2.28\ 10^{-7}$ for the Solar quadrupole moment and $R_0 \approx 6.96\ 10^{12} cm$ for the photospheric radius of the Sun. Moreover, we express the distance r in astronomical units, i.e., $r = n\ 1.496\ 10^{13} cm$, where n is a positive real number. Thus, we obtain

$$a_q = \frac{1}{n^4}\, 4.33\ 10^{-8} cm/s^2 . \tag{22}$$

So we see that at a distance of one astronomical unit, the order of magnitude of a_q coincides with a_P. Considering the uncertainty in the values of J_2, R_0 and m, one can show that a_q and a_P are still of the same order of magnitude at distances smaller than $n \approx 2.1$. In the interval between 20 and 60 AU's, where the anomalous acceleration has been measured, a_q is several orders of magnitude less than a_P.

CONCLUSIONS

In this work we have investigated the possibility of explaining Pioneer's anomalous acceleration by means of the Solar quadrupole moment. In the analysis of the hyperbolic motion of a test particle in the gravitational field of a mass with a negative quadrupole moment, we have shown that the complete trajectory becomes closer to the gravitational source, an effect that can be interpreted as due to the presence of an additional acceleration, a_q, generated by the quadrupole moment.

Although a_q is not a constant quantity, its value at distances very closed to one AU is of the same order of magnitude of a_P. The discrepancy between a_q and a_P increases drastically with distance, indicating that it is not possible to explain the anomalous acceleration at distances where it has been measured for Pioneer 10/11 spacecraft.

ACKNOWLEDGMENTS

This work was supported by DGAPA-UNAM grant IN112401 and CONACyT-Mexico grant 36581-E.

REFERENCES

1. J. D. Anderson et al., Phys. Rev. Lett. **81**, 2858 (1998); S. G. Turyshev et al., gr-qc/9903024.
2. J. D. Anderson et al., Phys. Rev. D **65**, 082004 (2002).
3. J. L. Rosales and J. L. Sánchez-Gomez, gr-qc/9810085.
4. V. Guruprasad, astro-ph/9907363, gr-qc/0005014, gr-qc/0005090.
5. D. Ostvang, gr-qc/9910054.
6. W. B. Belayev, gr-qc/9903016.
7. G. Modanese, Nucl. Phys. B **556**, 397 (1999).
8. R. Mansouri, F. Nasseri, and M. Khorammi, Phys. Lett. A **259**, 194 (1999).
9. J. P. Rozelot, S. Pireaux, S. Lefebvre, and T. Corbard, astro-ph/0403382.
10. G. Erez and N. Rosen, Bull. Res. Council of Israel **8F**, 47 (1959); Ya. B. Zeldovich and I. D. Novikov, *Relativistic Astrophysics* (University of Chicago Press, Chicago, 1971).
11. H. Quevedo, Fortschr. Phys. **38**, 733 (1990).
12. R. Geroch, J. Math. Phys. **11**, 2580 (1970); R. O. Hansen, J. Math. Phys. **15**, 46 (1974).
13. H. Quevedo and L. Parkes, Gen. Rel. Grav. **21**, 1047 (1989).

Noncommutative Theories, Seiberg-Witten Map, and Gravity

Victor O. Rivelles

Instituto de Física
Universidade de São Paulo
Caixa Postal 66318, 05315-970, São Paulo, SP, Brazil
E-mail: rivelles@fma.if.usp.br

Abstract. In this paper we review the connection between noncommutative field theories and gravity. When the noncommutativity is induced by the Moyal product we can use the Seiberg-Witten map in order to deal with ordinary fields. We then show that the effect of the noncommutativity is the same as a field dependent gravitational background. The gravitational background is that of a gravitational plane wave and the coupling is charge dependent. Uncharged fields couple more strongly than the charged ones. Deviations from the usual dispersion relations are discussed and we show that they are also charge dependent.

Noncommutative (NC) theories have a long history [1]. More recently they were found in a limit of string theory with D-branes in a constant NS-NS background B field [2]. In this limit gravity decouples but still leaves a trace in the emerging NC gauge theory. If point splitting regularization is used in the world-sheet then NC appears through the Moyal product which is defined as

$$A(x)\star B(x) = e^{\frac{i}{2}\theta^{\mu\nu}\partial^x_\mu\partial^y_\nu}A(x)B(y)|_{y\to x}, \tag{1}$$

where $\theta^{\mu\nu}$ is the NC parameter. If Pauli-Villars regularization is used instead then ordinary gauge symmetry is preserved and NC appears in higher dimension operators. Since the S matrix must be regularization independent these two descriptions are related by a space-time field redefinition known as the Seiberg-Witten (SW) map [2].

From the space-time point of view these two descriptions have quite different properties. When the Moyal product is used the ultraviolet structure of the theory is not modified [3] but new infrared divergences appear and get mixed with the ultraviolet ones [4]. This mixing of divergences turns the theory non-renormalizable except in some cases where supersymmetry is present [5]. An important property of NC theories induced by the Moyal product, which distinguishes them from the conventional ones, is that translations in the NC directions are equivalent to gauge transformations [6]. This can be seen even for the case of a real scalar field which has the NC gauge transformation $\delta\hat{\phi} = -i[\hat{\phi},\hat{\lambda}]_\star$, where $[A,B]_\star = A\star B - B\star A$ is the Moyal commutator. Under a global translation the scalar field transforms as $\delta_T\phi = \xi^\mu\partial_\mu\hat{\phi}$. Derivatives of the field can be rewritten using the Moyal commutator as $\partial_\mu\hat{\phi} = -i\theta^{-1}_{\mu\nu}[x^\nu,\hat{\phi}]_\star$ so that $\delta\hat{\phi} = \delta_T\hat{\phi}$ with gauge parameter $\hat{\lambda} = -\theta^{-1}_{\mu\nu}\xi^\mu x^\nu$. The only other field theory which has a similar property is general relativity where local translations are gauge transformations associated to

CP758, *Gravitation and Cosmology: 2nd Mexican Meeting on Mathematical and Experimental Physics*
edited by A. Macías, C. Läemmerzahl and D. Nuñez

general coordinate transformations. This remarkable property shows that, as in general relativity, there are no local gauge invariant observables in NC theories.

In the description obtained through the SW map the theory is presented as a series expansion in θ. In this way a local field theory is obtained at the expense of introducing a large number of non-renormalizable interactions [7]. At the classical level, on the other side, it is possible to understand very clearly the breakdown of Lorentz invariance induced by the noncommutativity. The dispersion relation for plane waves in a magnetic background gets modified so that photons do not move with the velocity of light [8]. However, the connection between translations and gauge transformations seems to be lost. A global translation on a commutative real scalar field $\delta_T \phi = \xi^\mu \partial_\mu \phi$ can no no longer be rewritten as a gauge transformation since $\delta\phi = 0$. We will discuss how other aspects concerning gravity emerges in the description which makes use of the SW map. In this case NC field theories can be interpreted as ordinary theories immersed in a gravitational background generated by the gauge field. We will show that the θ dependent terms in the commutative action can be interpreted as a gravitational background which depends on the gauge field. We then determine the metric which couples to real and complex scalar fields. We find that the uncharged field coupling is twice that of the charged one. So we can interpret the gauge coupling in NC theory as a particular gravitational coupling which depends on the charge of the field. We also determine the geodesics followed by a massless particle in this background. We find that its velocity differs from the velocity of light by an amount proportional to θ with the deviation for the uncharged case being twice that of the charged one. For the uncharged case the deviation is the same as that found for the the gauge theory in flat space-time [8, 9]. As a final check we derive these same velocities in a field theoretic context.

The action for the NC Abelian gauge theory in flat space-time is

$$S_A = -\frac{1}{4}\int d^4x\ \hat{F}^{\mu\nu} \star \hat{F}_{\mu\nu}, \tag{2}$$

where $\hat{F}_{\mu\nu} = \partial_\mu \hat{A}_\nu - \partial_\nu \hat{A}_\mu - i[\hat{A}_\mu, \hat{A}_\nu]_\star$. For a real scalar field in the adjoint representation of $U(1)$ the flat space-time action is

$$S_\varphi = \frac{1}{2}\int d^4x\ \hat{D}^\mu \hat{\varphi} \star \hat{D}_\mu \hat{\varphi}, \tag{3}$$

where $\hat{D}_\mu \hat{\varphi} = \partial_\mu \hat{\varphi} - i[\hat{A}_\mu, \hat{\varphi}]_\star$. On the other side, for a complex scalar field in the fundamental representation of $U(1)$ the action is

$$S_\phi = \int d^4x\ \hat{D}^\mu \hat{\phi} \star (\hat{D}_\mu \hat{\phi})^\dagger, \tag{4}$$

with $\hat{D}_\mu \hat{\phi} = \partial_\mu \hat{\phi} - i\hat{A}_\mu \star \hat{\phi}$. The gauge transformations which leave the above actions invariant are given by

$$\delta \hat{A}_\mu = \hat{D}_\mu \hat{\lambda}, \qquad \delta\hat{\varphi} = -i[\hat{\varphi}, \hat{\lambda}]_\star, \qquad \delta\hat{\phi} = i\hat{\lambda} \star \hat{\phi}, \qquad \delta\hat{\phi}^\dagger = -i\hat{\phi}^\dagger \star \hat{\lambda}. \tag{5}$$

To first order in θ the SW map is [2, 10]

$$\hat{A}_\mu \;=\; A_\mu - \frac{1}{2}\theta^{\alpha\beta} A_\alpha (\partial_\beta A_\mu + F_{\beta\mu}),$$

$$\hat{\varphi} = \varphi - \theta^{\alpha\beta} A_\alpha \partial_\beta \varphi,$$
$$\hat{\phi} = \phi - \frac{1}{2}\theta^{\alpha\beta} A_\alpha \partial_\beta \phi, \quad (6)$$

We can now expand the NC actions (2),(3) and (4) using (1) and apply the map (6) to get the corresponding commutative actions.

For the real scalar field we find, always to first order in θ,

$$S_\varphi = \frac{1}{2}\int d^4x \left[\partial^\mu \varphi \partial_\mu \varphi + 2\theta^{\mu\alpha} F_\alpha{}^\nu \left(-\partial_\mu \varphi \partial_\nu \varphi + \frac{1}{4}\eta_{\mu\nu}\partial^\rho \varphi \partial_\rho \varphi\right)\right]. \quad (7)$$

It is worth to remark that the tensor inside the parenthesis is traceless. If we now consider this same field coupled to a gravitational background

$$S_{g,\varphi} = \frac{1}{2}\int d^4x \sqrt{-g} g^{\mu\nu} \partial_\mu \varphi \partial_\nu \varphi, \quad (8)$$

and expand the metric $g_{\mu\nu}$ around the flat metric $\eta_{\mu\nu}$,

$$g_{\mu\nu} = \eta_{\mu\nu} + h_{\mu\nu} + \eta_{\mu\nu} h, \quad (9)$$

where $h_{\mu\nu}$ is traceless, we get

$$S_{g,\varphi} = \frac{1}{2}\int d^4x \left(\partial^\mu \varphi \partial_\mu \varphi - h^{\mu\nu}\partial_\mu \varphi \partial_\nu \varphi + h\partial^\rho \varphi \partial_\rho \varphi\right), \quad (10)$$

where indices are raised and lowered with the flat metric. Since both actions, (7) and (10), have the same structure we can identify a linearized background gravitational field

$$h^{\mu\nu} = \theta^{\mu\alpha} F_\alpha{}^\nu + \theta^{\nu\alpha} F_\alpha{}^\mu + \frac{1}{2}\eta^{\mu\nu}\theta^{\alpha\beta} F_{\alpha\beta},$$
$$h = 0. \quad (11)$$

Then, the effect of noncommutativity on the commutative scalar field is similar to a field dependent gravitational field [10].

The same procedure can be repeated for the complex scalar field [10]. We find the linearized metric

$$h^{\mu\nu} = \frac{1}{2}(\theta^{\mu\alpha} F_\alpha{}^\nu + \theta^{\nu\alpha} F_\alpha{}^\mu) + \frac{1}{4}\eta^{\mu\nu}\theta^{\alpha\beta} F_{\alpha\beta},$$
$$h = 0. \quad (12)$$

Then charged fields feel a gravitational background which is half of that felt by the uncharged ones. Therefore, the gravity coupling is now dependent on the charge of the field, being stronger for uncharged fields.

Notice that the gauge field has now a dual role, it couples minimally to the charged field and also as a gravitational background. As it is well known the SW map gives rise to the following action

$$S_A = -\frac{1}{4}\int d^4x \left[F^{\mu\nu} F_{\mu\nu} + 2\theta^{\mu\rho} F_\rho{}^\nu \left(F_\mu{}^\sigma F_{\sigma\nu} + \frac{1}{4}\eta_{\mu\nu} F^{\alpha\beta} F_{\alpha\beta}\right)\right]. \quad (13)$$

Again, the tensor inside the parenthesis is traceless. At this point we could be tempted to consider this action as some gravitational action build up from the metric (11) or (12). Since the field strength always appears multiplied by θ inside the metric, all invariants constructed with it will be of order θ. Hence, they can not give rise to (13), unless they appear in combinations involving the inverse of θ. If we insist in having an action which is polynomial in θ the best we can do is to regard the gauge field as having a double role again and couple it to gravitation as in the previous case. We then find that $h^{\mu\nu}$ is given by (12). Since the NC gauge field resembles a non-Abelian gauge field we expect that its commutative counterpart couple to the same gravitational field as the charged one. It should also be remarked that in this case the gravitational field can not be interpreted just as a fixed background since it depends on the dynamical gauge field.

Having determined the field dependent background metric we can now study its properties. A detailed analysis shows that it describes a plane gravitational wave [10].

We can now turn our attention to the behavior of a massless particle in this background. Its geodesics is described by

$$ds^2 = \left(1 + \frac{1}{4}\theta^{\alpha\beta}F_{\alpha\beta}\right) dx^\mu dx_\mu + \theta_{\mu\alpha}F^{\alpha}{}_{\nu} dx^\mu dx^\nu = 0. \tag{14}$$

If we consider the case where there is no noncommutativity between space and time, that is $\theta^{0i} = 0$, and calling $\theta^{ij} = \varepsilon^{ijk}\theta^k$, $F^{i0} = E^i$, and $F^{ij} = \varepsilon^{ijk}B^k$, we find to first order in θ that

$$(1 - \vec{v}^2)(1 - 2\vec{\theta}\cdot\vec{B}) - \vec{\theta}\cdot(\vec{v}\times\vec{E}) + \vec{v}^2\vec{\theta}\cdot\vec{B} - (\vec{B}\cdot\vec{v})(\vec{\theta}\cdot\vec{v}) = 0, \tag{15}$$

where $\vec{v}$ is the particle velocity. Then to zeroth order, the velocity $\vec{v}_0$ satisfies $\vec{v}_0^2 = 1$ as it should. We can now decompose all vectors into their transversal and longitudinal components with respect to $\vec{v}_0$, $\vec{E} = \vec{E}_T + \vec{v}_0 E_L$, $\vec{B} = \vec{B}_T + \vec{v}_0 B_L$ and $\vec{\theta} = \vec{\theta}_T + \vec{v}_0\theta_L$. We then find that the velocity is

$$\vec{v}^2 = 1 + \vec{\theta}_T\cdot(\vec{B}_T - \vec{v}_0\times\vec{E}_T). \tag{16}$$

Hence, a charged massless particle has its velocity changed with respect to the velocity of light by an amount which depends on θ. For an uncharged massless particle

$$\vec{v}^2 = 1 + 2\vec{\theta}_T\cdot(\vec{B}_T - \vec{v}_0\times\vec{E}_T), \tag{17}$$

and the correction due to the noncommutativity is twice that of a charged particle.

We can now check the consistency of these results by going back to the original action (7) and computing the group velocity for planes waves. Upon quantization they give the velocity of the particle associated to the respective field. For the uncharged scalar field we get the equation of motion

$$\left(1 - \frac{1}{2}\theta^{\mu\nu}F_{\mu\nu}\right)\Box\varphi - 2\theta^{\mu\alpha}F_{\alpha}{}^{\nu}\partial_\mu\partial_\nu\varphi = 0. \tag{18}$$

If the field strength is constant we can find a plane wave solution with the following dispersion relation

$$\left(1-\frac{1}{2}\theta^{\mu\nu}F_{\mu\nu}\right)k^2-2\theta^{\mu\alpha}F_\alpha{}^\nu k_\mu k_\nu=0, \tag{19}$$

and using the same conventions for vectors as before, it results in

$$\frac{\vec{k}^2}{\omega^2}=1-2\vec{\theta}_T\cdot(\vec{B}_T-\frac{\vec{k}}{\omega}\times\vec{E}_T), \tag{20}$$

where $k^\mu=(\omega,\vec{k})$. We then find that the phase and group velocities coincide and are given by (17) as expected. For the charged scalar field we have to turn off the gauge coupling in order to get a plane wave solution. In this case the equation of motion is

$$\left(1-\frac{1}{4}\theta^{\mu\nu}F_{\mu\nu}\right)\Box\phi-\theta^{\mu\alpha}F_\alpha{}^\nu\partial_\mu\partial_\nu\phi=0. \tag{21}$$

In a constant field strength background the dispersion relation for a plane wave reads as in (19) with θ replaced by $\theta/2$. Then we must perform the same replacement in the phase and group velocities and we get (16). Therefore, in both pictures, noncommutative and gravitational, we get the same results.

For the gauge field the situation is more subtle because of its double role. There is no clear way to split the action (13). What can be done is to break up the gauge field into a background plus a plane wave as in [8]. We then get the following dispersion relation

$$k^2-2\theta^{\mu\alpha}F_\alpha{}^\nu k_\mu k_\nu=0, \tag{22}$$

where $F_\alpha{}^\nu$ is now the constant background. This leads to (20), that is, the dispersion relation for the uncharged scalar field. It also reproduces the result in [8, 9] when the background is purely magnetic. This shows the dual role of the gauge field, since it couples to gravitation as a charged field but its dispersion relation is that of an uncharged field.

We have seen that it is possible to regard noncommutative theories as conventional theories embedded in a gravitational background produced by the gauge field. This brings a new connection between noncommutativity and gravitation. We could imagine that this is a peculiarity of the first order term in the θ expansion of the SW map but an analysis to all orders in θ was performed in [11].

ACKNOWLEDGMENTS

I would like to thank the organizers for the kind invitation to deliver this talk. This work was partially supported by FAPESP, CNPq and PRONEX under contract CNPq 66.2002/1998-99.

REFERENCES

1. For a review on NC field theories see: Szabo, R.J., Quantum field theory on noncommutative spaces, arXiv:hep-th/0109162. Douglas, M.R., and Nekrasov, N.A., *Rev. Mod. Phys.* **73**, 977 (2001). Rivelles, V.O., Noncommutative supersymmetric theories, arXiv:hep-th/0211169. Rivelles, V.O., *Nucl. Phys. Proc. Suppl.* **127**, 63 (2004).
2. Seiberg, N., and Witten, E., JHEP **9909**, 032 (1999).
3. Filk, T., *Phys. Lett.* B **376**, 53 (1996).
4. Minwalla, S., Van Raamsdonk, M., and Seiberg, N., *JHEP* **0002**, 020 (2000) .
5. Girotti, H.O., Gomes, M., Rivelles, V.O., and da Silva, A.J., *Nucl. Phys.* B **587**, 299 (2000). Girotti, H.O., Gomes, M., Rivelles V.O., and da Silva, A.J., *Int. J. Mod. Phys.* A **17**, 1503 (2002). Girotti, H.O., Gomes, M., Petrov, A.Y., Rivelles, V.O., and da Silva, A.J., *Phys. Lett.* B **521**, 119 (2001). Ferrari, A.F., Girotti, H.O., Gomes, M., Petrov, A.Y., Ribeiro, A.A., Rivelles, V.O., and da Silva, A.J., *Phys. Rev.* D **69**, 025008 (2004). Ferrari, A.F., Girotti, H.O., Gomes, M., Petrov, A.Y., Ribeiro, A.A., Rivelles, V.O., and da Silva, A.J., Towards a consistent noncommutative supersymmetric Yang-Mills theory: Superfield covariant analysis, arXiv:hep-th/0407040.
6. Gross D.J., and Nekrasov, N.A., *JHEP* **0010**, 021 (2000).
7. For a review see: Wulkenhaar, R., Quantum field theories on noncommutative R**4 versus Theta-expanded quantum field theories, arXiv:hep-th/0206018.
8. Guralnik, Z., Jackiw, R., Pi, S.Y., and Polychronakos, A.P., *Phys. Lett.* B **517**, 450 (2001).
9. Cai, R.G., *Phys. Lett.* B **517**, 457 (2001).
10. Rivelles, V.O., *Phys. Lett.* B **558**, 191 (2003).
11. Banerjee, R. and Yang, H.S., Exact Seiberg-Witten map, induced gravity and topological invariants in noncommutative field theories, arXiv:hep-th/0404064; Yang, H.S., Exact Seiberg-Witten map and induced gravity from noncommutativity, arXiv:hep-th/0402002; Banerjee, R., Lee, C.K., and Yang, H.S., Seiberg-Witten-type maps for currents and energy-momentum tensors in noncommutative gauge theories, arXiv:hep-th/0312103.

Towards the Witten-Isenberg-Yasskin-Green Correspondence in String Theory

Hugo García-Compeán*, Oscar Loaiza-Brito† and Norma Quiroz*

*Departamento de Física
Centro de Investigación y de Estudios Avanzados del IPN
Apdo. Postal 14-740, 07000, México D.F., México
†Physikalisches Institut, Universitat Bonn
Nussallee 12, 53115 Bonn, Germany

Abstract. Twistor methods are suitable to describe Yang-Mills instantons which receive corrections from string theory in powers of α'. The description of full fledged Yang-Mills fields can be also addressed through twistor methods. In this paper a stable brane configuration describing naturally the set up involved in the Witten-Isenberg-Yasskin-Green correspondence, is presented. This gives the first steps towards a description of this correspondence in string theory.

1. INTRODUCTION

Instantons in string theory are described by a (0,4) superconformal field theory (SCFT) represented by a (0,4) supersymmetric non-linear sigma model [1, 2]. From the target space point of view, instantons correspond to solitonic NS fivebranes in whose transverse four coordinates is located a Yang-Mills instanton. In the limit when the size of the instanton goes to zero a tube is created in the fivebrane in such a way that the conformal field theory becomes singular on this tube [3]. Moreover, the dilaton diverges linearly on the throat and nonperturbative quantum effects has to be included in order to give a sensible description of the small instanton. In [4] it was constructed a family of (0,4) supersymmetric linear sigma models. This family is constructed to be in the same universality class of the (0,4) SCFT, that means that this family of linear sigma models flows in the infrared (IR) to a (0,4) SCFT which corresponds to solution of string theory giving an instanton. This linear sigma model, termed the ADHM sigma model, is a field theory in two dimensions where the Yukawa couplings satisfy the ADHM equations of the ADHM construction of instantons [5]. This implies that the couplings of the linear sigma model provides the ADHM linear algebra data of the ADHM construction. The ADHM sigma model contains the heterotic five-brane soliton in the large scale limit. However in the limit of zero size instanton, it differs from the current string theory approach. In this limit the space of classical vacua develops two (asymptotic flat) branches connected by a wormhole in the region where the string effects are important [3, 4].

An explanation in terms of string theory of the involved physics when a instanton goes to zero size was given by Witten in [6]. In this paper he showed that the SO(32) heterotic string theory compactified on K3 leads to a six-dimensional theory with extra

CP758, *Gravitation and Cosmology: 2nd Mexican Meeting on Mathematical and Experimental Physics*
edited by A. Macías, C. Läemmerzahl and D. Nuñez

hypermultiplets arising in the limit when the instanton goes to zero size. In this same paper it was verified by using the S-duality between the SO(32) heterotic and the Type I theories. Thus on the Type I side the small instanton is mapped to a D-brane configuration: D5-D9. The hypermultiplet matter describes completely the ADHM data in a codimension four space transverse to the D5-brane. Thus the physical interpretation was clear: D-branes are fundamental objects which captures the string scale interactions. The gauge theory on the D-branes describe the low energy limit but the singularities of the moduli space still does contain information of the high energy theory and the ADHM data are high energy variables which become apparent in the singularities as these low energy modes. For further description of instantons using D-branes, see [7].

Later Douglas [8, 9] showed that configurations Dp-D$(p+4)$ in Type I and Type II string theories can be probed with D1 branes living at the worldvolume of the Dp-brane. The Higgs branch is identified with the moduli space of instantons and the two-dimensional theory living on the D1 worldsheet carries an ADHM linear sigma model of the Witten type [4].

Much effort has been done in the case of intantons both in the Atiyah-Ward correspondence and in the ADHM construction. It is well known that Atiyah-Ward correspondence admits a generalization involving Penrose twistor transform to construct non-self-dual Yang-Mills fields. This generalization was done by Witten [10] and independently by Isenberg, Yasskin and Green [11] (see also a different construction [12]). We will give an overview of this in Sec. 2. For details the reader can consult [13, 14, 15]. This construction establishes a one-one relation between gauge fields satisfying the full Yang-Mills equations and holomorphic vector bundles over the quadric on the Cartesian product of projective twistor spaces $\mathbf{PT} \times \mathbf{PT}^*$. We will call this correspondence the *WIYG correspondence*. This correspondence has not been yet related to string theory. Moreover, it was conjectured by Witten [16] the existence of an analog "ADHM" construction for gauge fields satisfying the full Yang-Mills equations. As far we know, this problem was never addressed. In the present contribution we will give a D-brane configuration which can be considered as a preliminary step towards the realization of WIYG correspondence in string theory (other application of twistors to string theory is given, for instance, in Ref. [17]).

The structure of this paper is as follows, in order to fix some notation and conventions, in Sec. 2, we give a brief overview of the WIYG correspondence. Sec. 3 is devoted to construct the D-brane configuration by using the technology of branes in angles.

2. WITTEN-ISENBERG-YASSKIN-GREEN CORRESPONDENCE

ADHM data are linear algebra data which gives all instantons over certain space $\mathscr{M}$ and with some suitable gauge group G. In string theory, we have seen that small instantons becomes described precisely by these ADHM data. However, one would like to see if the ADHM construction is also valid for full fledged Yang-Mills fields, not only their self-dual (or anti-self-dual counterpart). Due the full Yang-Mills fields (or non-self-dual) is not an integrable system, it seems to be a not easy problem. However, in vistas of the brane technology advances, it would be interesting to reexamine the problem in this context. In the present paper we will give evidence that such construction does exists. We will show that the set up of Witten's [10] and Isenberg-Yasskin-Green [11] correspondences arises in a natural way in string theory, in particular, in the context of intersecting D-branes and anti-D-branes. Before to do that we would like to remind this construction in order to introduce the

Ward's Correspondence.. There is a one-one correspondence between:

1. A $g\ell(N,\mathbf{C})$- valued gauge fields $A_{\alpha\dot{\alpha}}(x^{\beta\dot{\beta}})$ that obeys ASDYM equations on the Minkowski (or Euclidean) space $\mathscr{M}$.
2. Rank N holomorphic vector bundles E over $\mathbf{PT}(=\mathbf{CP}^3)$ such that E is a trivial holomorphic bundle over the holomorphic curve of degree one $\mathbf{CP}^1$ in the twistor space $\mathbf{PT}$.

The proof of this theorem requires from the Penrose twistor transform which transforms local differential geometric data on the spacetime and encodes it into global analytic data on the twistor space or on vector bundles over the twistor space.

Penrose twistor transform is beautifully expressed through the twistor equation given by

$$\mu_{\dot{\alpha}}+x_{\alpha\dot{\alpha}}\lambda^{\alpha}=0, \tag{1}$$

where the spacetime vector index i transforming with respect to the Lorentz group $SO(4)$, decomposes into a pair of spinorial indices $(\alpha,\dot{\alpha})$ transforming with respect to the Lorentz group written as the product $SU(2)_L\times SU(2)_R$. Twistor space $\mathbf{PT}$ is spanned by the four complex coordinates $(\mu_{\dot{\alpha}},\lambda^{\alpha})$. By fixing these coordinates the space of solutions of twistor equation is given by the space $\mathscr{M}$. A given point in twistor space corresponds to a null-plane $\mathscr{W}$ in $\mathscr{M}$, that means that over $\mathscr{W}$ all vector is null. These planes are called α-plane $\mathscr{W}_{\alpha}$. α-planes correspond with planes in spacetime where the curvature $F_{\alpha\dot{\alpha}\beta\dot{\beta}}$ of the connection $A_{\alpha\dot{\alpha}}(x^{\beta\dot{\beta}})$ vanishes and is self-dual on $\mathscr{M}$. On the other hand β-planes $\mathscr{W}_{\beta}$ are null planes where $F_{\alpha\dot{\alpha}\beta\dot{\beta}}$ vanishes and it is anti-self-dual on $\mathscr{M}$. Then according to the Penrose transform, the twistor space $\mathbf{PT}$ consists of points which are α-planes in spacetime $\mathscr{M}$. While the space of β-planes determines the *dual* twistor space $\mathbf{PT}^*$. On the other hand, for a given point $x^{\beta\dot{\beta}}\in\mathscr{M}$, the space of solutions of twistor equation is given by a holomorphic curve $\mathbf{CP}^1$ inside the twistor space $\mathbf{PT}$. As we mentioned before, there is a fibration, where for a given point of $\mathscr{M}$ in an α-plane $\mathscr{W}_{\alpha}$, the fiber corresponds to a curve $\mathbf{CP}^1$.

Now we describe the WIYG-construction [10, 11]. It is very well known that, in four dimensions Yang-Mills equations $D_{\mu}F^{\mu\nu}=0$ cannot be written as an integrability condition through a Lax pair just as the ASDYM equation does. However in eight dimensional space $\mathscr{K}=\mathscr{M}'\times\mathscr{M}''$, with local coordinates (y^i,z^i) with $i=1,\ldots,4$, this is possible. Our original four dimensional space $\mathscr{M}$ is obtained as the diagonal subspace of $\mathscr{K}$ consisting in all points satisfying $y^i=z^i=x^i$. Now consider a gauge field $A(y^i,z^i)$ defined on $\mathscr{K}$ whose field strength $F(y^i,z^i)$ satisfies the following equations

$$\left[\frac{D}{Dy^i},\frac{D}{Dy^j}\right]=\frac{1}{2}\varepsilon_{ijkl}\left[\frac{D}{Dy^k},\frac{D}{Dy^l}\right] \tag{2}$$

and

$$\left[\frac{D}{Dz^i},\frac{D}{Dz^j}\right] = -\frac{1}{2}\varepsilon_{ijkl}\left[\frac{D}{Dz^k},\frac{D}{Dz^l}\right] \tag{3}$$

such that

$$\left[\frac{D}{Dy^i},\frac{D}{Dz^j}\right] = 0. \tag{4}$$

These equations correspond with having an instanton gauge field over $\mathscr{M}'$, and an anti-instanton over $\mathscr{M}''$. In the diagonal space the Yang-Mills equation can be written as

$$\left[\frac{D}{Dx^i},\left[\frac{D}{Dx^j},\frac{D}{Dx^k}\right]\right] = 0. \tag{5}$$

By using the above system and the relation between coordinates $x^i = \frac{1}{2}(y^i + z^i)$ and $w^i = \frac{1}{2}(y^i - z^i)$ one can check that the full Yang-Mills equations are fulfilled. Conversely, if one assumes that full Yang-Mills are satisfied then the connection $A(y^i, z^i)$ on $\mathscr{K}$ which will satisfy our system is given at the second order in the coordinate w^i as follows

$$A_i^w = -\frac{1}{2}F_{ij}w_j - \frac{1}{3}(w^k\frac{D}{Dx^k})\widetilde{F}_{ij}w_j \tag{6}$$

and

$$A_i^x = A_i(w=0) - \widetilde{F}_{ij}w_j - \frac{1}{2}(w^k\frac{D}{Dx^k})F_{ij}w_j. \tag{7}$$

In terms of the Penrose twistor transform the WIYG correspondence establishes the following relation:

WIYG Correspondence.. There is a one-one correspondence between:

1. A $g\ell(N,\mathbf{C})$- valued gauge fields $A_{\alpha\dot{\alpha}}(x^{\beta\dot{\beta}})$ that obeys the full Yang-Mills equations on the Minkowski (or Euclidean) space $\mathscr{M}$.
2. Rank N holomorphic vector bundles E over a quadric in $\mathscr{A} = \mathbf{PT}\times\mathbf{PT}^*(=\mathbf{CP}^3\times\overline{\mathbf{CP}}^3)$, where $\mathscr{A} = \{(U^\alpha, V_\beta)\}$, such that E is trivial holomorphic bundle over the holomorphic curve $\mathbf{CP}^1\times\overline{\mathbf{CP}}^1$ in the twistor space $\mathbf{PT}\times\mathbf{PT}^*$.

Extension to the Super Yang-Mills theory formulated in the superspace also can be done following similar considerations [10].

In the present paper we will see that the WIYG correspondence can be realized in the context of an appropriate brane configuration involving also $\overline{\text{D}}p$-branes. Taking advantage that ADHM construction can be regarded also in this context, we propose a way to find a D-brane configuration which might be useful to construct an ADHM construction for full Yang-Mills fields.

3. THE BRANE CONFIGURATION

Let us consider a stationary configuration of two parallel D5-branes in Type I superstring theory. Suppose that we have chosen the coordinates of our spacetime so that the branes are parallel to the coordinate axes. The worldvolume of the D5-branes fill the directions $(x^0, x^1, x^2, x^3, x^4, x^5)$, where x^0 is assumed to be the time direction. We will consider the case in which one of the D5-branes is rotated relative to the other by an rotation angle θ in SO(8). To do that, we arrange the coordinates $(x^2, \ldots, x^9)$ in four complex planes with coordinates $z^a = x^{a+1} + ix^{a+5}$ with $a = 1, \ldots, 4$. We will concentrate in the particular case where the rotation is $\theta = (\frac{\pi}{2}, \frac{\pi}{2}, \frac{\pi}{2}, \frac{3\pi}{2})$ on the planes $z^1, \ldots, z^4$ respectively. The open string that is stretched between the two branes satisfies for every coordinate a Newmann (N) or Dirichlet (D) boundary conditions at each end of the string. Then NN denotes the number of coordinates for which both ends of the string satisfy N boundary condition. DD denotes the number of coordinates for which both ends satisfy D boundary conditions and finally, DN denotes the number of coordinates for which one end of the string has a D boundary condition and the other end a N boundary condition. Rotations by $\pi/2$ at each plane change N into D boundary conditions and vice versa. The extra rotation by π in the 5-9 plane turns the rotated D5-brane into a $\overline{\text{D5}}$ brane. Therefore, the final configuration is an anti-D5-brane, perpendicular to a D5-brane. The $\overline{\text{D5}}$-brane extends along the coordinate axes $(x^0, x^1, x^6, x^7, x^8, x^9)$.

Then, in our case the open string stretched between the D5-brane and the $\overline{\text{D5}}$-brane has 2 NN along the coordinates (x^0, x^1) and 8 ND along $(x^2, \ldots, x^9)$. The number NN+ND=10, since we will consider one of the superstring theories.

D5 $\perp$ $\overline{D5}$ in Type I Theory. It was probed in [6, 9] that the small instanton are D5-branes bound to D9-branes in the Type I theory. In order to see its collective coordinates, it is necessary to introduce a D1-brane as a probe brane [9] of the D9-D5 system. We are now interested into a system of instanton-anti instanton, living in configurations perpendicular one each other. For this reason we have to consider a system of a D9-D5 perpendicular to a D9-$\overline{D5}_\perp$ intersecting in directions (x^0, x^1) where precisely we place a probe D1-brane. In our discussion we will follow the notation introduced in Refs. [9, 18]. We consider an stack of N D9-branes filling all spacetime, k D5-branes extended along directions 012345 and an stack of k $\overline{\text{D5}}$-branes along 016789. A D1-brane is placed along the 01-plane. The supersymmetry and spectrum of this configuration will be discussed in detail in a forthcoming paper [19], where we have followed similar consideratios to Ref. [20]. In fact, the effective theory of the D1 world-sheet can be regarded as a sigma model with (0,8) supersymmetry. It can be decomposed as a direct sum of two non-interacting sigma models, each with a (0,4) supersymmetry [19].

For the D9-D5-$\overline{D5}_\perp$-D1 system the Lorentz group breaks as follows:

$$SO(1,9) \subset SO(1,1)_{01} \times SO(4)_{2345} \times SO(4)_{6789}. \tag{8}$$

The group $SO(4)_{2345} = SU(2)' \times \widetilde{SU(2)}'$ has doublet indices $(A', \tilde{A}')$ and $SO(4)_{6789} = SU(2)_L \times SU(2)_R$ has doublet indices (A, Y). In the following we classify the spectrum of the open strings with end points on the different D-branes of our systems. The states of the open string are labeled by index of the covering group.

D1-D1 sector. In this case, the space-time symmetry is $SO(8) \times SO(1,1)$. The D9-brane supersymmetry decompose as $16 = 8^+ \oplus 8'^-$, with the first factor unbroken by the D1-brane. Thus, the worldsheet theory will have a chiral (0,8) supersymmetry. The parameter of the supersymmetry can be decompose into two parameters of (0,4) supersymmetry, $\eta_+^{AA'}$ and $\overline{\eta}_+^{\tilde{A}'Y}$.

The massless states in the NS sector are $(-1)^F = 1$ and Ω invariant states. They are scalars $\psi^i(x^0,x^1)|0\rangle$ and $\psi^j(x^0,x^1)|0\rangle$, where $i = 2,3,4,5$ and $j = 6,7,8,9$. According how they transform under the covering group we call these fields $b^{A'\tilde{A}'}$ and b^{AY} respectively. In the R sector, states are written in terms of the spinor representation of the group $SO(1,1)\times SO(8)$. GSO projection and Ω invariance leaves only states with $s_0 = -\frac{1}{2}$. The spinors are decomposed into two sectors, we call them $\varphi_-^{A'Y}$ (with $s_1 = s_2$, $s_3 = -s_4$) and $\varphi^{\tilde{A}'A}$ (with $s_1 = -s_2$, $s_3 = s_4$). With respect to the two (0,4) supersymmetries, the bosons and fermions are related as:

$$
\begin{aligned}
\delta b^{A'\tilde{A}'} &= i\varepsilon_{AB}\eta_+^{A'A}\varphi_-^{B\tilde{A}'}, \\
\delta b^{AY} &= i\varepsilon_{A'B'}\eta_+^{A'A}\varphi_-^{B'Y}
\end{aligned} \tag{9}
$$

and

$$
\begin{aligned}
\delta b^{A'\tilde{A}'} &= i\varepsilon_{YZ}\overline{\eta}_+^{\tilde{A}'Y}\varphi_-^{A'Z}, \\
\delta b^{AY} &= i\varepsilon_{\tilde{A}'\tilde{B}'}\overline{\eta}_+^{\tilde{A}'Y}\varphi_-^{A\tilde{B}'}.
\end{aligned} \tag{10}
$$

D1-D5, D1-$\overline{D5}_\perp$. The D1-D5 string in the NS sector has zero modes along the ND directions. Such states are labeled by the $SO(4)_{2345}$ spinor weight $|s_1,s_2\rangle$. GSO projection with $(-1)^F = 1$ leaves only states with $s_1 = s_2$. We denote this state by $\phi^{A'm}$, where m is an index counting the number of D5-branes. The R sector has zero modes along the NN and DD directions. The states are labeled by the $SO(1,1)\times SO(4)_{6789}$ spinor weight $|s_0,s_3,s_4\rangle$. GSO projection requires an odd number of minus signs. Physical states come into two classes, χ_-^{Am} (with $s_0 = -\frac{1}{2}, s_3 = s_4$) and χ_+^{Ym} (with $s_0 = \frac{1}{2}, s_3 = -s_4$). Only χ_-^{Am} is related to $\phi^{A'm}$ by

$$
\delta\phi^{A'm} = i\varepsilon_{AB}\eta_+^{A'A}\chi_-^{Bm}, \tag{11}
$$

and the most general supersymmetry transformation for χ_+^{Ym} is given by

$$
\delta\chi_+^{Ym} = i\eta_+^{AA'}C_{AA'}^{Ym}. \tag{12}
$$

The D1-$\overline{D5}_\perp$ has 4 DN directions along the directions 6789. The NS sector has zero modes which can be written in the spinor representation of the $SO(4)_{6789}$ group, that is $|s_3,s_4\rangle$. In this case the GSO projection has the opposite sign. Then, GSO invariant states are $\overline{\phi}^{Ym}$ (with $s_3 = -s_4$). The R sector has now zero modes along the directions 012345. In the spinor representation of $SO(1,1)\times SO(4)_{2345}$ they are $|s_0,s_2,s_3\rangle$. Again GSO has the opposite sign and the invariant states are two states $\overline{\chi}_-^{\tilde{A}'m}$ (with $s_0 = -\frac{1}{2}, s_2 = -s_3$)

and two states $\overline{\chi}_{+}^{A'm}$ ($s_0 = \frac{1}{2}, s_2 = s_3$). Only $\overline{\chi}_{-}^{\tilde{A}'m}$ is related to $\bar{\phi}^{Ym}$ by

$$\delta\overline{\phi}^{Ym} = i\varepsilon_{\tilde{A}'\tilde{B}'}\overline{\eta}_{+}^{Y\tilde{A}'}\overline{\chi}_{-}^{\tilde{B}'m} \tag{13}$$

and

$$\delta\overline{\chi}_{+}^{A'm} = i\overline{\eta}_{+}^{\tilde{A}'Y}\overline{C}_{\tilde{A}'Y}^{A'm} \tag{14}$$

where $C_{AA'}^{Ym}$ and $\overline{C}_{\tilde{A}Y}^{A'm}$ are functions to determine.

D1-D9. There are no massless states in the NS sector for the D1-D9 string, since the eight ND directions gives a zero point energy equal to $\frac{1}{2}$. In the R sector the massless states are given by 2 NN directions. In terms of the $SO(1,1)$ spinor representation we have $|s_0\rangle$. GSO projections leaves only the positive value. We denote this state by λ_{+}^{M} with M an index for the D9-branes. The supersymmetric transformations for λ_{+}^{M} are respectively

$$\begin{aligned} \delta\lambda_{+}^{M} &= \eta_{+}^{AA'}C_{AA'}^{M}, & (15)\\ \delta\lambda_{+}^{M} &= \overline{\eta}_{+}^{\tilde{A}'Y}\overline{C}_{\tilde{A}'Y}^{A'm} & (16) \end{aligned}$$

D5-D9 and $\overline{D5}_{\perp}$-D9. In in NS sector there are zero modes along 6789 with ND boundary conditions. In terms of the spinor representation of $SO(4)_{6789}$ we have the states $|s_3, s_4\rangle$. The GSO invariant state is denoted by h_M^{Am} (with $s_3 = s_4$). In the R sector there are zero modes along the directions with NN boundary conditions. The GSO invariant states are $\vartheta_M^{\tilde{A}'m} = |s_0, s_1, s_2\rangle$ in the spinor representation of $SO(1,1)\times SO(4)_{2345}$ with $s_1 = -s_2$.

In the sector $\overline{D5}_{\perp}$-D9, $(-1)^F = -1$. NS states come from the ND directions. They are labeled by $\overline{h}_M^{\tilde{A}'m} = |s_2, s_1\rangle$ with respect to the covering group of $SO(4)_{2345}$. GSO leaves states with $s_1 = -s_2$. States in the R sector are $\overline{\vartheta}_M^{Am} = |s_0, s_3, s_4\rangle$ of $SO(1,1)\times SO(4)_{6789}$ with $s_3 = s_4$.

Finally, the analysis for D5-D5 and $\overline{D5}_{\perp}$-$\overline{D5}_{\perp}$ can be computed in the standard way.

ACKNOWLEDGMENTS

H. G.-C. wish to thank the organizers of the Second Mexican Meeting on Mathematical and Experimental Physics, for invitation. This work was partially supported by the CONACyT grant No. 41993.

REFERENCES

1. Strominger, A., *Nucl. Phys. B*, **343**, 167 (1990).
2. Callan,Jr, C.G., Harvey, J.A., and Strominger, A., *Nucl. Phys. B*, **359**, 611 (1991); *Nucl. Phys. B*, **367**, 60 (1991).
3. Witten, E., "Some comments on String Dynamics," arXiv:hep-th/9507121.

4. Witten, E., *J. Geom. Phys.* , **15**, 215 (1995) [arXiv:hep-th/9410052].
5. Atiyah, M.F., Hitchin, N.J., Drinfeld,V.G., and Manin, Y.I., *Phys. Lett. A*, **65**, 185 (1978).
6. Witten, E., *Nucl. Phys. B*, **460**, 541 (1996) [arXiv:hep-th/9511030].
7. Park, J.S., *Nucl. Phys. B*, **493**, 198 (1997) [arXiv:hep-th/9612096].
8. Douglas, M.R., "Branes within branes," arXiv:hep-th/9512077.
9. Douglas, M.R., *J. Geom. Phys.* , **28**, 255 (1998)
10. Witten, E., *Phys. Lett. B*, **77**, 394 (1978).
11. Isenberg, J., Yasskin, P.B., and Green, P.S., *Phys. Lett. B* , **78**, 462 (1978).
12. Ivanov, E.A., *JETP Lett.*, **30**, 422 (1979).
13. Ward, R.S., and Wells, R.O., *Twistor Geometry And Field Theory*, "Cambridge, UK: Univ. Pr. (1990) 520 p"
14. Manin, Y.I., *Gauge Field Theory And Complex Geometry*, Berlin, Germany: Springer (1988) 295 p.
15. Witten, E., *J. Geom. Phys.* , **8**, 327 (1992).
16. Witten, E., "Some Comments On The Recent Twistor Space Constructions," in *Complex Manifold Techniques in Theoretical Physics*, Eds. D.E. Lerner and P.D. Sommers, Pitman Advanced Publishing Program, San Francisco (1979) pp. 207-218.
17. Witten, E., *Nucl. Phys. B* , **266**, 245 (1986).
18. Johnson, C.V., "D-brane Primer," arXiv:hep-th/0007170.
19. García-Compeán, H., Loaiza-Brito,O., and Quiroz, N., "Non-Self-dual Gauge Fields from Intersecting Branes" to appear (2004).
20. Epple, F., and Lust, D., *Fortsch. Phys.* , **52**, 367 (2004) [arXiv:hep-th/0311182].

ASTROPHYSICS AND COSMOLOGY

Very Massive Stars and their Cosmological Consequences

Jaime Klapp*, David Bahena†, Manuel G. Corona-Galindo** and Heinz Dehnen‡

*Instituto Nacional de Investigaciones Nucleares, Km. 36.5 Carretera México-Toluca, 52045 Estado de México, México
†Astronomical Institute, Charles University, Prague, Czech Republic
**Instituto Nacional de Astrofísica, Optica y Electrónica, A. P. 256, Tonanzintla, Puebla, México
‡Konstanz University, Germany

Abstract. In recent years there has been increasing interest in understanding the role that Very Massive Stars (VMS) could play in the cosmological scenario. Star formation and accretion calculations suggest that the proposed first stars are VMS rather than lower mass stars. However, recent nucleosynthesis results of very massive supernovae (SN) explosions suggest that the so called strong VMS hypothesis is inconsistent with data from Extremely Metal Poor (EMP) stars and that the observations are better explained by energetic supernovae or hypernovae connected to low redshift gamma-ray bursts. We propose an alternative scenario where the first stars are born very massive but their mass reduced to the Massive Star (MS) range during their quasi-static hydrogen and helium burning phases. In this work we present Zero Age Main Sequence (ZAMS) models and evolutionary calculations of mass losing VMS and show that for high enough mass loss rates, the mass is reduced to the MS range. Our results also show that, if as predicted by reionization models, a fraction of $\sim 10^{-3}$ of the baryons in the Universe is incorporated into VMS first stars, the amount of CNO elements returned to the interstellar medium through radiation driven winds is consistent with observations of EMP stars.

1. INTRODUCTION

Within the context of the so-called cold or tepid models of the Universe, the existence of a pregalactic generation of VMS was proposed as an alternative to the standard Hot Big Bang model of the Universe. In the former models, part or all of the microwave background and the observed helium abundance is not primordial and has to be generated by other mechanisms such as VMS. However, the idea of the existence of VMS was abandoned about two decades ago for the following reasons: if a large proportion of the baryons in the Universe is incorporated into a generation of VMS, these stars are quite capable of producing the observed helium abundance, but at the same time they will overproduce heavy elements [10, 11, 31]; stars with masses greater or equal to $60M_{\odot}$ are vibrationally unstable to nuclear energizing pulsations that could induce eruptions and/or very strong winds [14, 37]; VMS have a luminosity close to the Eddington luminosity, consequently a strong radiation pressure is present, inducing strong winds which reduces the mass of the VMS to the MS range during the quasi-static phases [10, 11]; the existence of the Humphreys-Davidson limit; and finally, the minimum Jeans mass obtained from opacity limited calculations was much less than $1M_{\odot}$ [29, 15, 30].

CP758, *Gravitation and Cosmology: 2nd Mexican Meeting on Mathematical and Experimental Physics*
edited by A. Macías, C. Läemmerzahl and D. Nuñez

Recently, Heger, Baraffe, Fryer & Woosley found that stars with a mass less or equal than $500 M_\odot$ are far less vibrationally unstable than previously thought, and may reach the end of their evolution with a massive helium core that encounter the pair-creation instability [9]. Baraffe, Heger & Woosley studied the stability of metal-free and metal-enriched massive stars and found that metal-free stars with masses in the range $120 M_\odot \leq M \leq 500 M_\odot$ become stable after about one third of the hydrogen burning lifetime [5].

However, all these calculations have limited significance because of the use of the so called "flux-freezing convection theory" that assumes that the perturbation in the luminosity is only given by the perturbation in the radiative luminosity. The stability properties of the models could change strongly if convection is properly taken into account. Other effects that are normally neglected are rotation and magnetic fields that could also affect the stability. Thus, the issue of whether VMS could be vibrationally unstable to pulsations and induce pulsationally driven winds is still open.

In the context of this controversy, we suggest the following scenario: ZAMS "First Stars" are composed of VMS but during the hydrogen and helium burning phases lose mass polluting the intergalactic medium with hydrogen, helium, and some metals so that at the end of the quasi-static burning phases, the mass of the star has been reduced to the MS range.

In section 2, we briefly summarize the relevant statements about the mass of Very Massive Objects, in section 3 we discuss some recent conflicting ideas about the role VMS could play in the cosmological scenario and propose an alternative scenario. In section 4, ZAMS and stellar evolution calculations are presented, and in section 5 the results are discussed.

2. THE VMS FIRST STARS HYPOTHESIS

In the present section, the different propositions of the typical mass of the pristine stars are discussed. Supermassive Stars (SMS) have a mass greater than about $10^5 M_\odot$, the mass of VMS lies between 10^2 to $10^5 M_\odot$ and the mass range of MS is $4 \leq M/M_\odot \leq 100$ [8]. With respect to metallicity, Pop III stars, also called EMP, have a metallicity of about $Z \sim 10^{-5} - 10^{-6}$, pregalactic stars that could be referred to as population IV stars (Pop IV) (see section 4) have a metallicity of $Z \sim 10^{-9} - 10^{-10}$ [10, 11] and produce the abundances observed in Pop III stars.

It has been suggested that the Initial Mass Function (IMF) of pregalactic stars may be in the VMS range [12].

Nakamura & Umemura studied the gravitational contraction and fragmentation of filamentary gas clouds and found that Pop III stars must had been of the order of $\sim 3 M_\odot$, which may grow up to $\sim 16 M_\odot$ by accreting the diffuse envelope [16]. For Pop I stars, in a subsequent paper, they found fragment masses in the range $1 - 2M_\odot$ to $\sim 100 M_\odot$ [17]. Including nonequilibrium processes for hydrogen molecule formation, the same authors predicted a bimodal IMF for Pop III stars [18] and Primordial Galaxies [19] with peaks at $100 M_\odot$ and $\sim 1 M_\odot$, and $100 M_\odot$ and $\sim 10 M_\odot$, respectively.

Assuming the star formation efficiency used by Padoan [26], Abel et al. predicted very first stellar systems to have a mass of $\sim 50M_\odot$ [1]. Bromm and collaborators have investigated the fragmentation properties of a metal-free gas and found that the gas dissipatively settles into a filamentary morphology and concluded that Pop III stars might have been quite massive, perhaps even "very massive" ($M_{\rm char} > 100M_\odot$), if one assumes that a Jeans mass clump does not undergo significant further fragmentation [6]. Later, Abel et al. followed the collapse of primordial molecular clouds and their subsequent fragmentation and found $\sim 200M_\odot$ for a protostellar cloud core [2], whereas the distribution of clump masses in Bromm et al. suggests somewhat higher masses, perhaps up to $1000M_\odot$, are possible [7]. Larson and Bromm [13] reported that Abel, Bryan and Norman have argued that the stars probably had masses not greater than $300M_\odot$, and their work suggests that masses as high as $1000M_\odot$ might have been possible. Both predictions might be valid in different circumstances [13]. Norman et al. discuss the results of 3D simulations of first structure formation and estimate that the first stars must be massive, of the order of the gravitationally unstable core which forms them, i.e. $\sim 200M_\odot$. The Bonnor-Ebert mass is $240M_\odot$, close to their simulated core mass. Thereby, the cold gas in the centers of primordial low mass halos, fragments into cores of this mass, each of which form a MS ($10-100M_\odot$), or possibly a binary [22].

Nishi et al. investigated the formation condition of primordial stars and estimate that the lower limit of Pop III luminous objects must be of several times $10^7M_\odot$, since smaller objects may be disrupted by supernovae (SN) explosions before they become luminous [20].

By performing a study of a full self-consistent three dimensional hydrodynamical simulation about primigene stars in the Universe, Abel et al. found a core of $10^2M_\odot$, which undergoes a rapid contraction [3]. Later, assuming a primordial gas composed of H, H^+, H^-, e^-, He, He^+, He^{++}, H_2, H_2^+, Abel et al. have found by three-dimensional numerical simulation an uncertain final mass, but when the dense central parts of the cold gas became self-gravitating, then a dense core of $\sim 100M_\odot$ undergoes a rapid contraction [3].

Omukai and Palla have investigated the process of star formation from metal-free gas following the evolution of accreting protostars with emphasis on the properties of massive objects. Their calculations show that star formation depends sensitively on the mass accretion rate and its time variation. The maximum mass is set by consideration of stellar lifetimes rather than by protostellar evolution. They suggest that the upper limit can be as high as $\sim 600M_\odot$ [24]. Previously, Omukai and Inutsuka found that accretion would stop at $M \sim 460M_\odot$ [25]. This limit is similar to that imposed by the amount of mass that can be accreted by the star during the entire main-sequence lifetime, i.e, $600M_\odot$ [3].

3. AN ALTERNATIVE SCENARIO

Recently, it has been distinguished between what has been called the strong and the weak VMS hypothesis. In the former, the first generation of stars was completely composed of VMS, whereas in the latter the first generation included VMS in addition to stars with masses less or equal to $\sim 40M_\odot$ [33]. The weak VMS hypothesis is in fact an almost normal star formation with the addition of some VMS.

Wasserburg & Qian [36] suggested that the wide dispersion in Eu to Ba abundances at [Fe/H] ~ -3 indicates unrelated sources of Fe-peak and r-elements, so that a prompt (P) inventory of Fe production was proposed with an initial population with large Fe-peak yields but little or no r production. The initial inventory ceased at [Fe/H] ~ -3 and the star formation changed from VMS to normal. These authors showed that the yields of a few hundred solar masses SN models, are Fe-rich and r-poor [27, 28]. However, a closer scrutiny shows the P-inventory does not match all the observations and that the P-inventory model is incompatible with the strong VMS hypothesis, because VMS have no-significant post-He nuclear burning and therefore produce no r-elements. If VMS produce all the Fe up to [Fe/H] ~ -3, the r-elements should be absent instead of appearing in EMP stars. An alternative explanation to EMP abundances has been developed by Umeda & Nomoto [34, 35] and Nomoto et al. [21]. They suggest that energetic supernovae or hypernovae ($E_{51} = 10 - 100$, $E_{51} = E/10^{51}$ erg) connected to some low redshift gamma-ray bursts could be relevant to early nucleosynthesis at low metallicity. Since such hypernovae have been found at lower redshift, the authors argue that these objects are less speculative than VMS.

However, the VMS hypothesis and the hypernovae scenario are not necessarily inconsistent if VMS have considerable mass loss during their quasi-static evolutionary phases. In recent papers, it has always been assumed that VMS evolve with no mass loss, so that they reach the SN stage with almost their initial main sequence mass [33]. However, the SN mass could be much smaller than the zero-age main sequence mass [4], and in order to keep the VMS hypothesis, we suggest a new scenario with the following two cornerstones: VMS are formed but their masses reduced to the MS range by winds during the hydrogen and helium burning phases [4], and that a fraction of $\sim 10^{-3}$ of the baryons in the Universe are incorporated into VMS [32]. The consequences of these assumptions are the following: through winds, VMS produce at least part of the CNO elements observed in EMP stars [4], the VMS evolve to MS during the H- and He-burning phase and ends as hypernovae producing the Fe-rich and r-poor elements observed in EMP stars [34, 35, 21] and could be connected to low redshift gamma-ray bursts [34, 35, 21] and the reionization of the Universe [32, 33, 4].

In this work we study the evolution of mass losing very massive stars with different initial compositions and mass loss rates and show how through radiation driven winds, VMS evolve to MS.

4. VMS ZAMS AND EVOLUTIONARY CALCULATIONS

In this section we present ZAMS models of stars from 100 to $10000 M_{\odot}$ and evolutionary calculations for a $250 M_{\odot}$ Pop III star. A more complete set of models will be presented in Bahena et al. [4]. The models are assumed to have the following initial compositions: For Pop I $(X,Z) = (0.74, 10^{-2})$, for Pop II $(X,Z) = (0.749, 10^{-4})$, and for Pop III $(X,Z) = (0.765, 10^{-6})$. We also define the Population IV (Pop IV) composition $(X,Z) = (0.765, 10^{-9})$, and the Pregalactic population (Pop P) with $(X,Z) = (1.0, 10^{-10})$. For the mass loss rate we take the empirical relation $\dot{M} = NL/c^2$, where N is a constant, L the

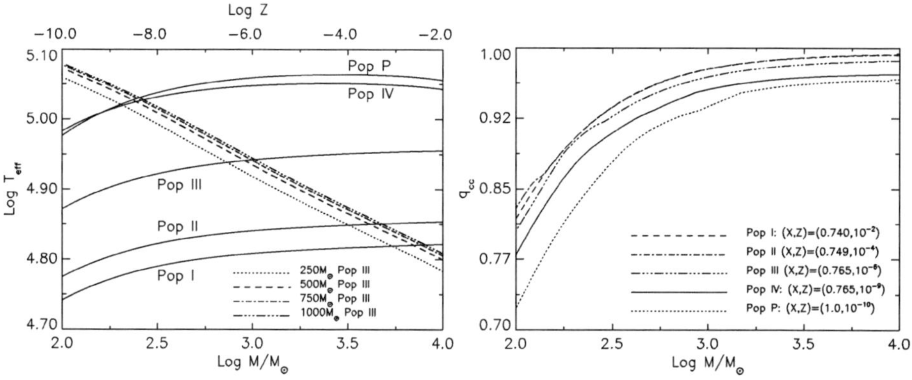

FIGURE 1. Effective temperature $T_{\rm eff}$ (ascending curves, left panel) and ratio q_{cc} of the convective core mass to the total mass of the star (right panel), both as a function of the ZAMS initial mass. The descending curves in the left panel shows the variation of the effective temperature with respect to the metallicity Z for stars of masses $250M_{\odot}, 500M_{\odot}, 750M_{\odot}$ and $1000M_{\odot}$ and for the Pop III composition.

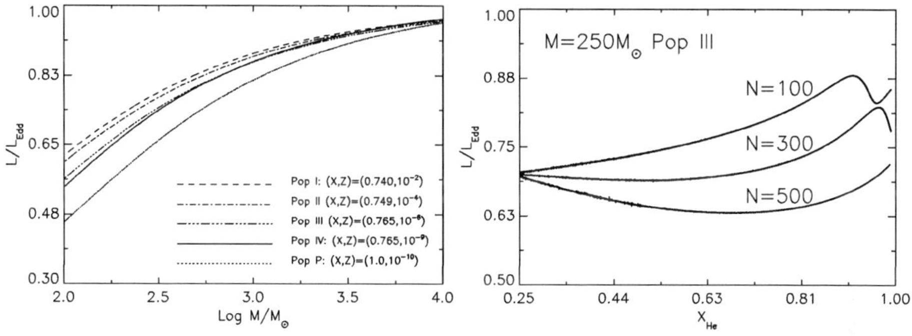

FIGURE 2. Ratio of the luminosity L to the modified Eddington Luminosity $L_{\rm Edd}$ (left panel) as a function of the ZAMS initial mass. In the right panel we show the same ratio for the hydrogen burning main sequence as a function of the central He-abundance for the Pop III composition and for various values of the mass loss parameter N.

luminosity and c the speed of light [11]. For MS, N is in the range 100-300, for VMS we expect to have higher values, perhaps up to 500 or 1000 [4].

The characteristics of the ZAMS models depends upon the mass of the star and its metallicity. In Fig. 1 (left panel), we show that the effective temperature increases with mass and depends strongly upon the metallicity. For the VMS range, the effective temperature is of about 60000 K, 67000 K, 84000 K, 108000 K and 112000 K for the compositions Pop I, Pop II, Pop III, Pop IV and Pop P, respectively. Hence, for low metallicity, VMS have very high effective temperatures which are translated into high ratios of radiation to total gas pressure, but as shown in Fig. 2 (left panel), the ratio of the luminosity to the Eddington luminosity diminishes with decreasing metallicity, since the opacity gets smaller. The ratio of the luminosity to the Eddington luminosity is given by

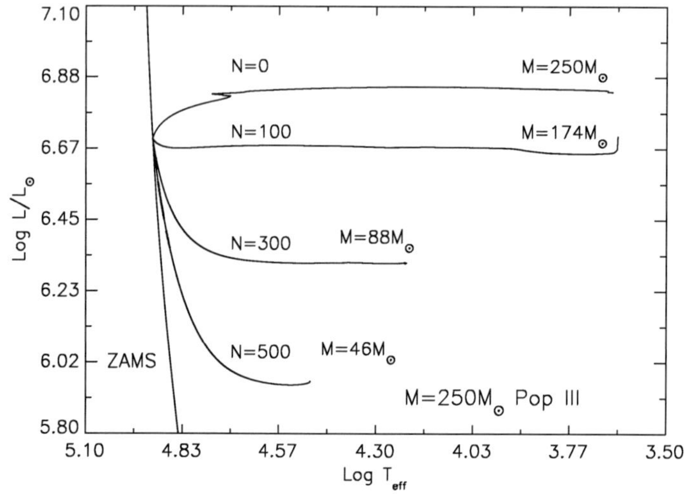

FIGURE 3. HR-diagram for the evolution of mass losing Pop III VMS. We also show ZAMS models for the same composition.

the expression $L/L_{\rm Edd} = \kappa_\nu L/4\pi cGM$, where κ_ν is the real mass absorption coefficient and the other symbols have their usual meaning. Therefore, our L_{Edd} corresponds to the modified Eddington luminosity. The high values of $L/L_{\rm Edd}$ for stars in the VMS range and for almost all compositions suggest that they have strong radiation driven winds.

In Fig.1 (right panel), we show the convective core size $q_{cc} = M_{cc}/M_*$ as a function of the mass of the star. M_{cc} is the mass of the central convective region and M_* the total mass of the star. We notice that all the models have large central convective regions which increase with mass and the models become almost fully convective for high enough masses, with a weak dependence upon metallicity.

As an example of the evolution of mass losing VMS, we describe in this paper the evolution of a $250M_\odot$ Pop III star with mass loss parameters $N = 0$, 100, 300 and 500. For a description of a full set of models see Bahena et al. [4]. The evolution in the HR-diagram of a $250M_\odot$ Pop III star for various values of the mass loss parameter N is shown in Fig. 3. The constant mass evolution ($N = 0$) is similar to that obtained by Baraffe et al. [5]. However, with the introduction of mass loss, the evolutionary tracks changes strongly. For the $N = 100$ case, the star evolves with an almost constant luminosity, increasing the $L/L_{\rm Edd}$ ratio from 0.75 to 0.82 by the end of the hydrogen burning phase (see Fig. 2, right panel), whereas the convective core size is reduced from about 0.9 to 0.5. During the H-burning phase, the ratio of radiation to total pressure increases from 0.75 to 0.81. For this evolution, the mass is reduced to $174M_\odot$ at the end of the He-burning phase which is still in the VMS range.

For the $N = 300$ case, which is more appropriate for VMS, the luminosity during the hydrogen burning phase decreases strongly mainly due to the reduction of the mass of the star. The $L/L_{\rm Edd}$ ratio remains almost constant but the convective core size behaves in much the same way as the N=100 case. By the end of the $H-$burning phase, the mass has been reduced to $88M_\odot$ which is almost in the MS range and so this evolution was stopped at the end of the $H-$burning phase. The evolution for the $N = 300$ case shows an even more drastic reduction of the luminosity which is also due to the reduction of mass that by the end of the H-burning phase reaches a mass of $46M_\odot$. For this case, the L/L_{Edd} ratio is reduced from 0.75 to about 0.63 and the ratio of radiation to total pressure is also reduced from 0.75 to about 0.63 which is expected to reduce the mass loss rate. The mass losing evolutions in this paper have been performed for a constant value of N. This is in general a common assumption in stellar evolution calculations, however, it would be more realistic to consider evolution with variable N which becomes smaller as the mass of the star is reduced from the VMS to the MS range.

A detailed description of the metal yields will be communicated in Bahena et al. [4], in here we just mention that as a result of the evolution, VMS return to the interstellar medium through radiation driven winds about 2-3% of their initial mass in the form of CNO elements, mostly carbon and oxygen.

5. DISCUSSION

In order to resolve the recent controversy in the literature with respect to the characteristic mass of the first stars, we propose a new scenario in which the first stars are born very massive but transformed into massive stars through radiation driven winds during their quasi-static burning phases. We have studied ZAMS models from 100 to $10000M_\odot$, and as an example we describe in this paper the evolution of a $250M_\odot$ Pop III star with mass loss parameters $N = 0$, 100, 300 and 500. The ZAMS models indicates that VMS have very high effective temperatures, a large ratio of radiation to total pressure and a luminosity close to the Eddington luminosity, and so we expect them to have strong radiation driven winds. The models for the $250M_\odot$ Pop III stars indicate that the mass of the star is reduced to the MS range if the mass loss parameter N is equal or greater than ~ 300. A description of a full set of models will be presented in Bahena et al. [4], where it is shown that through radiation driven winds, VMS lose about $\sim 2\%$ of their initial mass in the form of CNO elements, mostly carbon and oxygen. Hence, if, as predicted by Tegmark et al. [32], a fraction of $\sim 10^{-3}$ of the baryons in the Universe is incorporated into a generation of pregalactic VMS, then the enrichment generated by these stars and their hypernovae remnants, will be consistent with abundance determinations of EMP stars.

ACKNOWLEDGMENTS

The calculations of this paper were performed using the Xeon based Linux cluster of the Instituto Nacional de Investigaciones Nucleares (ININ) of Mexico. This work was partially supported by the mexican Consejo Nacional de Ciencia y Tecnología (CONACyT) under contract number U43534-R and J200.476/2004, and the Deutsche Forshungsgemainshaft (DFG) and the Deutscher Akademischer Austausch Dienst (DAAD) of Germany. In particular, one of us (D. B.) wishes to thank CONACYT for a postgraduate grant.

REFERENCES

1. Abel, T., Anninos, P., Norman, M.L., and Zhang, Y., *Ap. J.* **508**, 518-529 (1998).
2. Abel, T., Bryan, G. L., and Norman, M.L., *Ap. J.* **540**, 39-44 (2000).
3. Abel, T., Bryan, G.L., and Norman, M.L., *Science* **295**, 93-98 (2002).
4. Bahena, D., Klapp, J., Corona-Galindo, M. G., and Dehnen, H., *A & A*, in preparation.
5. Baraffe, I., Heger, A., and Woosley, S.E., *Ap. J.* **550**, 890-896 (2001).
6. Broom, V., Coppi, P.S., and Larson, R.B., *Ap. J.* **527**, L5-L8 (1999).
7. Bromm, V., Coppi, P.S., and Larson, R.B., *Ap. J.* **564**, 23-51 (2002).
8. Carr, B. J., Bond, J. R., and Arnett, W. D., *Ap. J*, **277**, 445-469 (1984).
9. Heger, A., Baraffe, I., Fryer, C.L., and Woosley, S. E., *Lighthouses of the Universe: The Most Luminous Celestial Objects and Their Use for Cosmology*, Proceedings of the MPA/ESO, Springer-Verlag (2000), pp 369.
10. Klapp, J., *Astrophy. & Space Sci.* **93**, 313-345 (1983).
11. Klapp, J., *Astrophy. & Space Sci.* **106**, 215-229 (1984).
12. Larson, B. L., *MNRAS* **301**, 569-581 (1998).
13. Larson, B. L., and Broom, V., *Sc. Am.* **12** 64-71 (2001).
14. Ledoux, P., *Ap. J.* **94**, 537-547 (1941).
15. Low, C., and Lynden-Bell, D., *MNRAS* **176**, 367-390 (1976).
16. Nakamura F., and Umemura M., *Ap. J.* **515**, 239-248 (1999).
17. Nakamura, F., and Umemura, M., *The First Stars*, Proceedings of the MPA/ESO, Weiss, Abel and Hill (Eds.), Springer-Verlag (2000), pp. 263-267.
18. Nakamura, F., and Umemura, M., *Ap. J.* **548**, 19-32 (2001).
19. Nakamura, F., and Umemura, M., *Ap. J.* **569**, 549-557 (2002).
20. Nishi, R., Susa, H., and Omukai, K., *The First Stars*, Proceedings of the MPA/ESO, Weiss, Abel and Hill (Eds.), Springer-Verlag (2000), pp. 268-272.
21. Nomoto, K., Maeda, K., Umeda, H., Tominaga, N., Ohkubo, T., Deng, J., and Mazzali, P.A., *Origin and Evolution of the Elements*, in Carnegie Obs. Astrophys. Ser. 4, A. McWilliam & M. Rauch (Eds.) (Pasadena: Carnegie Obs.), pp. 1.
22. Norman, M. L., Abel, T., and Bryan, G., *The First Stars*, Proceedings of the MPA/ESO, Weiss, Abel and Hill (Eds.), Springer-Verlag (2000), pp. 250-260.
23. Norman, M. L., O'Shea, B. W., and Paschos, P., *Ap. J.* **601**, L115-L118 (2004).
24. Omukai, K., and Palla, P., *Ap. J.* **589**, 677-687 (2003).
25. Omukai, K., and Inutzuka, S., *MNRAS* **332**, 59-64 (2002).
26. Padoan, P., Jimenez, R., and Jones, B. J. T., *MNRAS* **285**, 711-717 (1997).
27. Qian, Y.-Z., and Wasserburg, G. J., *Ap J.* **559**, 925-941 (2001).
28. Qian, Y.-Z., and Wasserburg, G. J., *Ap J.* **567**, 515-531 (2002).
29. Rees, M. J., *MNRAS*, **176**, 483-486 (1976).
30. Silk, J., *Ap. J.* **211**, 638-648 (1977).
31. Tarbet, P. W., and Rowan-Robinson, M., *Nature* **298**, 711-715 (1982).
32. Tegmark, M., Silk, J., Rees, M. J., Blanchard, A., Abel, T., and Palla, F., *Ap J.* **474**, 1-12 (1997).
33. Tumlinson, J., Venkatesan, A., and Shull, J. M., *Ap J.* **612**, 602-614 (2004).
34. Umeda, H., and Nomoto,K., *Ap J.* **565**, 385-404 (2002).
35. Umeda, H., and Nomoto,K., *Nature* **422**, 871-873 (2003).
36. Wasserburg, G. J., and Qian, Y.-Z., *Ap J.* **529**, L21-L24 (2000).
37. Ziebarth, K., *Ap. J.* **162**, 947-962 (1970).

Non Cuspy Galactic Halos from Scalar Field Dark Matter

Argelia Bernal* and Tonatiuh Matos*

*Departamento de Física, Centro de Investigación y de Estudios Avanzados del IPN, AP 14-740, 07000 D.F., México

Abstract. The Lambda Cold Dark Matter model has recently shown an enormous predictive power. Nevertheless, the CDM paradigm predicts a density profile which corresponds to the Navarro-Frenk-White profile which is cusp and singular in the center of the galaxies. However, observations in LSB galaxies show a density profile which seems to follow an almost constant density profile. In this work we show that a flat central density profile naturally arises within the scalar field dark matter hypothesis. This adds evidence in favor that the cold dark matter in galaxies can be identified with the scalar field.

1. INTRODUCTION

Dark matter in the universe seems to be one of the most important puzzles science has to face in this moment. These component account for the 23% of the total material content of the universe [1, 2] and it existence has been only detected trough gravitational interactions. The fact that the dark matter does not interact with the rest of the matter components of the universe implies an important puzzles for ths Standard Model of Particles (SMP) as well.

Addressing the dark matter problem The Lambda Cold Dark Matter model (LCDM) is very successful at cosmological level. Nevertheless, some problems of this model at galactic level have arisen, (Moore [3], Burkert [4] and Tyson et al [5]). The CDM paradigm predicts a density profile which corresponds to the Navarro-Frenk-White (NFW) profile [6] given by

$$\rho_{\rm NFW} = \frac{\rho_0}{\frac{r}{r_0}\left(\frac{r}{r_0}+1\right)^2}. \tag{1}$$

This cusp density profile seems to have some differences with the observed profiles of LSB galaxies, see de Blok et al [7, 8, 9] and Simon et al [10]. The evidence points to the fact that, in the central regions, galaxies prefer to follow an almost flat density profile. In this work we review the main feature of the Scalar Field Dark Matter hypothesis (SFDM) which recover the success of CDM at cosmological level. We also review some analytical solutions of the flat and weak field limit of the Einstein- Klein-Gordon equations within the SFDM model. With this solutions we show that its scalar field density profile corresponds to a halo with an almost flat central density and that this halo coincides with the CDM model in a large outer region. Such a result could solve the

CP758, *Gravitation and Cosmology: 2nd Mexican Meeting on Mathematical and Experimental Physics*
edited by A. Macías, C. Läemmerzahl and D. Nuñez

problem of the cusp dark matter halo in galaxies without extra hypothesis, adding to the viability of the SFDM model.

2. THE SCALAR FIELD DARK MATTER MODEL

We work within the specific context of the so–called 'strong self–interacting scalar field dark matter' hypothesis that has recently been developed by several authors, Matos and Ureña-López [11, 12, 13], Matos and Guzmán [14], Alcubierre et al [15, 16], Ureña-López [17] (see also Peebles [18]).

The SFDM scenario can be explained as follows. We suppose that the dark matter responsible for structure formation in the Universe is a real scalar field, Φ, minimally coupled to Einstein gravity with self-interaction parameterized by a potential energy. Therefore, we propose the following effective Lagrangian for the SFDM model,

$$\begin{aligned} \mathscr{L}_{\text{L-SFDM}} &= \mathscr{L}_{\text{GR}} + \mathscr{L}_{\text{B}} + \mathscr{L}_{\Lambda} - \sqrt{-g}\left[\Phi^{,\mu}\Phi_{,\mu} + 2V(\Phi)\right], \\ V(\Phi) &= \frac{m_\Phi^2}{8\pi G\lambda^2}\left[\cosh\left(\sqrt{8\pi G}\lambda\Phi\right) - 1\right] \end{aligned} \quad (2)$$

where λ and m_Φ are the free parameters of the model; the latter is recognized as the mass of the scalar field. The values of λ and m_Φ can be fixed by imposing that the SFDM model should reproduce the success of CDM at cosmological scales [11, 12], and then

$$\lambda \sim 20, \qquad m_\Phi \sim 10^{-23}\,\text{eV}. \quad (3)$$

The behavior of the SFDM obtained from the lagrangian (2) is the following. After and during inflation, the universe is dominated by the radiation component, the scalar potential is exponential and is far from the minimum. The large value of λ makes the scalar matter sub dominant and behave as part of the radiation fluid. After inflation, the universe has cooled and the scalar field approaches the minimum of the potential, ceases to follow the radiation fluid, and begins to behave as a dust fluid (see M. S. Turner [19]). From this time on, the evolution of the homogeneous and isotropic universe proceeds exactly as in the LCDM model.

The latter is also true for the development of scalar fluctuations, which grow as in the CDM case, except that all structure on scales smaller than $\lambda_\Phi = m_\Phi^{-1} \sim 10\,\text{pc}$ is suppressed. This fact eventually prevents the formation of too much substructure like dwarf galaxies in clusters. At the other hand, the collapse of the scalar field form bound objects of the size of a galaxy. Numerical simulations for the collapse of a single scalar fluctuation, which ends as a self-gravitating scalar object called *oscillaton*[20, 15, 16, 21] have shown that this objects avoid the formation of gravitationally bounded objects with a cusp density profile, the energy density profile of oscillatons is regular everywhere. Oscillatons are stable scalar configurations if their mass is below the critical value[16]

$$M_{\text{crit}} \sim 0.6\frac{m_{\text{Pl}}^2}{m_\Phi}. \quad (4)$$

For the scalar mass value (3), we get $M_{\rm crit} \sim 10^{12} M_\odot$, which is roughly the mass content of a typical galaxy. Also, numerical studies within the weak-gravity regime have shown that a galaxy scalar fluctuation rapidly virializes[22], which seems to be in accord with observations of the oldest galaxies in the universe[23].

We want to underline the fact that the scalar field has no interaction with the rest of the matter, thus, it does not follow the standard lines of reasoning for the particle-like candidates for dark matter. The scalar field was not thermalized, that is, the scalar field forms a Bose condensate, and thus behaves strictly as cold dark matter from the beginning (see J. Lee and I. Koh [24]). In what follows we briefly describe the SFDM at galactic level.

3. DENSITY PROFILES FROM A FLAT SPACE-TIME

For the scenario of galactic formation with the SFDM hypothesis, when the scalar field fluctuations reaches the non-linear regime, the scalar field collapses in a different way as the standard CDM. In a normal dust collapse, as for example in CDM, there is in principle nothing to avoid that the dust matter collapse all the time. In the scalar field paradigm this collapse is different. The radial and angular pressures are two natural components of the scalar field which stops the collapse, avoiding the cusp density profiles in the centers of the collapsed objects. The pressures play an important roll in the SFDM equilibrium [25].

Galaxies have a weak gravitational field. In this way, their space-time is almost flat. We will suppose first that the space-time of the galaxy is flat. This is a crude approximation but it shows us with great clarity the physics taking place in the oscillatons. In the next section we perform the weak field limit to the Einstein-Klein-Gordon (EKG) equations and show that the density profile remains almost flat and regular in the center of the galaxy.

Imposing spherical symmetry, we work with the line element

$$ds^2 = -e^{2\nu} dt^2 + e^{2\mu} dr^2 + r^2 d\Omega^2\,, \tag{5}$$

with $\mu = \mu(r,t)$ and $\nu = \nu(r,t)$, being this last function the Newtonian potential. The energy momentum tensor of the scalar field is

$$T_{\mu\nu} = \Phi_{,\mu}\Phi_{,\nu} - \frac{g_{\mu\nu}}{2}\left[\Phi^{,\alpha}\Phi_{,\alpha} + 2V(\Phi)\right]. \tag{6}$$

In the Minkowski background $\mu \sim \nu \sim 0$ and making the approximation of the potential (2) to $V = \frac{m_\Phi^2}{8\pi G\lambda^2}\left[\cosh\left(\sqrt{8\pi G}\lambda\Phi\right) - 1\right] \sim \frac{1}{2} m_\Phi^2 \Phi^2$ (we will see later that it is not a so bad approximation neither in the flat or weak field cases) the Klein-Gordon equation $\mathscr{D}\Phi - dV/d\Phi = 0$, where $\mathscr{D}$ stands for the DAlambertian, reads

$$\Phi'' + \frac{2}{r}\Phi' - m_\Phi^2 \Phi = \ddot{\Phi} \tag{7}$$

where over-dot denotes $\partial/\partial t$ and prime denotes $\partial/\partial r$. The exact general solution for the scalar field Φ is

$$\Phi(t,r) = \frac{e^{\pm ikr}}{r} e^{\pm i\omega t} \tag{8}$$

where we obtain the dispersion relation $k^2 = \omega^2 - m_\Phi^2$. For $\omega > m_\Phi$ the solution is non-singular and vanishes at infinity. We will restrict ourselves to this case. It is more convenient to use trigonometric functions and to write the particular solution in the form

$$\Phi(t,x) = \Phi_0 \frac{\sin(x)}{x} \cos(\omega t) \tag{9}$$

where $x = kr$.

Even though it is not a solution to the Einstein equations as we are neglecting the gravitational force provoked by the scalar field, the solution is analytic and it helps us to understand some features that appear in the non-flat oscillatons. The scalar field oscillates in harmonic manner in time. and it spreads over all space, *i.e.*, it is not confined to a finite region, as we are neglecting the gravitational interactions. In Ureña-López, [17], and Ureña-López, Matos and Becerril [26] has been shown that when the gravitational interaction is taken into account, the oscillaton gets confined in a finite region. The analytic expression for the scalar field energy density derived from Eq. (9) is

$$\rho_\Phi = \frac{1}{2}{\Phi_0}^2 \left(\left(\frac{x\cos(x) - \sin(x)}{x^2} \right)^2 - \frac{\sin^2(x)}{x^4} \right) k^4 \cos^2(\omega t) \tag{10}$$

$$+\frac{1}{2}{\Phi_0}^2 \frac{\omega^2 k^2 \sin^2(x)}{x^2} \tag{11}$$

which oscillates with a frequency 2ω. Observe that close to the central regions of the object, the density of the oscillaton behaves like

$$\rho_\Phi \sim \frac{1}{2}{\Phi_0}^2 k^2 \left[\omega^2 - k^2\cos^2(\omega t)\right] + O(x^2) \tag{12}$$

which implies that the density is almost constant in the central regions, i.e. when $x \to 0$ the central density oscillates around a fixed value. Recall that this is an exact solution of the Klein Gordon field equation, so this behaviors arise naturally.

On the other hand, the asymptotic behavior when $x \to \infty$, is such that $\rho_\Phi \sim 1/x^2$, *i.e.* far away from the center, in this approximation, the flat oscillaton density profile behaves like the isothermal one. Nevertheless, if the gravitational interaction is taken into account, the object must be confined (see Ureña-López, [17], and Ureña-López, Matos and Becerril [26]) and a more realistic profile should change this behavior. Obviously, the mass function oscillates around $M \sim x$. In this approximation the integrated mass of the scalar field gives an infinite value.

In order to understand what is happening within the object, observe that the KG equation can be rewritten in a more convenient form in terms of the energy density, as

$$\frac{\partial \rho_\Phi}{\partial t} - \frac{1}{r^2}\frac{\partial}{\partial r}\left(r^2 \mathscr{P}_\Phi\right) = 0. \tag{13}$$

where $\mathscr{P}_\Phi$ is the momentum density T_{01}. This last equation has a clear interpretation: Since its form looks like the conservation equation, $\dot{\rho} + \nabla \cdot \vec{J} = 0$, equation (13) represents the conservation of the scalar field energy. It also tells us that there is a scalar field current given by

$$\begin{aligned} \vec{J}_\Phi &= -\mathscr{P}_\Phi \vec{r} \\ &= \Phi_0^2 \frac{k\omega}{2} \left[x\cos(x) - \sin(x) \right] \frac{\sin(x)\sin(2\omega t)}{x^3} \vec{r}. \end{aligned}$$

Although the flux of scalar radiation at large distances does not vanish, there is not a net flux of energy, as it can be seen by averaging the scalar current on a period of a scalar oscillation. We also see that the only transformation process is that of the scalar field energy density into the momentum density, and vice versa. For the realistic values this transfer is very small.

In the next section we will see that if the gravitational force is taken into account, the oscillaton is more confined (see also Ureña-López, [17], Ureña-López, Matos and Becerril [26] and Alcubierre et al [16])

4. DENSITY PROFILES FROM THE WEAK-FIELD LIMIT

In this section we make the analysis of the oscillaton behavior in the presence of a weak gravitational field. In the weak field limit the space-time metric can be written as

$$g_{\alpha\beta} = \eta_{\alpha\beta} + h_{\alpha\beta} \tag{14}$$

were $\eta_{\alpha\beta}$ is the Minkowski metric and $|h_{\alpha\beta}| \ll 1$ is a small correction to it. Considering spherical symmetry (5) the non zero perturbations will be

$$h_{00} = -2\nu(r,t) \tag{15}$$
$$h_{11} = 2\mu(r,t) \tag{16}$$

In this case the dynamics of the scalar field is given by the linearized EKG equations

$$R_{\alpha\beta} = \kappa_0 S_{\lambda\beta} \tag{17}$$
$$\mathscr{D}\Phi - m_\Phi^2 \Phi = 0 \tag{18}$$

where $R_{\alpha\beta}$ is the first order Ricci tensor in $h_{\alpha\beta}$ and $\kappa_0 = 8\pi G$ the equation (18) is the same as in the flat space (see S. Weinberg [27]), then the solution of (18) will be in the form (9). Following Ureña-López [17] and Ureña-López, Matos and Becerril [26] we propose for μ and ν

$$\mu(t,r) = \mu_0(r) + \mu_2(r)\cos(2\omega t) \tag{19}$$
$$\nu(t,r) = \nu_0(r) + \nu_2(r)\cos(2\omega t) \tag{20}$$

Replacing this anzats we obtain a closed system of linear equations for μ_0, μ_2, ν_0, ν_2 and Φ. It is convenient to introduce the scaled variables

$$x = m_\Phi r, \quad \tau = m_\Phi t, \quad \Omega = \frac{\omega}{m_\Phi} \tag{21}$$

In terms of these new variables the solutions for the scalar field and the r–dependent functions of the perturbations to the metric are

$$\Phi = \frac{\sigma_0}{\sqrt{\kappa_0}}\frac{\sin(x\sqrt{\Omega^2-1})}{x}\cos(\Omega\tau) \tag{22}$$

$$\mu_2 = \frac{1}{8}\sigma_0^2\left[-\frac{\sin^2(\sqrt{x\Omega^2-1})}{x^2}+\frac{1}{2}\frac{\sqrt{\Omega^2-1}\sin(2x\sqrt{\Omega^2-1})}{x}\right] \tag{23}$$

$$\mu_0 = \frac{1}{8}\sigma_0^2\left[-\frac{1}{2x^2}+\frac{1}{2}\frac{\cos(2x\sqrt{\Omega^2-1})}{x^2}-\frac{1}{2}\frac{\sin(2x\sqrt{\Omega^2-1})}{x\sqrt{\Omega^2-1}}+\Omega^2\right]+\frac{C_1}{x} \tag{24}$$

$$\nu_2 = \frac{1}{8}\sigma_0^2\left[\frac{1}{2}\frac{\sqrt{\Omega^2-1}\sin(2x\sqrt{\Omega^2-1})}{x}+Ci\left(2x\sqrt{\Omega^2-1}\right)-\ln(2x\sqrt{\Omega^2-1})\right]+C_2 \tag{25}$$

$$\nu_0 = \frac{1}{8}\sigma_0^2(2\Omega^2-1)\left[\frac{\sin(2x\sqrt{\Omega^2-1})}{2\sqrt{\Omega^2-1}x}-Ci\left(2x\sqrt{\Omega^2-1}\right)+\ln(2x\sqrt{\Omega^2-1})\right]+C3 \tag{26}$$

Observe that the final solution for the space-time metric and the scalar field depend only on two parameters: the scalar field amplitude σ_0 and the quotient $\Omega = \omega/m_\Phi$. If we require a non-singular and asymptotically flat solution for the x-dependent scalar field part we have to restrict $\Omega > 1$. On the other hand remember that the weak field limit is maintained only if $|h_{\alpha\beta}| \ll 1$. Both $\mu_0(x)$ and $\mu_2(x)$ are bounded and have an oscillatory nature. On the contrary $\nu_0(x)$ and $\nu_2(x)$ are increasing unbounded functions however it is possible to find a space range, starting from the origin, where $|h_{00}| \ll 1$ and $|h_{11}| \ll 1$ (see A. Bernal,T. Matos and D. Nuñez [25] and C. Misner, k.Thorne and J. Wheeler [28]). Those conditions are satisfied if

$$\Omega \approx 1 \tag{27}$$

$$\sigma_0^2 \sim 10^{-6} \tag{28}$$

The analytic expression for the scalar field energy density is given by $\rho_\Phi = -T^0{}_0$. In our case it reads as

$$\begin{aligned}\rho_\Phi &= \frac{1}{2}M^2(1+2\nu)^{-1}\sigma^2\Omega^2 \\ &+\frac{1}{2}M^2\left[-(1+2\nu)^{-1}\sigma^2\Omega^2+(1+2\mu)^{-1}\sigma'^2+\sigma^2\right]\cos^2(\Omega\tau)\end{aligned} \tag{29}$$

where we have defined $M^2 \equiv m_\Phi^2/\kappa_0$, σ is the spatial part of the scalar field and prime denote $\partial/\partial x$. As in the flat space case when $x \to 0$ the central density oscillates around a fixed value, this is because both $\nu(x)$ and $\mu(x)$ are regular at the origin.

In the figure 1 we compare the density profile from the weak field which is analytical, and the density profile from a relativistic oscillaton obtained with the numerical

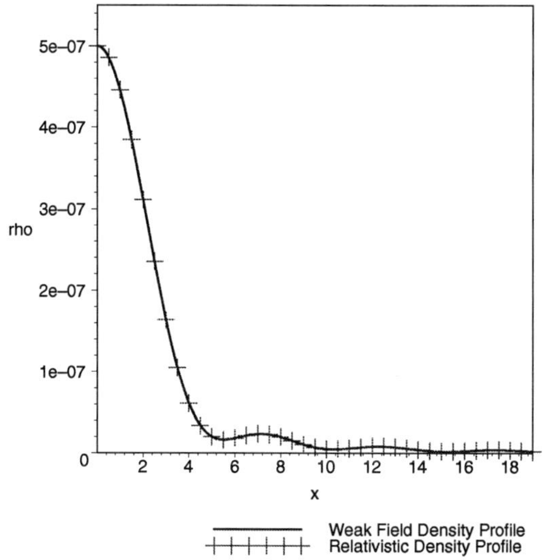

FIGURE 1. The input values for the relativistic oscillaton are specified in A. Bernal, T. Matos and D. Nuñez [25]. The output value for Ω is 1.18100, given this, the value for σ_0 is $1.59156e^{-3}$. The rescaled x does not have units because the change of variables (21) and the density's, ρ, units are in terms of M^2, for this graphic $M^2 = 1$. For these values of σ_0 and Ω the flat density profile is equal to the weak field density profile.

solutions in Ureña-López, Matos and Becerril [26]. Ω is an output value of the relativistic oscillaton. Taking this Ω, in the weak field, the lastly free parameter σ_0 is obtained scaling the central scalar field amplitude to the central scalar field amplitude for the relativistic oscillaton (see [25]).

It so important to note that because the change of variables made (21) the units and the magnitude of the energy–momentum density (29) are in terms of $M^2 \equiv m_\Phi^2/\kappa_0$. In A. Bernal, T. Matos and D. Nuñez [25] the scalar field mass is taken as (3) and the fits with some LSB galaxies density profiles are made. Finally we have to point that we have been working with an approximation of the SFDM potential (2) however we found that in the weak field limit the argument of the cosh goes like $\sqrt{\kappa_0}\lambda\Phi \sim 10^{-2}$ so it is valid to take just a quadratic potential in this limit. Further work account for study the complete potential.

5. DISCUSSION

There are many theoretical proposals that go beyond the standard model of particles, among which the most successful are the superstring theory and the brane models. A common feature of all of these super-theories is the existence of *scalar fields*: the so called dilatons, radions, etc., which are key elements by themselves. The presence of scalar fields in these models is surprising, since no fundamental scalar field has ever been detected, nor even the so-needed Higgs particle of the SMP. There are only four fundamental interactions in Nature: gravitational, electromagnetic, strong and weak. Gravitation is a spin-2 interaction, and the others are spin-1. As there is not evidence for the existence of a fundamental scalar field nor of a spin-0 interaction, it seems that our super-theories have to have an internal mechanism to suppress their own scalar fields. But, the evidence is not at all conclusive and we should ask: *did Nature forget the most simple interaction?*

On one hand, we have to face the theoretical existence of fundamental scalar fields; on the other hand, we have to face the problem of the missing matter in the universe. A deal seems possible: could it be possible that Nature did include the spin-0 interaction and made it the dominant one as part of what we called dark energy and dark matter?

We have seen in this work how the SFDM scenario offers the same results as the concordance LCDM model at large scales but it is different at galactic level. The SFDM provides a new paradigm for galaxy formation that seems to be in accord with actual observations. If this model is realistic, it could give us a solution of the dark matter problem and give us a connection between our supertheories, the simple spin-0, long range interaction and the actual observations of the universe.

ACKNOWLEDGMENTS

We thanks the helpful conversations with Luis Ureña-López and Francisco S. Guzmán. TM thanks the fanny conversations with Dario Núñez, Roberto Sussman and Luis Cabral. This work was partly supported by CONACyT México, under grants 32138-E and 42748. We also thank the support of the Germany-México bilateral project DFG-CONACYT 444 MEX-13/17/0-1.

REFERENCES

1. LAMBDA, Legacy archive for microwave background analysis (2004), URL `http://lambda.gsfc.nasa.gov/`.
2. SCP, Supernova cosmology project (2004), URL `http://supernova.lbl.gov/`.
3. B.Moore, *Nature*, **370**, 629 (1994).
4. A.Burkert, *Astrophys. J.*, **477**, L25 (1995).
5. J.A.Tyson, G.P.Kochanski, and I.P.Dell'Antonio, *Astrophys. J.*, **498**, L107 (1998).
6. J.Navarro, C.S.Frenk, and S.S.M.White, *Astrophys. J.*, **490**, 493 (1997).
7. Blok, W., S.S.MacGaugh, A.Bosma, and V.C.Rubin, *Astrophys. J.*, **552**, L23 (2001).
8. S.S MacGaugh, V. R., and de Blok, W., *Astrophys. J.*, **122**, 2381 (2001).
9. W.J.G de Blok, S. M., and Rubin, V., *Astrophys. J.*, **122**, 2396 (2001).
10. J.D.Simon, A.D.Bolatto, A.Leroy, and L.Blitz, *Astrophys. J.*, **596**, 957 (2003).
11. T.Matos, and L.A.Ureña-López, *Class. Quantum Grav.*, **17**, L75 (2000).
12. T.Matos, and L.A.Ureña-López, *Phys. Rev. D*, **63**, 063506 (2001).
13. T.Matos, and L.A.Ureña-López, *Phys. Lett. B*, **538**, 246 (2002).
14. T.Matos, and F.S.Guzman, *Class. Quantum Grav.*, **18**, 5055 (2001).
15. M.Alcubierre, F.S.Guzmán, T.Matos, D.Núñez, L.A.Ureña-López, and P.Wiederhold, *Class. Quantum Grav.*, **19**, 5017 (2002).
16. M.Alcubierre, R.Becerril, F.S.Guzmán, T.Matos, D.Núñez, and L.A.Ureña-López, *Class. Quantum Grav.*, **20**, 2883 (2003).

17. L.A.Ureña-López, *Class. Quantum Grav.*, **19**, 2617 (2002).
18. P.J.E.Peebles, Fluid dark matter (2000), astro-ph/0002495.
19. M.S.Turner, *Phys. Rev. D*, **28**, 1243 (1983).
20. E.Seidel, and W-M.Suen, *Phys. Rev. Lett.*, **66**, 1659 (1991).
21. D.Page, *Phys. Rev. D*, **70**, 023002 (2004).
22. F.S.Guzmán, and L.A.Ureña-López, *Phys. Rev. D*, **68**, 024023 (2003).
23. K.Glazebrook, R.G.Abraham, P.J.McCarthy, S.Savaglio, H-W.Chen, D.Crampton, R.Murowinski, I.Jørgensen, K.Roth, I.Hook, R.O.Marzke, and R.G.Carlberg, *Nature*, **430**, 181 (2004).
24. J.Lee, and I.Koh, *Phys. Rev. D*, **53**, 2236 (1996).
25. A.Bernal, T.Matos, and D.Nuñez, Evolution of the schrödinger–newton system for a self–gravitating scalar field (2003), to be published.
26. L.A.Ureña-López, T.Matos, and R.Becerril, *Class. Quantum Grav.*, **19**, 1 (2002).
27. Weinberg, S., *Gravitation and Cosmology:Principles and Applications of the General Theory of Relativity*, Jhon Wiley & Sons Inc., New York, 1972, ISBN 0-201-54199-8.
28. C.Misner, k.Thorne, and Wheeler, J., *Gravitation*, W. H. Freeman and Company, San Francisco, 1973, ISBN 0-201-54199-8.

On a parity issue of the thermal Sunyaev-Zel'dovich effect

A. Sandoval-Villalbazo* and L.S. García-Colín†**

*Departamento de Física y Matemáticas, Universidad Iberoamericana
Lomas de Santa Fe 01210 México D.F., México
E-Mail: alfredo.sandoval@uia.mx
†Departamento de Física, Universidad Autónoma Metropolitana
México D.F., 09340 México
**El Colegio Nacional, Centro Histórico 06020
México D.F., México
E-Mail: lgcs@xanum.uam.mx

Abstract. The thermal Sunyaev-Zel'dovich (SZ) effect is probed to be described by means of convolution integrals involving direct scattering laws and its moments. It is suggested that the distorted SZ spectrum may contain a new term in its description based on derivatives of the undistorted intensity. Possible implications of this fact are discussed

1. INTRODUCTION

The Sunyaev-Zel'dovich (SZ) effect is a signature left in the cosmic microwave background (CMB) by collapsed structures. The thermal Sunyaev-Zeldovich effect arises from the frequency shift of CMB photons that are scattered off the hot electrons contained in galaxy clusters. For a given line of sight, The frequency dependence of this effect results in a temperature decrement in the Rayleigh-Jeans region of the CMB spectrum, and to a temperature excess at high frequencies. The effect has been detected in observations of some rich, X-ray luminous clusters.

2. SUMMARY OF PREVIOUS WORK

We begin this paper by recalling the approach followed previously to discuss the SZ effect in terms of a scattering law. This means that the distorted radiation spectrum which originates from inverse Compton scattering between electrons and photons in a certain cluster is given by:

$$I(\nu) = \int_{-\infty}^{\infty} I_o(\nu) G_s(\bar{\nu}, \nu) d\bar{\nu} \tag{1}$$

Here, $I(\nu)$ is the scattered radiation off the plasma, $I_o(\nu)$ the undistorted spectrum and $G_s(\bar{\nu}, \nu)$ the scattering law. For the thermal non-relativistic effect we have argued that

CP758, *Gravitation and Cosmology: 2nd Mexican Meeting on Mathematical and Experimental Physics*
edited by A. Macías, C. Läemmerzahl and D. Nuñez

the form $G_s(\bar{\nu},\nu)$ is given by

$$G_s(\bar{\nu},\nu) = (1-\tau)\delta(\bar{\nu}-\nu) + \tau G(\bar{\nu},\nu) \tag{2}$$

where τ is the optical depth for a thin plasma, and

$$G(\bar{\nu},\nu) = \frac{1}{\sqrt{2\pi}\sigma(\nu)} e^{-\left(\frac{\bar{\nu}-(1-2z)\nu}{\sqrt{2}\sigma(\nu)}\right)^2}. \tag{3}$$

Here, $\sigma^2(\nu) = 2\frac{kT_e}{m_e c^2}\nu^2 = 2z\nu^2$ is the square of the width of the spectral line at frequency ν, and m_e and T_e are the mass and temperature of an electron and the electron gas, respectively.

The first term in Eq. (1) is self explanatory, it represents the probability that a photon traverses the plasma unscattered. The second term is a Gaussian probability function describing the scattering of a photon with an incoming frequency $\bar{\nu}$ by an electron whose average kinetic energy is $\frac{1}{2}kT_e$. The reason of why the outgoing frequency is shifted by a factor $2z\nu$ arises precisely form the fact that the average frequency of a photon scattered by electrons with speeds $u = \beta c$ will exhibit a temperature blue shift given by $2z\nu^2$ [8].

Inserting Eqs. (2) and (3) into (1), and upon integration [11], one is lead to the result that

$$\frac{\Delta I(\nu)}{\tau} = -2z\nu\frac{\partial I_o}{\partial \nu} + z\nu^2\frac{\partial^2 I_o}{\partial \nu^2} + 2z^2\nu^2\frac{\partial^2 I_o}{\partial \nu^2} - 2z^2\nu^3\frac{\partial^3 I_o}{\partial \nu^3} + \frac{z^2\nu^4}{2}\frac{\partial^4 I_o}{\partial \nu^4} \tag{4}$$

to second order in z and first order in τ. Eq. (4) is a useful result in order to discuss the SZ thermal effect. Nevertheless, several characteristics of this result must be underlined. Here, of course, $\Delta I(\nu) = I(\nu) - I_o(\nu)$.

Setting aside the fact that to first order in z Eq. (4) is identical to the diffusion approximation described by the Kompaneets equation [1] [2] [7], a fact that has been analyzed in Ref.[12], we have other interesting features. The first one is the presence of a third order derivative term in the right hand side of Eq. (4) and the other one is that the last term leads to the somewhat surprising fact that the Wien side of the distortion curve is practically identical to its relativistic counterpart [11]. This is clearly exhibited in Fig. 1 for $kT_e = 7.5$ keV.

A careful examination of the analytic expression for the relativistic thermal SZ effect derived by Sazonov and Sunyaev shows the absence of a third order derivative term. Indeed, Eq.(12) of Ref. [10] for a static cluster ($V = 0$) is equivalent to the expression:

$$\frac{\Delta I_{SS}}{\tau} = \left(-2z + \frac{17}{5}z^2\right)\nu\frac{\partial I_o}{\partial \nu} + \left(z - \frac{17}{10}z^2\right)\nu^2\frac{\partial^2 I_o}{\partial \nu^2} + \frac{7}{10}z^2\frac{\partial^4 I_o}{\partial \nu^4} \tag{5}$$

Before attempting to analyze whether a third order derivative term should arise in the distortion curve, let us go back one step and ask how is it possible to obtain an equation for $\Delta I(\nu)/\tau$ without the third order derivative term. We don't challenge the first derivative term, since it appears already in the diffusion approximation. This can be achieved by writing the scattering law for G_s as follows:

$$G_s = (1-\tau)\delta((\bar{\nu} - (1-2z\tau)\nu) + \tau G'(\bar{\nu},\nu) \tag{6}$$

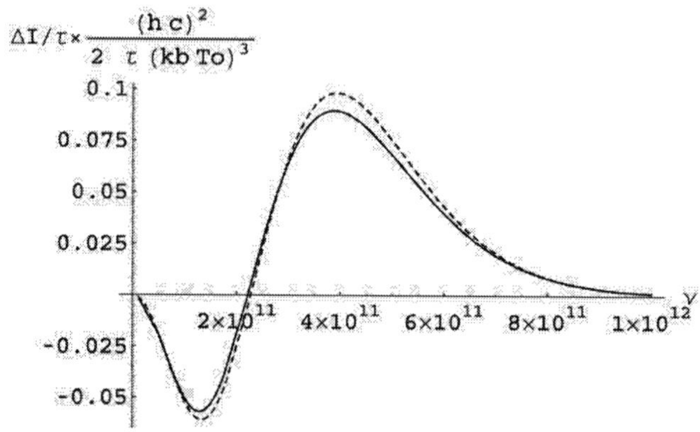

FIGURE 1. CMB distortion for the case of Eq. (4) (dashed line) compared with the relativistic curve, Eq. (5) (solid line). Here, $kT_e = 7.5$ keV, the frequency is given in Hz and$\Delta I/\tau$ is given in units of $2(kT_{\text{CMB}})^3(hc)^{-2}$.

where

$$G'(\bar{\nu},\nu) = \frac{1}{\sqrt{2\pi}\sigma} e^{-\left(\frac{\bar{\nu}-\nu}{\sqrt{2}\sigma}\right)^2} \tag{7}$$

Physically, this implies that $G'(\bar{\nu},\nu)$ is a purely Gaussian function peaked at $\bar{\nu} = \nu$ and that the central limit theorem governs the scattering of photons by electrons and a blue shift in the frequency is experienced by the unscattered photons. Although we believe that this is physically incorrect, substitution of Eqs. (6) and (7) into Eq. (1) leads to an interesting result, namely,

$$\frac{\Delta I(\nu)_a}{\tau} = -2z\nu\frac{\partial I_o}{\partial \nu} + z\nu^2\frac{\partial^2 I_o}{\partial \nu^2} + \frac{1}{2}z^2\nu^4\frac{\partial^4 I_o}{\partial \nu^4} \tag{8}$$

In Fig. (2) we exhibit the effects of this result compared with with the full relativistic curve of Sazonov and Sunyaev [10] for $kT = 10KeV$, the curve arising form Eq. (4) is also included. In Wien's region, where the effects of these z^2 terms are important, is not so easy to decide which is the most satisfactory. The reader may wonder why all this fuss about the higher order in z corrections but we believe that the underlying physics is basic for understanding the effect and, as we shall see below, also to grasp the nature of the relativistic corrections to G_s.

Thus, we are brought back to the original question, namely, if there is a third order derivative present in Eq. (4). To obtain a guide for replying it we went back to a detailed examination of two papers, one by Stebbins written six years ago [14] and the publication of Sazonov and Sunyaev on relativistic corrections to both the thermal and kinematic SZ effects. Although based on different approaches, both led to results which turn out to be identical for the thermal SZ effect. Indeed, Eq.(10) of Ref. [14] is essentially an expression for the distorted spectra written in terms of the occupation

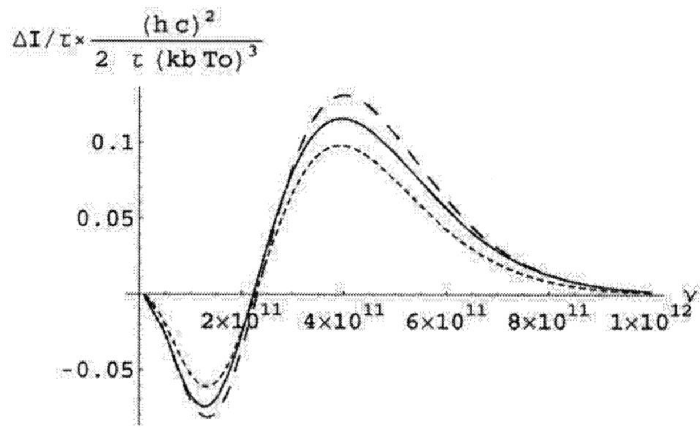

FIGURE 2. CMB distortion for the case of Eq. (8) (long-dashed line) compared with the relativistic curve, Eq. (5) (solid line). The curve resulting from Eq. (4) is included as a short-dashed line. Now, $kT_e = 10$ KeV, the frequency is given in Hz and $\Delta I/\tau$ is given in units of $2(kT_{CMB})^3(hc)^{-2}$.

number $n_\gamma = \left(\exp\frac{h\nu}{kT} - 1\right)^{-1}$ instead of the intensity $I(\nu)$, noticing that

$$\frac{\Delta n}{n_o \tau} = \frac{\Delta I}{I_o \tau} \tag{9}$$

where τ is the optical depth, identifying $\varepsilon = h\nu$ and assuming that $h\nu \ll mc^2$ in that equation. The expansion of all terms of Eq.(10) of Ref. [14] is a laborious but straightforward calculation that shows that the third order derivative in I_o disappears and that the resulting expression is identical to Eq.(5). The surprising result of the comparison between the non-relativistic limit of Eq.(5) and Eq. (4) is that the third order derivative in I_o ought to be absent. In our context, this would imply that, for low temperature clusters $kT_e \simeq 3$ keV, Eq. (8) should be essentially correct, which we object because it is physically incorrect.

This leads us to a different test, namely the value of the critical frequency ν_c defined by $\Delta I(\nu_c) = 0$. For $kT_e = 3$ keV one gets, by the Kompaneets equation that $\nu_c =$ 220.027 GHz, a value which according to the relativistic correction increases only to 221.554 GHz, a 0.68 percent correction. However, if the third order derivative is present in Eq.(4), as we assert, $\nu_c = 223.761$ GHz, giving a more significant correction of 1.66 percent. The simple physics behind Eq.(4) is quite attractive but merely theoretical up to this point. Notice also that from the observation of the results, it is really hard to see which is in agreement with observations, all follow the same pattern, specially in the Wien region.

Finally, a comment on the last term of Eq.(4). Our method, which is non-relativistic, yields a numerical factor of $\frac{5}{10}$ instead of the $\frac{7}{10}$ factor obtained in references [10] and [14] using explicit relativistic corrections. The question is if this difference and possible improvements can be obtained using the scattering law approach in the relativistic case. We turn to this question in the following section.

3. RELATIVISTIC SCATTERING LAW

From the discussion of the previous section it is rather clear that in order to introduce relativistic corrections into the scattering Kernel $G_s(\bar{\nu},\nu)$ two facts must be considered. First, the energy of an electron has to be written in its relativistic form $E = mc^2(\gamma - 1)$ and second, their velocity distribution is no longer given by Maxwell's distribution, but by its relativistic extension. Nevertheless, since the relativistic parameter z is small even at high temperature clusters, $kT_e = 15$ KeV, we may simply generalize the scattering law, Eqs. (2) - (3), allowing for z^2 corrections for the peak shift and $\sigma(\nu)$, namely

$$G_r(\bar{\nu},\nu) = (1-\tau)\delta(\bar{\nu}-\nu) + \frac{1}{\sqrt{2}N\sigma(\nu)} H\left(\frac{\bar{\nu}-(1-\varepsilon)\nu}{\sqrt{2}\sigma(\nu)}\right) \tag{10}$$

In this case, the evaluation of the expression $I(\nu) = \int_{-\infty}^{\infty} I_o(\nu) G_r(\bar{\nu},\nu) d\bar{\nu}$ leads to a generic Kompaneets type equation. This is achieved performing the change of variable:

$$\alpha = \frac{\bar{\nu}-(1-\varepsilon)\nu}{\sqrt{2}\sigma(\nu)} \tag{11}$$

with normalization factor

$$N = \int_{-\infty}^{\infty} H(\alpha)d\alpha \tag{12}$$

and identifying

$$\Delta\nu(\alpha) = \sqrt{2}\sigma(\nu)\alpha - \varepsilon\nu \tag{13}$$

Direct integration leads to

$$\begin{aligned}\frac{\Delta I_r}{\tau} &= -\varepsilon\nu\frac{\partial I_o}{\partial \nu} + \left(\frac{\varepsilon^2}{2}\right)\nu^2\frac{\partial^2 I_o}{\partial \nu^2} + \left(\frac{\sigma^2}{N}\right)\frac{\partial^2 I_o}{\partial \nu^2}\int_{-\infty}^{\infty}\alpha^2 H(\alpha)d\alpha \\ &+ \left(\frac{\sigma^2\varepsilon\nu}{N}\right)\frac{\partial^3 I_o}{\partial \nu^3}\int_{-\infty}^{\infty}\alpha^2 H(\alpha)d\alpha + \left(\frac{\sigma^4}{6N}\right)\frac{\partial^4 I_o}{\partial \nu^4}\int_{-\infty}^{\infty}\alpha^4 H(\alpha)d\alpha \end{aligned} \tag{14}$$

In this case, the change of variable is $\alpha = \dfrac{\bar{\nu}-\nu}{\sqrt{2}\sigma(\nu)}$. Thus, while the scattering law is able to reproduce the standard expression, it would need to include the frequency shift in the Delta function.

4. DISCUSSION

The present paper shows firstly that the ν^3 term discussed is essentially enough to account for the relativistic features if the SZ thermal effect without entering into long and sophisticated calculations, and secondly, it provides a bridge between the convolution integral approaches to the thermal SZ effect and the expansions of derivatives of the undistorted intensity-number occupation. The calculation of convolution integrals in the logarithmic space prevents the establishment of expressions that could relate the approaches, while the calculation of the convolution in the frequency space fills this gap. Similar techniques applied to CMB distortions have been discussed in other contexts [13].

ACKNOWLEDGMENTS

This work has been supported by CONACyT (México), project $41081-F$.

REFERENCES

1. Y.B. Zel'dovich and R.A. Sunyaev, Astrophys. Space Sci. **4**, 301 (1969).
2. R.A. Sunyaev and Ya. B. Zel'dovich. Comm. Astrophys. Space Phys. **4**, 173 (1972).
3. Y. Rephaeli, Astrophys. J. **445**, 33 (1995).
4. A.D. Dolgov, S.H. Hansen, D.V. Semikoz and S. Pastor, Astrophys. J. **554** 74 (2001) [astro-ph/0010412].
5. P.J.E. Peebles, Physical Cosmology (Princeton Univ. Press, Princeton, 2nd. Ed. 1993).
6. R.A. Sunyaev and Ya. B. Zel'dovich; Ann. Rev. Astrom. Astrophys. **18**, 537 (1980).
7. A.S. Kompaneets; JETP. **4**, 730 (1957).
8. A. Sandoval-Villalbazo and L.S. García-Colín; J.Phys. A: Math and Gen. **36** 4641 (2003) [astro-ph/0208440].
9. Rybicki, G.B., and Lightman, A.P., Radiative Processes in Astrophysics (Wiley, New York 1980).
10. S.Y. Sazonov and R.A. Sunyaev, Astrophys. J. **508**, 1 (1998) [astro-ph/9804125 (1998)].
11. A. Sandoval-Villalbazo and L.S. García-Colín; Gen. Rel. Grav. 2005 (in press) [astro-ph/0310465].
12. A. Sandoval-Villalbazo and L.S. García-Colín; [astro-ph/0305144].
13. R. Di Stefano, L.H. Ford, H. Yu and D.J. Fixsen [astro-ph/0107001].
14. A. Stebbins [astro-ph/9709065].

Cosmic censor violation, Big Crunch, or none of the above?

M. Alcubierre*, J.A. González†, M. Salgado* and D. Sudarsky*

*Instituto de Ciencias Nucleares, Universidad Nacional Autónoma de México, A.P. 70-543, México D.F. 04510, México.
†Institute for Gravitational Physics and Geometry. Penn State University, University Park, PA 16802.

Abstract. A generic set of initial data recently proposed by Hertog, Horowitz and Maeda [1], within the asymptotically anti deSitter, spherically symmetric, class of spacetimes in Einstein scalar field theory, has been argued will evolve toward a singularity that can not be contained within a black hole. We will argue that there is a logical loophole in the argument associated with the supposition that the end state needs to be a stationary spacetime. We then describe a physically reasonable possibility for the evolution of the data that would allow the singularity to be contained within a black hole.

1. INTRODUCTION

One of the most famous conjectures of gravitational physics and perhaps of all physics is the so called *cosmic censor conjecture* (CCC) [2]. The physical formulation of this conjecture states that, for physically reasonable initial data, space-time cannot evolve toward a naked singularity. That is, if a singularity forms, it will be covered by an event horizon (*i.e.* it will be contained within a black hole), indicating that far away observers will not be influenced by it. The importance of the conjecture resides in the fact that its validity will prevent the complete collapse of the predictive power of physical laws (a naked singularity will eventually influence the rest of the space-time), as far as asymptotic observers are concerned (among which we count ourselves). On the other hand, singularity theorems predict the formation of singularities starting from certain regular initial data when appropriate energy conditions are satisfied. However, up to now, an despite the efforts of many relativists, it has not been possible to prove that such singularities will always be contained within a black hole (the conjecture resists to be promoted to the status of a theorem).

Most of the reasonable matter that undergoes gravitational collapse ends up in the formation of a black hole with a completely regular domain of outer communication, a fact that produces strong supporting evidence for the validity of the CCC. On the other hand, there are indeed "counterexamples" that clearly violate the CCC. However, these are not generic in the sense that the initial conditions leading to the formation of a naked singularity are fine-tuned [3]. Since it is thought to be physically impossible to prepare a system with such precise initial conditions, those counterexamples are considered to be rather artificial, and the CCC is nowadays considered to apply only in the generic sense.

CP758, *Gravitation and Cosmology: 2nd Mexican Meeting on Mathematical and Experimental Physics*
edited by A. Macías, C. Läemmerzahl and D. Nuñez

Recently, however, Hertog, Horowitz and Maeda (HHM) [1], reported having found a generic counterexample to the CCC. To that end, they construct an open set of initial data within an Einstein-Higgs system in asymptotically anti deSitter (AdS), spherically symmetric, space-times with a scalar-field potential $V(\phi)$ which is not positive semi-definite. This situation has been considered by Dafermos, who has shown that under general conditions that cover the example by HHM, the singularity, if not enclosed within an event horizon would have to extend all the way to asymptotic infinity leading to a Big Crunch type scenario rather than to a naked singularity. Furthermore, in a revision of the initial claims HHM have argued that there might be a further loophole that calls into question even the formation of the singularity [4]. The CCC is thus clearly unaffected by the original situation described by HHM. Nevertheless it is still interesting to consider if indeed it would be impossible for the singularity, if it did form, to be contained within an event horizon.

In this report, we show that the HHM arguments have a generic loophole that would allow the formation of a very large black hole (so that one could not argue that it will not be capable of containing the singularity), despite the fact that the initially available mass would seem to be insufficient for the formation of large black holes in the asymptotically AdS spacetime. We start by briefly reviewing the model and arguments of HHM and then proceed to discuss the loophole that we have encountered.

2. THE MODEL

In their proposal HHM consider a scalar field minimally coupled to gravity and having a self interaction potential with two local minima having negative values: $V(0) = -3V_0$ and $V(\phi_1) = -3V_1$ with $0 < V_1 < V_0$. The potential is chosen to have a small positive barrier between these minima and is required to satisfy the positive energy condition among the configurations that are asymptotically AdS corresponding to the false vacuum ϕ_1 (i.e. the effective cosmological constant being $\Lambda_{\text{eff}} = -V_1$ in units where $8\pi G = 1$). That is, we consider initial data consisting of the three metric γ_{ab}, the extrinsic curvature K_{ab}, the scalar field ϕ and its conjugate momentum Π, which satisfy the Hamiltonian and momentum constraints and which correspond to an asymptotically AdS geometry. One then adjusts the details of the potential so that, within this set of data, the ADM Mass is ensured to be positive definite.

Within the spherical symmetric situation one chooses to parametrize the 3-metric as

$$d\sigma^2 = \left(1 - \frac{2m(r)}{r} + V_1 r^2\right)^{-1} dr^2 + r^2 d\Omega^2 \tag{1}$$

The first thing we must note is that such parametrization is appropriate as long as no “bag of gold" type configuration needs to be considered. For any configuration having the asymptotic parametrization given above one uses the notion of “ADM" mass given by

$$M_{\text{ADM}} = \lim_{r \to \infty} m(r) \tag{2}$$

One considers now the set of initial data $C[R_1] = \{$*the set of configurations that correspond to* $\phi = \phi_1$ *for all* $r > R_1$ *and* $\phi(0) = 0\}$. Within this set one considers the instantaneously static subset consisting of configurations having $K_{ab} = 0$ and $\Pi = 0$. The mass of such a configuration can be expressed by use of the Hamiltonian constraint as:

$$M_{\text{ADM}} = \frac{V_1}{2}R_1^3 + \tilde{M}_V R_1^3 + \tilde{M}_\partial R_1 \quad (3)$$

where M_V and M_∂ are expressions (the first related to the contribution of the scalar field potential and the second related only to the scalar field gradients) that depend only on the scalar field configuration expressed in terms of a rescaled radial coordinate $y = r/R_1$ and thus do not depend explicitly on R_1. Concretely the functionals are given by

$$\begin{aligned} \tilde{M}_V[\phi(y)] &\equiv \frac{1}{2}\int_0^1 e^{-\int_y^1 \frac{\hat{y}}{2}(\partial_y\phi)^2 d\hat{y}} V(\phi) y^2 dy \\ \tilde{M}_\partial[\phi(y)] &\equiv \frac{1}{2}\int_0^1 e^{-\int_y^1 \frac{\hat{y}}{2}(\partial_y\phi)^2 d\hat{y}} \frac{1}{2}(\partial_y\phi)^2 y^2 dy \end{aligned}$$

Thus, if one keeps the rescaled configuration fixed (i.e. one keeps $\phi(y)$ fixed) the scaling properties of the mass are given by the expression (3). The situation considered in [1] involves adjusting the parameters of the theory so that for the configuration that minimizes $\tilde{M}_V$ (denoted by $\phi_0(y)$) the terms proportional to R_1^3 cancel out, i.e. $V_1 = -2\tilde{M}_V[\phi_0(y)]$. The argument for cosmic censorship violation is the following: The configuration $\phi_0(y)$ has a value $y_0 < 1$ such that for $y < y_0$, the configuration is very close to the true vacuum $\phi = 0$, i.e. $\phi_0(y) < \varepsilon$ for a small ε. The corresponding region is essentially a region of radius $r_0 = R_1 y_0$ of AdS spacetime with a cosmological constant given by V_0 and a scalar field slightly removed from the true minimum. Such a configuration is known to evolve to a singularity. The ADM mass of this configuration is $M_0(R_1) = \tilde{M}_\partial[\phi_0]R_1$. Let's assume that a black hole develops and that it has at asymptotically late times a radius R_{BH}. Assuming it is a standard AdS black hole corresponding to the asymptotic value of the effective cosmological constant, its ADM mass is given by $M_{BH} = (1/2)(R_{BH} + V_1 R_{BH}^3)$. Its radius therefore must satisfy $(1/2)(R_{BH} + V_1 R_{BH}^3) < M_0(R_1)$. In particular $R_{BH} < (2M_0(R_1)/V_1)^{1/3} = CR_1^{1/3}$. As the matter fields in this theory satisfy the null energy condition, the area of the event horizon has to be an increasing quantity in the sense that given two Cauchy hypersurfaces the intersection of the event horizon with the later one has larger area than its intersection with the earlier one. Thus the intersection of the initial data hypersurface with the event horizon has to occur (if at all) at $r_{\text{intersection}} < R_{BH} < CR_1^{1/3}$. Clearly one can choose R_1 sufficiently large so that $r_0 = R_1 y_0$ be much larger than $CR_1^{1/3}$ given the fact that the former scales like R_1 while the latter scales like $R_1^{1/3}$. In this situation, it is argued that the black hole can not encompass the region that evolves into a singularity. Thus the singularity that results from the evolution should be naked.

3. THE ALTERNATIVE POSSIBILITY

The loophole that we want to discuss lies in the assumption that the spacetime will evolve toward a stationary configuration at late times. In fact, it is quite possible in principle for the following alternative scenario to develop: The outermost region corresponding to the asymptotic regime of the SAdS solution with $\Lambda_{\rm eff} = -V_1$ is continuously shrinking (from the inside) due to a domain wall which expands outward and which corresponds to the scalar field interpolating between the two minima. The inner region becomes a SAdS spacetime with a large black hole that encloses the singularity and which would, if extending infinitely outward, have a mass that exceeds the initial mass of the space-time. The incompatibility that one might presume would prevent the joining of the inner an outer regions disappears when we recognize that here we are dealing with two very different notions of mass corresponding to two different values of the cosmological constant for the initial and final AdS space-times one would be considering ($\Lambda^a = -V_0$ and $\Lambda^b = -V_1 = \Lambda_{\rm eff}$). Of course, in reality the interior AdS region never covers the whole space-time. The wall would behave as a solitonic type solution (such as a boosted "kink" in Sine-Gordon system) in that, once initially set it would always keep moving outward (in our case it could, of course, be changing its form and needs not be a true soliton). The apparent paradoxical situation which would allow the generation of a large black hole which will be present in any late time hypersurface is explained as due to the large negative energy which, relative to AdS with cosmological constant Λ^b, represents the large inner region with cosmological constant Λ_a.

In [4] it is argued that "*It is easy to show that in our case this could not occur*" because"*the region where ϕ is close to the global minimum (of the potential) cannot expand without increasing the total energy. This is because the initial profile $\phi(y)$ is already chosen to minimize the potential contribution to the energy... Any other shape for the wall will have higher energy*". There is a serious flaw in this argument: The initial profile was chosen to minimize the value of $\tilde{M}_V$, and **not that of the total energy:** $M_{\rm ADM}$. In fact the configuration $\phi_0(y)$ could in principle have a total energy that is quite higher that the minimum within $C[R_1]$. If one were to choose as initial configuration the true minimum of the total energy in $C[R_1]$, the scaling properties (as one changes the value of R_1) would certainly differ from those (3) above. Thus, there is a clear possibility, as suggested in [5], that the configuration will evolve into a barrier that expands indefinitely (i.e. R_1 would expand), while the energy at every "instant" of the corresponding frozen configuration (i.e. one where the kinetic terms are made zero by hand) decreases, the term $\tilde{M}_V$ increasing slowly and the term $\tilde{M}_\partial$ decreasing faster (as needed). The true total energy of the configuration ($M^{\rm true}_{\rm ADM}$) being of course conserved, after adding the kinetic terms in the actual solution ($M^{\rm true}_{\rm ADM} = M^{\rm frozen}_{\rm ADM} + M^{\rm kinetic}$). What is known is therefore that $\tilde{M}_V$ has to increase relative to its initial value and that $\tilde{M}_\partial$ has to be positive definite. The issue is then if one can envision an evolution (with some appropriate time parameter t labeling the spacetime foliation) such that the instantaneous configuration $\phi(r,t)$ differs from ϕ_1 only for values of $r < R_1(t)$, and where $R_1(t)$ increases without bound. Energetically all we need is to show that it is possible for the corresponding ADM mass of the instantaneously frozen configuration to decrease. It is easy to construct an example compatible with what is known about the functionals $\tilde{M}_V$ and $\tilde{M}_\partial$: Let us take as parameter the value of $R = R_1(t)$ rather than t. So the instantaneous configuration will be described by $\phi(y,R)$. As a result, through the dependence of the configuration on R, the functionals become, when evaluated on $\phi(y,R)$, functions of R. About these functions we know, in principle, only that $\tilde{M}_V[R]$ increases relative to its value at $R_0 = R_1(t=0)$, and that $\tilde{M}_\partial[R]$ is positive definite. Now one can give a simple example of functions satisfying the required behavior:

$$\tilde{M}_V[R] = \tilde{M}_V[R_0] + A(R - R_0)R^{-9/2} \tag{4}$$

where $A = \tilde{M}_\partial[R_0]R_0^{3/2}$, and

$$\tilde{M}_\partial[R] = \tilde{M}_\partial[R_0]\left(\frac{R_0}{R}\right)^{5/2} \tag{5}$$

In this way

$$M_{\text{ADM}}^{\text{frozen}}[R] = V_1 R^3 + \tilde{M}_V R^3 + \tilde{M}_\partial R = \tilde{M}_\partial(R_0)\frac{R_0^{3/2}}{R^{1/2}} \tag{6}$$

which is clearly a decreasing function of R. As mentioned above the true mass of the evolving configuration will be increased relative to the value above by the kinetic terms, leading to a mass that is conserved through out the evolution. Note that within this possibility one should include the case in which the wall expands indefinitelly but asymptots a certain finite value of its radius. Therefore, and in contrast with the claims made in [4] and seconded by [6], these energetic arguments can not be used to exclude the possibility that a domain wall will developed in the situation that has been proposed in [1]. Furthermore, as the arguments in [1] are supposed to refer to a generic situation (i.e. they are presented as evidence of a generic "violation of Cosmic Censorship"), the discussion of the issue at hand can not (if the generic nature of the argument is to be preserved) be based on the detailed properties of a specific configuration (i.e. the absolute minimum of $\tilde{M}_V$).

It is worth pointing that in a recent work [4], the original proponents of the generic cosmic censor violation have argued that the connection between the almost homogeneous region and a singularity is in doubt due to possible influences of the arbitrarily far away regions in finite time in AdS spacetimes. This is further concern regarding the arguments originally put forward by HHM.

Further support against the formation of a naked singularity like the one suggested by HHM is provided by the numerical analysis of Garfinkle [8], which indicates that the initial data devised by HHM evolves to a small stationary BH rather than to a Big Crunch type naked singularity.

4. CONCLUSION

In conclusion, there are at this point various issues that cast doubts on the argument of [1], that in their example a singularity would form that could not be contained within a Black Hole: The one raised here, and the one raised in [4]. However the main issue that is raised as the result of [1], whether or not a Big Crunch singularity [7] arises in the *original* proposed situation, remains at this time an important and open question.

ACKNOWLEDGMENTS

The authors wish to thank G. Horowitz for useful communications. This work was supported in part by DGAPA-UNAM through grants IN112401, IN122002 and IN108103-3, and by NSF grant PHY-02-18750.

REFERENCES

1. T. Hertog, G.T. Horowitz and K. Maeda, *Phys. Rev. Lett.* **92** 131101 (2004), gr-qc/0307102.
2. R. Penrose, *General Relativy, an Einstein Centenary Survey*. Ed. S.W. Hawking and W. Israel, Cambridge University Press, (1979).
3. M.W. Choptuik, *Phys. Rev. Lett.* **70** 9 (1993).
4. T. Hertog, G.T. Horowitz and K. Maeda, gr-qc/0405050.
5. M. Alcubierre, J.A. Gonzalez, M. Salgado and D. Sudarsky, gr-qc/0402045.
6. V. E. Hubeny, X. Liu, M. Rangamani and S. Shenker, hep-th/0403198.
7. M. Dafermos, gr-qc/0403033.
8. D. Garfinkle, gr-qc/0408064.

Cosmological Radion Problem in Models with Extra Dimensions

Abdel Pérez-Lorenzana

*Departamento de Física, Centro de Investigación y de Estudios Avanzados del I.P.N.
Apdo. Post. 14-740, 07000, México, D.F., México*

Abstract. Moduli problem is a known common feature of models that involve extra compact dimensions, as the Kaluza-Klein theories, the String theory, and the large extra dimension models. Moduli are scalar fields that emerge when the configuration of the compact space is perturbed. Usually, these fields feel no potential and represent run away modes which makes it difficult to stabilize the shape of the compact manifold. The radion is a particular example of this type of fields, which is associated to the variations of the total volume of the compact space. Radions usually couple to all other fields, affecting the definition of coupling constants and gravity strength. Thus, they are a potential threat for cosmology of the early Universe. Here we discuss some aspects of radion dynamics during inflation, and show that a heavy radion ($m \gg H$) seems to be required for a successful baryogenesis.

1. INTRODUCTION

Models that involve more than four dimensions have been under consideration for about eighty years already. The first ideas date back to the works of Kaluza and Klein in the 1920's [1], and continued all along the twenty century, well up to the present. In modern terms, extra dimensions are a known fundamental ingredient for String Theory, which needs at least six extra spatial dimensions (seven if there is M-theory) to be consistent. Although it was conventional to assume that such extra dimensions were compact, with very small sizes of the order of the Planck length, $\ell_P \sim 10^{-33}$ cm, recent developments, originally based on the studies of the the non-perturbative regime of the $E_8 \times E_8$ theory by Witten and Horava [2], have suggested that some, if not all, of these extra dimensions could rather be larger than ℓ_P. It has also been pointed out that the extra dimensions might even be as large as some hundred microns, and yet scape to detection [3]. This would be possible if our observable world is constrained to live on a four dimensional hypersurface (the brane) embedded in a higher dimensional space (the bulk), such that the extra dimensions can only be tested by gravity, a set up that resembles D-brane theory constructions. This simplified model has been the basis for a large number of theoretical studies that appeared on the last few years, most of them built up on a bottom-up approach, using an effective field theory point of view, with almost no any real string theory calculations. In spite of being quite speculative, this models might provide some insight to the low energy manifestations of the fundamental theory, since it is likely that the excited modes of the string could appear on the experiments way before any quantum gravity effect, in which case the effective field theory approach would be acceptable.

CP758, *Gravitation and Cosmology: 2nd Mexican Meeting on Mathematical and Experimental Physics*
edited by A. Macías, C. Läemmerzahl and D. Nuñez

If there are n extra spatial dimensions compatified in a flat manifold, the Newton constant, G_N, and thus, the Plack scale, defined by

$$M_P\, c^2 = \left[\frac{\hbar c^5}{8\pi G_N}\right]^{1/2} \sim 2.4\times 10^{18}\ \text{GeV} \tag{1}$$

are not fundamental but rather effective quantities. Instead, the fundamental quantum gravity or string scale, M_*, would be given as

$$M_*\, c^2 = \left[\frac{\hbar^{1+n} c^{5+n}}{8\pi G_*}\right]^{1/(2+n)}, \tag{2}$$

where G_* is the true gravitational coupling that gives gravity strength in the complete $4+n$ dimensional theory. Both the couplings are related to each other by the volume of the compact space, V_n, through out the expression

$$G_N = G_*/V_n\ , \tag{3}$$

which in natural units where $c=\hbar=1$, that we use hereon, implies $M_P^2 = M_*^{n+2} V_n$ [3]. The actual value of M_* is unknown, it can lay anywhere within few TeV and M_P.

In a model where the space is dynamical, as expected for cosmology, the actual value of our effective 4D gravitational coupling could be jeopardized by changes on V_n. Variations on gravity strength are unwelcome during baryogenesis [4], thus, any dynamics of the compact space that changes its total volume should have been gradely ameliorated by the time the Universe was about one second old. However, this seems potentially troublesome if one considers that, during inflation, our four dimensions were expanding exponentially, which makes difficult to believe that the internal manifold could remain stable all along this process. Associated to the variations on the volume of the extra dimensions one can identify a scalar field, here called the radion, which couples to all other fields, including the inflaton. The radion, as any other modulus field, suffers from run away problems. It has no natural potential, and so, any perturbation may push it away, meaning the compact space volume could indefinitely expands (or contracts). To avoid this problem a stabilization potential must be provided by some other source. The same potential would also be responsible for providing a mass to the radion, explaining why the associated long range scalar interaction had not been so far detected. Although the specific form of this potential is required to make any quantitative analysis for the radion dynamics, some qualitative results can be obtained by the use of toy model potentials [5]. Along the present discussion we will assumed a minimal squared mass term for the radion, and we shall show that, by coupling the radion to the inflaton, the dynamics may indeed quickly stabilize the radion field during inflation, such that the expansion would proceed on as in four dimensions. However, as we will discuss, in the early stages the volume of the compact space typically grows larger than V_n, and then, slowly shrinks down as 4D inflation continues, finally becoming V_n only when Hubble parameter H had became smaller than the radion mass. Hence we conclude that having a standard expansion rate around BBN era seems to prefer a heavy radion.

The present work is organized as follows. First, we introduce the radion field by dimensionally reducing the gravity action on a cosmological background, then, we discuss radion couplings, particularly to the inflaton field. Next, by assuming a simple toy model potential, we show how the radion is dynamically stabilized during the very first stages of 4D inflation, fixing the volume of the extra space to a value that in general differs from the V_n involved in the definition of the effective M_P. Finally, we further study the dynamical evolution of the radion field as 4D inflation proceeds, and show that as H becomes smaller, the dynamics drives the radion field to zero, and thus, the physical volume of the extra space gradually shrinks to its final stable value.

2. DIMENSIONAL REDUCTION AND THE RADION FIELD

Let us consider the following simple model. We assume to have n extra dimensions compactified in the orbifold T^n/Z_2, with Z_2 the identification $y^a \to -y^a$, such that the extra coordinates y^a take values in the interval $[0,1]$, with our Universe located on a 4D surface or brane at $y^a = 0$. We then consider the background metric

$$ds^2 = g_{AB}dx^A dx^B = g_{\mu\nu}dx^\mu dx^\nu - h_{ab}dy^a dy^b\,, \tag{4}$$

where $g_{\mu\nu}$ has to be conformally consistent with the brane cosmological principle. Note that since y^a is dimensionless, h_{ab} has length dimension two. Also notice that we are not considering vector-like connection A^a_μ pieces, which are common in Kaluza Klein theories, since we shall be only interested in the zero mode part of the metric where these terms vanish due to the Z_2 identification of points.

Next, let us reduce the Einstein-Hilbert action

$$S = \frac{1}{2k_*^2}\int d^4x d^n y \sqrt{|g_{(4+n)}|}\,R_{(4+n)} \tag{5}$$

to four dimensions, considering only the zero mode level. Here, $1/k_*^2 = M_*^{2+n}$, and $R_{(4+n)}$ obviously stands for the $4+n$ dimensional scalar curvature, whereas the determinant of the metric $|g_{(4+n)}| = -g_{(4)}\cdot h$. One then obtains

$$S = \frac{1}{2k^2}\int d^4x\sqrt{|g_{(4)}|}\,\frac{\sqrt{h}}{V_n}\left\{R_{(4)} - \frac{1}{4}\partial_\mu h^{ab}\,\partial^\mu h_{ab} - \frac{1}{4}h^{ab}\partial_\mu h_{ab}\cdot h^{cd}\partial^\mu h_{cd}\right\}, \tag{6}$$

where $1/k^2 = M_P^2$, and V_n stands for the volume of the extra space (at the desired stable configuration) as defined above in Eq. (3). In order to obtain the 4 dimensional scalar curvature term in a canonical form, we perform a conformal transformation on the 4D part of the metric,

$$g_{\mu\nu} \to \frac{V_n}{\sqrt{h}}\,g_{\mu\nu}\,, \tag{7}$$

The action in Eq. (6) is then transformed into

$$S = \frac{1}{2k^2}\int d^4x\sqrt{-g_{(4)}}\left\{R_{(4)} - \frac{1}{4}\partial_\mu h^{ab}\,\partial^\mu h_{ab} + \frac{1}{8}h^{ab}\partial_\mu h_{ab}\cdot h^{cd}\partial^\mu h_{cd}\right\}. \tag{8}$$

For $g_{\mu\nu}$ we can now assume the standard Friedmann-Robertson-Walker (FRW) metric with a flat geometry, i.e.

$$g_{\mu\nu} = \mathrm{diag}(1, -a(t), -a(t), -a(t)), \tag{9}$$

for an isotropic and homogeneous (brane) Universe, whereas we consider a diagonal form for the h part of the metric:

$$h_{ab} = b(t)^2 \delta_{ab} \, . \tag{10}$$

Obviously, the true physical volume of the extra space is given as $vol_{\rm phys} = \sqrt{h} = b^n(t)$. If the bulk were in its stable configuration, the physical size of the extra dimension would given by the identification $b = R$, and thus the physical volume would just become $vol_{\rm phys} = V_n = R^n$, as expected.

The action (8) can now be simplified by introducing the radion field

$$\sigma(t) = M_P \sqrt{\frac{n(n+2)}{2}} \ln\left(\frac{b(t)}{R}\right) , \tag{11}$$

which has the straightforward physical interpretation of being related to the variation of the physical size of the volume. Notice that with this definition the radion is set to be zero when the stabilized volume is reached. Any other value would mean that $v_{phys} \neq V_n$. In this terms, one gets the effective 4D action

$$S = \frac{1}{2k^2} \int d^4x \sqrt{-g_{(4)}} R_{(4)} + \frac{1}{2} \int d^4x \sqrt{-g_{(4)}} \left(\partial^\mu \sigma\right) \left(\partial_\mu \sigma\right) \, . \tag{12}$$

Clearly, the very first term corresponds to the 4D gravity action of the FRW model. On the other hand, the last term is identified as the action of a run away mode. Without potential, the radion field can take any value. Furthermore, under any perturbation, the volume of the extra space can expand or contract without control. This is what is called the radion stabilization problem, and it is a particular case of the more general moduli problem. Understanding the stability of the volume of the compact space can be seeing as finding the mechanism that provides the force that keeps the radion fixed at its zero value. Thus, in order to have a stable bulk volume, there has to be a potential, $U(\sigma)$, which provides such a force. Some ideas on the possible origin of this potential can be found in the literature, see for instance [6, 7, 8, 9]. Here we will only assume that this potential exist.

3. RADION COUPLINGS

As already mentioned, radion couples to almost all matter fields. This is regardless whether those are brane or bulk fields. To get the couplings we proceed in general by dimensionally reducing the $(4+n)D$ action, including the same conformal transformation that we performed on the metric above. For instance, for the scalar field we get the effective 4D action at the zero mode level [5]

$$S_\phi = \int d^4x \sqrt{-g_{(4)}} \left[\frac{1}{2}\partial^\mu \phi\, \partial_\mu \phi - e^{-\alpha\sigma/M_P} V(\phi)\right] , \tag{13}$$

where the coupling constant is given by $\alpha = \sqrt{2/n(n+2)}$. Notice that the radion coupling is exponential, just as expected for any moduli field. It is this coupling and its consequences we will study in the next section. One can also calculate the effective radion coupling to brane fermion fields,

$$S_\psi = \int d^4x \sqrt{-g_{(4)}}\, e^{-(3\alpha/2)\sigma/M_P} \left(-i\bar{\psi} D_\mu \gamma_a \psi\right) e^a_\nu\, g^{\mu\nu} . \tag{14}$$

This last action shows that when the radion is away from its zero point value it changes the fermion dispersion relations, and thus it may affect the late evolution of the Universe if the volume were not stabilized before BBN. Notice also that the radion does not couple to massless gauge fields.

4. THE COSMOLOGICAL RADION PROBLEM

Cosmology of models with extra dimensions has received a great deal of attention, from building inflationary models [10, 11, 12, 13, 14, 15] to studying moduli problem [16] thermal history [17], and baryogenesis [18]. However, less attention has been paid to moduli stabilization [5, 12]. This is yet an important issue on these models. By looking up on Eq. (12), considering the addition of a radion potential, we can immediately see that in a model where the extra volume is stabilized at very early times, that is, where σ is fixed to zero in a time scale smaller than the required time for inflation ($t_{stab} \leq t_{inf}$), the rest of Universe history should proceed as in the standard 4D. To understand how this may happened, let us consider the following simple model. Let us assume that inflation is being drive by a bulk scalar field [19]. Thus Eq. (13) very well describes the action for the inflaton. As usual, we assume that the observed Universe was initially a small patch with a characteristic length of order $\sim M_*{}^{-1}$. Besides, one has to make some assumption on the initial size of the extra compact dimensions. Naturalness suggest that they also were of same order, $b(0) \sim M_*{}^{-1}$. If we insert this value in Eq. (11), we get that the natural initial value for the radion was large and negative, $\sigma(0) \approx -M_P\sqrt{(n+2)/n}\, \ln(M_p/M_*)$. Notice the sign reflects the fact that we are starting with a much smaller volume than V_n.

In the small rolling regime of the inflaton, the equations that govern the dynamics of the fields are given in a good approximation by

$$3H\dot{\sigma} \approx -\frac{\alpha}{M_p} e^{-\alpha\sigma/M_P} V(\phi) + \frac{\partial U(\sigma)}{\partial \sigma} , \tag{15}$$

$$3H\dot{\phi} \approx e^{-\alpha\sigma/M_p} \frac{\partial V(\phi)}{\partial \phi} , \tag{16}$$

$$H^2 \approx \frac{1}{3M_p^2}\left[e^{-\alpha\sigma/M_P} V(\phi) + U(\sigma)\right] . \tag{17}$$

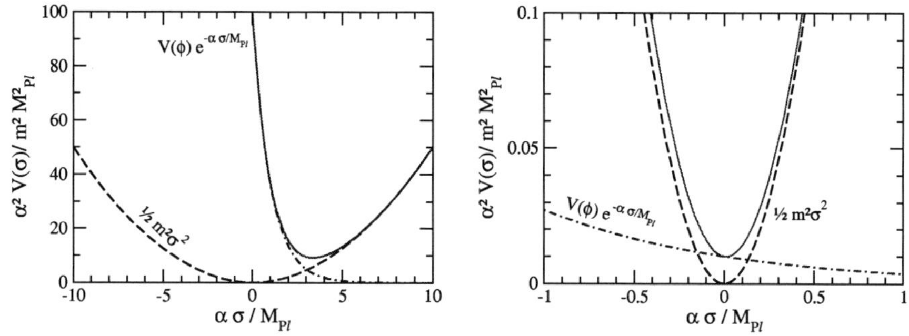

FIGURE 1. Total radion potential (continuous curve) for $\alpha M_I/m = 10$ (left) and $\alpha M_I/m = 0.1$ (right) respectively. Dashed (dash-dotted) curve corresponds to the radion mass potential (inflaton term).

First of above equations indicates that, within first bounds of expansion, the radion would be quickly driven towards its local minimum, since the negative term on the RHS would be dominating due to the exponential enhancement on the inflaton contribution. Nevertheless, it may happen that the actual local minimum, due to the presence of the inflaton, does not correspond to $\sigma = 0$, as one would expect. Indeed, even though the potential of the radion-inflaton system, $V(\sigma,\phi) = e^{-\alpha\sigma/M_P}V(\phi) + U(\sigma)$ has a global minimum at $\sigma = 0$, it is easy to see that during the phase of inflation driven by the inflaton potential the radion is trapped in a a false (effective) minimum. Taking $V(\phi)$ as constant, the minimization condition for the total radion potential reads

$$\frac{\partial U(\sigma)}{\partial \sigma} = \frac{\alpha}{M_P} e^{-\alpha\sigma/M_P} V(\phi) . \tag{18}$$

Clearly, $\sigma = 0$ is not longer a solution, unless $V(\phi)$ were null. Thus, the minimum for the radion is displaced [5] to a position which strongly depends on the actual form of the radion potential. We should notice, however, that this effective minimum is always located on the positive range of σ values, which means that within a Hubble time the bulk has grown larger than its expected value V_n. This is because we need to have a positive slope for $U(\sigma)$ in order to catch up with the now suppressed inflaton potential, a condition that is expected to be fulfilled only for positives values of σ.

This point is very well illustrated if one takes for instance the potential $U(\sigma) = m^2\sigma^2$, and draw a plot for $V(\sigma)$. This is shown in Fig. 1, where the radion mass is taken to be lighter than the inflaton scale, $M_I = \sqrt{V(\phi)}/M_P$. Here we have taken $\alpha M_I/m = 10$ for the plot on the left. From the figure one can easily see why and where the minimum gets displaced. The inflaton part of the potential drops very fast due to the exponential suppression of a growing radion, however, this is not fast enough as to catch with the radion potential which, being quite flat, grows much more slowly, and therefore, the minimum is close to the point where both the potentials are almost equally relevant. This effect is more dramatic when the hierarchy among m and H_I is larger. In any case, both the inflaton and the radion fields are driving inflation together at this stage.

The system will remain in the false vacuum as long as the inflaton potential remains constant. Inflaton was, nevertheless, slow rolling right before the radion field settles down to the false vacuum. As inflation proceeds, the inflaton slowly takes smaller vevs, as does the energy density residing in its potential. For small values of the scale of the $V(\phi)$ potential, M_I, the minimum gets shifted to smaller values of σ. It would look like if the radion were slowly rolling down the effective potential

$$V(\sigma) = m^2\sigma\left[\frac{1}{2}\sigma + \frac{M_P}{\alpha}\right], \tag{19}$$

which is obtained by replacing the minimization condition back into the total potential $V(\sigma,\phi)$. Notice also that in the RHS of Eq. (17) the exponential suppression on the inflaton potential does contribute for a very slow evolution of the inflaton field, which behaves as if it were effectively frozen. All these would result in a large period of inflation, that may last plenty of e-foldings.

As inflation continues in 4D, inflaton scale gradually becomes smaller, while the radion rolls down and the extra volume now shrinks. If inflation last long enough as to diminish M_I well below the radion mass, then the picture will dramatically change. If we now draw a plot for the potential for $\alpha M_I/m = 0.1$, as also depicted in Fig. 1, we can see that the effective minimum would now be located consistently at $\sigma \approx 0$. This means that if one starts on a scenario where $m \ll M_I$, inflation should last for a very long time in order to correct the over growing of the internal volume. In other words, phenomenology constraints the Universe to exit inflation only on the regime where $M_I \ll m$, otherwise, the radion would end having a still large amplitude that will be reflected on large variations on the effective 4D gravity coupling due to the oscillations of the radion field. This of course is a problem for models with large extra dimensions, where the radion mass is expected to be very light (of the order of few eV), since they would mean exiting inflation with a very unrealistic Hubble scale (of few eV at most). Notice that in the last stage, when radion is almost gone, the expansion is dominated only by the inflaton, so from there on Universe evolution is expected to be standard.

5. CONCLUDING REMARKS

Models that consider extra dimensions as one of their ingredients suffer in general of moduli problems. Here we have explored some of the features of one of those modular fields, the radion, on a cosmological setup. We have qualitatively explored the dynamics of the radion field when coupled to an inflaton. We have shown that, including the stabilization potential, the total radion potential has a local minimum that correspond to an oversized compact volume, meaning that in the very first stages of expansion the internal volume grows much beyond what is supposed, and then, as inflation continues the extra volume now contracts towards its definite (stable) size, V_n. Nevertheless, if inflation is ended before the V_n is reached, the radion field will end having a Planckian amplitude, which means that the internal volume will also oscillate with a large amplitude (of few factors of V_n), inducing unwanted time variations on the gravity coupling. The problem may be solved if the radion stabilization potential provides a large mass for the radion. Thus, a complete solution of the radion cosmological problem strongly depends on a better understanding of the radion potential.

ACKNOWLEDGMENTS

This work was supported in part by CONACyT Mexico, under grant 44596.

REFERENCES

1. Th. Kaluza, Sitzungober. Preuss. Akad. Wiss. Berlin, p. 966 (1921); O. Klein, Z. Phys. **37** (1926) 895.
2. E. Witten, Nucl. Phys. **B 471** (1996) 135; P. Horava and E. Witten, Nucl. Phys. **B460** (1996) 506; *idem* **B475**(1996) 94.
3. N. Arkani-Hamed, S. Dimopoulos and G. Dvali, Phys. Lett.**B429** (1998) 263; I. Antoniadis, *et al.*, Phys. Lett.**B436** (1998) 257; I. Antoniadis, S. Dimopoulos, G. Dvali, Nucl. Phys. **B516** (1998) 70.
4. C. M. Will, *Theory and Experiment in Gravitational Physics*, Cambridge University Press (1993).
5. A. Mazumdar, R.N. Mohapatra and A. Pérez-Lorenzana, JCAP **06** (2004) 004.
6. L. Amendola, E. W. Kolb, M. Litterio, F. Occhionero, Phys. Rev. D**42**, 1944 (1990).
7. S. Tsujikawa, JHEP **0007** (2000) 024.
8. A. R. Frey and A. Mazumdar, Phys. Rev. D **67** (2003) 046006
9. S. B. Giddings, S. Kachru and J. Polchinski, Phys. Rev. D **66** (2002) 106006.
10. D. Lyth, Phys. Lett **B 448**, 191 (1999); G. Dvali, S. H. H. Tye, Phys. Lett. **B450**, 72 (1999); N. Kaloper and A. Linde, Phys. Rev **D 59**, 101303 (1999); A. Mazumdar, Phys. Lett. **B 469**, 55 (1999); N. Arkani-Hamed, *et al.*, Nucl. Phys. **B567**, 189 (2000); J. Cline, Phys. Rev. D **61**, 023513 (2000).
11. R.N. Mohapatra, A. Pérez-Lorenzana and C.A. de S. Pires, Phys. Rev. **D 62**, 105030 (1999).
12. A. Mazumdar and A. Perez-Lorenzana, Phys. Lett. B **508**, 340 (2001).
13. A. M. Green and A. Mazumdar, Phys. Rev. D **65**, 105022 (2002).
14. T. Matsuda, Phys. Rev. **D 66**, 107301 (2002).
15. M. Bastero-Gil, V. Di Clemente and S. F. King, hep-ph/0211011; hep-ph/0211012.
16. C. Csaki, M. Graesser and J. Terning, Phys. Lett. B **456**, 16 (1999).
17. K. Benakli and S. Davidson, Phys. Rev. D**60**, 025004 (1999); L. J. Hall and D. Smith. Phys. Rev. D **60**, 085008 (1999).
18. A. Mazumdar and A. Perez-Lorenzana, Phys. Rev. D **65**, 107301 (2002); R. Allahverdi, K. Enqvist, A. Mazumdar and A. Perez-Lorenzana, Nucl. Phys. B **618**, 277 (2001); D. J. Chung and T. Dent, Phys. Rev. D **66**, 023501 (2002).
19. R.N. Mohapatra, A. Pérez-Lorenzana and C.A. de S. Pires, Phys. Rev. **D 62**, 105030 (1999).

MATHEMATICAL FORMALISM, NUMERICAL RELATIVITY AND EXACT SOLUTIONS

Brief Introduction to Numerical Relativity

Miguel Alcubierre

Instituto de Ciencias Nucleares, Universidad Nacional Autónoma de México, A.P. 70-543, México D.F. 04510, México

Abstract. In this manuscript I give a brief introduction to the fundamental concepts behind the field of numerical relativity. I start by introducing the Cauchy problem and the 3+1 formalism. I discuss the constraint and evolution equations, gauge conditions and the problem of finding initial data. I also introduce some basic notions of finite differencing techniques for solving partial differential equations numerically. Finally I discuss some applications of numerical relativity to the study of black hole and neutron star spacetimes.

1. INTRODUCTION

General relativity is a highly successful theory. It has radically modified our vision of space and time and possesses an enormous predictive power. Its more important achievements include the prediction of exotic objects such as black holes and neutron stars, and the cosmological model of the Big Bang. In addition, it predicts the existence of gravitational waves, which will probably be detected for the first time before the end of this decade. Still, the field equations of general relativity are extremely complex. These equations form a system of ten coupled non-linear partial differential equations in four dimensions. When fully expanded in a general coordinate system they have thousands of terms. Because of this, exact solutions to these equations usually involve high degrees of symmetry, either in space or in time: spherical or axially symmetry, homogeneity, isotropy, staticity, stationarity, etc. When one is interested in studying highly dynamic systems that have little or no symmetry it is simply impossible to solve the equations exactly and one must fall back on numerical solutions.

Numerical relativity started in the 1960's with the pioneering work of Hahn and Lindquist [1], but it wasn't until the mid 1970's that the first successful simulations were performed by Smarr and Eppley in the context of the head-on collision of two black holes [2, 3]. Since then numerical relativity has come a long way. The availability of powerful super-computers, together with an improved understanding of the fundamental theoretical issues that underlie the field, have finally allowed simulations of fully three-dimensional (3D) systems with strong and highly dynamical gravitational fields. All this activity has arrived at the right time, as the first generation of advanced gravitational wave detectors are finally coming on line [4, 5, 6, 7]. We are living a very exciting stage in the development of this field.

In this manuscript I will present a very brief introduction to numerical relativity. For readers interested in going deeper into the field I have tried to give an extensive list of references at the end.

CP758, *Gravitation and Cosmology: 2nd Mexican Meeting on Mathematical and Experimental Physics*
edited by A. Macías, C. Läemmerzahl and D. Nuñez

2. THE 3+1 DECOMPOSITION

The Einstein field equations are usually written in fully covariant form, with no distinction between space and time. This is elegant and mathematically powerful, but it is not very useful when one is interested in studying the evolution in time of a gravitational system starting from some appropriate initial data, the so called "initial value problem".

It is well known from the seminal work of Choquet-Bruhat [8] that general relativity does allow an initial value formulation. Today there are three main procedures in which one can obtain such a formulation: the "Cauchy" or "3+1" formalism, the "conformal" formalism, and the "characteristic" formalism. The 3+1 and conformal approaches share the property of splitting spacetime into a foliation of spacelike hypersurfaces. The main difference lies in the fact that in the 3+1 approach these hypersurfaces reach spatial infinity, while in the conformal approach the hypersurfaces are hyperboloidal and reach future null infinity instead. This allows the hypersurfaces to be conformally compactified, permitting a rigorous treatment of fields at infinity and eliminating the need for the boundary conditions at a finite distance that have to be introduced in the standard 3+1 approach. The characteristic formalism, on the other hand, foliates spacetime using null cones emanating from a central world tube and also allows a compactification that brings null infinity to a finite coordinate distance.

Each approach has its strong points, but to date most work in numerical relativity has been done using the 3+1 formalism, so here I will concentrate on this approach. However, interested readers are referred to the excellent review of Winicour on the characteristic approach [9] and the articles of Friedrich, Fraundiener and Husa on the conformal approach [10, 11, 12].

In the 3+1 approach one introduces a global time function t whose levels sets are the hypersurfaces defining the foliation. One then defines three main ingredients (see Fig. 1): 1) The three-dimensional metric γ_{ij} $(i,j=1,2,3)$ that measures distances within a given hypersurface, 2) the "lapse" function α that measures proper time between adjacent hypersurfaces, and 3) the "shift vector" β^i that measures the relative speed between observers moving along the normal direction to the hypersurfaces, and those keeping constant spatial coordinates. One then writes the four dimensional metric as

$$ds^2 = \left(-\alpha^2 + \beta_i\beta^i\right) dt^2 + 2\,\beta_i dt dx^i + \gamma_{ij}\, dx^i dx^j \,, \tag{1}$$

where $\beta_i := \gamma_{ij}\,\beta^j$.

Having split the metric, one introduces the "extrinsic curvature" tensor K_{ij} (also known as the "second fundamental form") that measures how the spatial hypersurfaces are immersed in spacetime. This tensor is given by the Lie derivative of γ_{ij} along the time lines, or in 3+1 language:

$$\partial_t \gamma_{ij} = -2\alpha K_{ij} + D_i\beta_j + D_j\beta_i \,, \tag{2}$$

with D_i the covariant derivative associated with γ_{ij}.

The next step is to decompose the Einstein equations. Doing this one finds that they naturally split in two groups. One group involves no time derivatives and represents constraints that must be satisfied at all times. The "Hamiltonian constraint" is given by

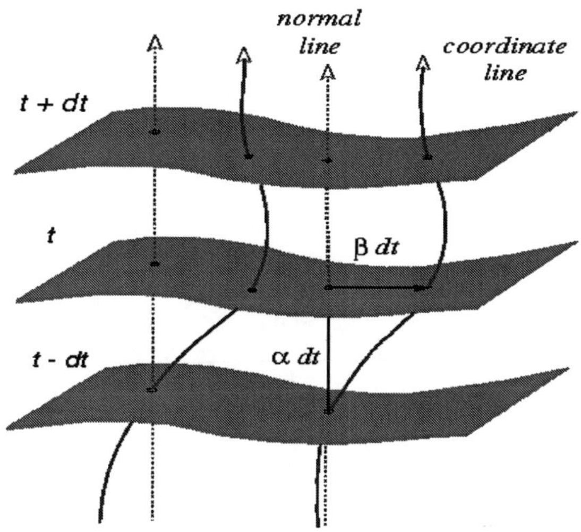

FIGURE 1. 3+1 foliation of spacetime.

(in units such that $G = c = 1$)

$$R + (\mathrm{tr}\,\mathrm{K})^2 - K_{ij}K^{ij} = 16\pi\rho\,, \tag{3}$$

with R the scalar curvature of the spatial geometry, $\mathrm{tr}\,\mathrm{K} \equiv \gamma^{ij}\mathrm{K}_{ij}$ the trace of the extrinsic curvature and ρ the energy density of matter measured by the normal observers. The "momentum" constraints take the form

$$D_j\left[K^{ij} - \gamma^{ij}\mathrm{tr}\,\mathrm{K}\right] = 8\pi j^i\,, \tag{4}$$

with j^i the momentum flux of matter measured by the normal observers. The existence of the constraints implies, in particular, that one is not free to specify the 12 dynamical quantities $\{\gamma_{ij}, K_{ij}\}$ as arbitrary initial conditions. The remaining 6 Einstein equations contain the dynamics of the system:

$$\begin{aligned} \partial_t K_{ij} &= \beta^a D_a K_{ij} + K_{ia} D_j \beta^a + K_{ja} D_i \beta^a - D_i D_j \alpha \\ &+ \alpha\left[R_{ij} - 2K_{ia}K^a_j + K_{ij}\,\mathrm{tr}\,\mathrm{K}\right] + 4\pi\alpha\left[\gamma_{ij}\left(\mathrm{tr}\,\mathrm{S} - \rho\right) - 2S_{ij}\right]\,, \end{aligned} \tag{5}$$

with S_{ij} the stress tensor of matter.

It is important to mention that the Bianchi identities imply that the evolution equations preserve the constraints, that is, if they are satisfied initially they will remain satisfied at subsequent times. The equations just described are know as the Arnowitt-Deser-Misner equations, or ADM for short. They represent the starting point of practically all of 3+1 numerical relativity. The reader interested in seeing how these equations are derived is referred to the original ADM article [13] or the classic review by York [14].

The ADM evolution equations introduced above are in fact non-unique. One can easily add arbitrary multiples of the constraints to them, *i.e.* multiples of zero, without affecting their physical solutions. Of course, solutions that do not satisfy the constraints will certainly change. Constraint violating solutions, though not of interest physically, are inevitable in numerical simulations where the constraints are never satisfied exactly. The non-uniqueness of the evolution equations is well known: The original equations of ADM [13] differ from those of York [14] precisely by the addition of a multiple of the Hamiltonian constraint.

When considering different formulations of the evolution equations one has to worry about their well-posedness, by which one understands that solutions exist and are stable with respect to small changes in the initial data. One usually looks for symmetric or strongly hyperbolic systems of equations as such systems are well posed under very general conditions [15]. In general relativity, the problem of well-posedness was studied by Choquet-Bruhat in the 1950's [8]. By the mid 1980's a number of hyperbolic reductions were known [16, 17, 18, 19] (an review of the subject can be found in [20]). In the early 1990's, Bona and Masso studied hyperbolic formulations for numerical relativity [21, 22, 23], and this effort was later followed by a number of people [24, 25, 26, 27]. However, a generalized interest of the numerical community in hyperbolicity had to wait until 1999, when Baumgarte and Shapiro [28] showed that a reformulation of the ADM equations proposed by Nakamura, Oohara and Kojima [29], and Shibata and Nakamura [30], had far superior numerical stability properties than ADM. This empirical observation was put on firmer ground in [31, 32, 33], and today it is understood that ADM is only weakly hyperbolic and thus not well posed [34], whereas the new formulation, which is now known as BSSN, is strongly hyperbolic [32, 33]. The search for more general hyperbolic reductions of the Einstein evolution equations has continued, and today many such formulations exist [35, 36].

3. GAUGE

The Einstein equations provide us with evolution equations for the spatial metric γ_{ij} and extrinsic curvature K_{ij}. However, they say nothing about the evolution of the lapse α and shift β^i. This is as it should be, as the lapse and shift represent our freedom in choosing the coordinate system, *i.e.* they are "gauge" functions. Choosing a good gauge is crucial when one solves the Einstein equations numerically, and one must choose carefully if one wants to avoid coordinate (and physical) singularities and to cover the interesting regions of spacetime. The lapse and shift must be chosen dynamically as functions of the evolving geometry, that is, we choose the coordinates as we go.

3.1. Slicing conditions

Consider first the choice of the lapse α. This choice determines the proper time between one hypersurface and the next, that is, it fixes the spacetime foliation. Because of this the choice of lapse is called a "slicing condition".

There is an infinite number of ways in which one can choose the lapse function, but typically the different choices can be classified in the following way: 1) Prescribed slicings, where the lapse is specified as an *a priori* known function of space and time, 2) algebraic slicings, where the lapse itself or its time derivative is specified as some function of the geometric variables (metric and extrinsic curvature) at each hyper-surface, 3) elliptic slicings, where the lapse is obtained by solving an elliptic differential equation that typically enforces some geometric condition on the spatial hyper-surfaces.

The most obvious choice one can make is to ask for coordinate time and proper time to coincide, which corresponds to $\alpha = 1$. In this case one can show that observers moving in the normal direction are in free fall, so this choice is therefore know as "geodesic" slicing. Geodesic slicing is in fact a very bad choice. This can be understood if one thinks for a moment about the motion of freely falling particles in a non-uniform and dynamic gravitational field: the particles will accelerate in different directions and will very easily collide with one another. But our coordinates are tied to these particles, so when the particles collide our coordinate system breaks down.

A better slicing condition should stop the normal observers from approaching each other by, for example, asking for the volume elements associated with them to remain fixed. From the ADM equations one finds that this implies $\mathrm{trK} = \partial_t \mathrm{trK} = 0$, which in turn means that the lapse must satisfy the following elliptic equation:

$$D^2 \alpha = \alpha K_{ij} K^{ij} \,, \tag{6}$$

with D^2 the spatial Laplace operator. This condition is know as "maximal" slicing [37, 38] and is very robust, producing smooth evolutions whenever it can be applied. It also has the property of being "singularity avoiding", which means that when a singularity is approached the slicing reacts by making the lapse collapse to zero in that region, thus halting the evolution there while allowing it to continue in regions away from the singularity (see Fig. 2). Singularity avoiding slicings, however, develop a problem known as "slice stretching", where the radial metric components grow rapidly as the slices are stretched between the frozen inner regions and the evolving outer ones.

The drive towards hyperbolicity and computational efficiency has made algebraic slicing conditions increasingly popular in recent years. In particular, the Bona-Masso family of slicing conditions [23] allows one to construct strongly hyperbolic systems of equations that include the gauge. This condition has the form of the following evolution equation for the lapse

$$\partial_t \alpha - \beta^i \partial_i \alpha = -\alpha^2 f(\alpha) K \,, \tag{7}$$

where $f(\alpha)$ is a positive, but otherwise arbitrary function of the lapse. The Bona-Masso condition includes some well known particular cases. For $f = 1$ the condition reduces to a very old choice know as "harmonic slicing", for which the time coordinate t can be shown to obey a wave equation: $\Box t = 0$. This condition also implies that $\alpha = h(x^i)\gamma^{1/2}$, with γ the determinant of the 3-metric (*i.e.* the volume element), and $h(x^i)$ a time independent function. This means that for harmonic slicing the "densitized" lapse $q := \alpha\gamma^{-1/2}$ does not evolve in time, which would seem to imply that perhaps this variable is more important than the lapse itself [39].

Another important sub-family of the Bona-Masso condition is known as "1+log" slicing, and is obtained when one chooses $f(\alpha) = N/\alpha$, with N a positive constant. The name comes from the fact that in this case one finds that $\alpha = h(x^i) + \ln(\gamma^{N/2})$.

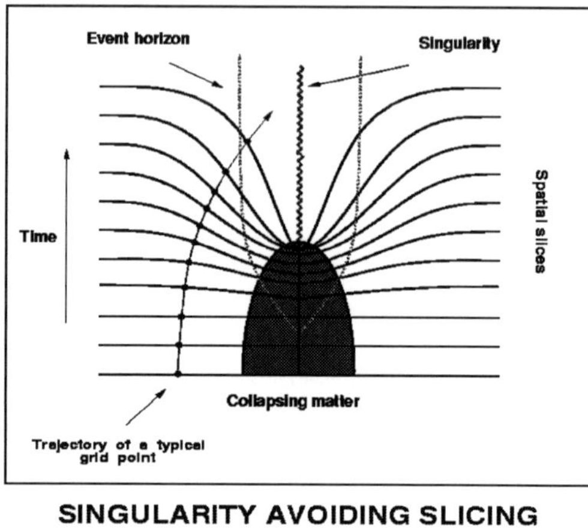

FIGURE 2. Singularity avoiding slicing condition.

This family mimics maximal slicing in its singularity avoiding properties. In practice, the choice $N = 2$ has been found to be particularly well behaved [40], and is commonly used in numerical simulations of both black hole and neutron star spacetimes.

3.2. Shift conditions

Considerably less in known about shift conditions. The main reason is that taking a zero shift vector works well in many cases. Still, there are some important proposals about shift choices that can be used in different circumstances. Moreover, is has become clear that 3D simulations of rotating and/or orbiting compact objects require non-trivial shift vectors to avoid large shears in the metric and allow long lived simulations.

The classic shift condition is known as "minimal distortion". It was suggested by Smarr and York in 1978 [37] and is still in use today in one way or another. This condition was designed to minimize the distortion in the volume elements over the hypersurface. One starts by considering the distortion tensor $\Sigma_{ij} = \frac{1}{2}\gamma^{1/2}\partial_t\tilde{\gamma}_{ij}$, where $\tilde{\gamma}_{ij} = \gamma^{-1/3}\gamma_{ij}$ is the conformal metric with unit determinant. The minimal distortion condition then asks for the integral of the square of the distortion tensor $\Sigma_{ij}\Sigma^{ij}$ to be minimized over a given hypersurface. The variation principle then yields:

$$D^j\Sigma_{ij} = 0 \qquad \rightarrow \qquad D^j\left(\hat{L}\beta\right)_{ij} = 2\,D^j\left[\alpha\left(K_{ij} - \frac{1}{3}\gamma_{ij}\mathbf{tr}K\right)\right], \tag{8}$$

where $\hat{L}$ is a differential operator defined as:

$$(\hat{L}W)_{ij} = \tilde{D}_i W_j + \tilde{D}_j W_i - \frac{2}{3}\tilde{\gamma}_{ij}\tilde{D}_k W^k, \tag{9}$$

with $\tilde{D}_i$ the covariant derivative associated with the conformal metric. The minimal distortion condition given by Eq. (8) is a system of three coupled second order elliptic differential equations for the shift vector components. Because of this, it is difficult to solve in 3D (though not impossible), and is therefore not often used in practice.

Shift conditions related to minimal distortion, but much easier to solve, have been proposed in the last few years. The main development has been the introduction of hyperbolic shift conditions associated with the BSSN formulation. This formulation introduces as auxiliary variables the contracted conformal Christoffel symbols $\tilde{\Gamma}^i := \gamma^{jk}\tilde{\Gamma}^i_{jk}$, and shift conditions have been designed that make the second time derivatives of the shift proportional to the time derivatives of the $\tilde{\Gamma}^i$: $\partial_t^2 \beta^i \sim \partial_t \tilde{\Gamma}^i$. It turns out that the principal part of the operator on the right hand side of this equation is identical with that of the minimal distortion condition, so these "Gamma driver" conditions can be seen as a hyperbolic version of minimal distortion. These type of shift conditions are extremely easy to implement numerically, in contrast with the difficulty involved in solving the minimal distortion elliptic equations. Moreover, they have proven to be extremely robust in practice, controlling the slice stretching associated with singularity avoiding slicings, and allowing very long term evolutions of black hole spacetimes [41]. Because of this, such shift conditions are now being used by a large fraction of existing 3D codes. Recently, shift conditions of this type have been shown to yield well-posed systems when considered together with Einstein's evolution equations [42, 43].

As a final comment, it is important to mention that when one is interested in evolving systems with angular momentum, one has to deal with the phenomenon know as "dragging of inertial frames", which means that observers close to the rotating object are dragged around it by the spacetime. The use of co-rotating coordinates using a shift that approaches a rigid rotation far away has been found to be crucial for evolving these type of systems. In particular, such co-rotating shifts have recently allowed long lived simulations of black hole binaries to be achieved [44, 45].

4. INITIAL DATA

Before starting a numerical simulation we need initial data that represents the physical situation we want to evolve. The existence of the constraint equations implies that one needs to solve a system of coupled elliptic equations to find such initial data.

The classic method for finding initial data is the York-Lichnerowicz conformal decomposition [46, 14], which starts by factoring out the spatial metric as:

$$\gamma_{ij} = \phi^4 \tilde{\gamma}_{ij} \,. \tag{10}$$

The extrinsic curvature is also decomposed by first separating out its trace free part

$$A_{ij} := K_{ij} - \frac{1}{3}\gamma_{ij}\, \mathrm{trK}, \tag{11}$$

and making a conformal transformation of A_{ij}

$$A_{ij} = \phi^{-10}\tilde{A}_{ij} \,. \tag{12}$$

One uses now the fact that any trace-free symmetric tensor can be split in two parts in the following way:

$$\tilde{A}_{ij} = \tilde{A}^*_{ij} + (\hat{L}W)_{ij}, \tag{13}$$

where $\tilde{A}^*_{ij}$ has zero divergence, W_i is a vector, and $\hat{L}$ is the operator defined in Eq. (9).

In order to find initial data one assumes that the conformal metric $\tilde{\gamma}_{ij}$, the trace of the extrinsic curvature tr$K$ and its divergence-less trace-free part $\tilde{A}^*_{ij}$ are given, and uses the constraints to solve for $\{\phi, W^i\}$. The Hamiltonian constraint gives an equation for ϕ:

$$8\tilde{D}^2\phi - \tilde{R}\phi + \phi^{-7}\left(\tilde{A}_{ij}\tilde{A}^{ij}\right) - \frac{2}{3}\phi^5(\mathrm{trK})^2 + 16\pi\tilde{\phi}^{-3}\rho = 0, \tag{14}$$

while the momentum constraint gives equations for W^i:

$$\tilde{D}^2 W^i - \frac{2}{3}\phi^6\tilde{D}^i\mathrm{trK} - 8\pi\tilde{\mathrm{j}}^{\mathrm{i}} = 0, \tag{15}$$

The problem can be considerably simplified if we choose $\mathrm{trK} = 0$ and also assume that the conformal metric is flat: $\tilde{\gamma}_{ij} = \delta_{ij}$. The constraints then reduce (in vacuum) to

$$8D^2_{\mathrm{flat}}\phi + \phi^{-7}\left(\tilde{A}_{ij}\tilde{A}^{ij}\right) = 0, \tag{16}$$

$$D^2_{\mathrm{flat}}W^i = 0, \tag{17}$$

with D^2_{flat} the ordinary flat space Laplacian. The second equation is linear and can be solved analytically in many cases.

As an example we can construct the vacuum solution corresponding to a conformally flat metric at a moment of time symmetry, that is, $\tilde{\gamma}_{ij} = \delta_{ij}$ and $K_{ij} = 0$. The above equations then reduce simply to

$$D^2_{\mathrm{flat}}\phi = 0, \tag{18}$$

which is just Laplace's equation. Boundary conditions for an asymptotically flat space imply that far away $\phi = 1$. The simplest non-trivial solution is then

$$\phi = 1 + k/r, \tag{19}$$

with k an arbitrary constant. The spatial metric then takes the final form

$$dl^2 = (1 + k/r)^4\left[dr^2 + r^2 d\Omega^2\right]. \tag{20}$$

Identifying $k = M/2$ we see that this is just Schwarzschild solution in isotropic coordinates. Now, Laplace's equation is linear, so we can superpose solutions to find new solutions. For example, we can take

$$\phi = 1 + \sum_{i=1}^{N}\frac{M_i}{2|r - r_i|}. \tag{21}$$

This represents N black holes momentarily at rest and is known as Brill-Lindquist data.

Brill-Lindquist data can be shown to correspond to a topological construction where each black hole is represented by a wormhole (Einstein-Rosen bridge) joining one universe with a series of disjoint multiple other universes. One can also construct data that joins two universes by multiple wormholes, corresponding to Misner type multiple black hole data. When the black holes have momentum, as in the case of orbiting black holes, exact solutions do not exist but numerical solutions are not difficult to find. In 1994 Cook found Misner type data for black holes in quasi-circular orbits using an effective potential method [47]. Later, in 1997 Brandt and Bruegmann showed that Brill-Lindquist type data could be obtained more easily by factoring out analytically the poles representing each black hole [48], and in 2000 this "puncture" method was used by Baumgarte to construct Brill-Lindquist type data for orbiting black holes [49].

More recently, an alternative method to construct initial data has been proposed by York [50]. This is known as the "thin sandwich" method, and gives initial data in the form of the 3-metric and its time derivative. The thin sandwich approach has been used to construct binary black hole initial data that is more "astrophysically motivated". The idea comes from the fact that black holes in orbit have an approximate helical symmetry as they rotate. The symmetry is not exact since the system emits gravitational waves, but for sufficiently separated holes this happens very slowly. One can then exploit the fact that an approximate helical Killing field exists to obtain black hole initial data in a co-rotating frame. Grandclement, Gourgoulhon and Bonazzola have used this approach to construct binary black hole initial data of the Misner type [51]. This data seems to agree closer with high order post-Newtonian predictions than the data based on the effective potential method. Tichy and Bruegmann have also applied this approach to puncture type data [52], and Yo *et al.* have done the same for Kerr-Schild type data [53].

5. FINITE DIFFERENCING

The are several different ways to solve partial differential equations numerically. The most popular methods are three: "finite differences" [54], "finite elements" [55] and "spectral methods". By far the most common method in numerical relativity has been finite differences, though spectral methods have been gaining popularity [56, 57, 58].

The basic idea of finite difference approximations is to substitute the continuous space-time for a discrete set of points. For simplicity one usually assumes that those points are equally spaced (see Fig. 3). One then substitutes the original differential equations for a system of algebraic equations by using Taylor series to compute approximations to derivatives that involve neighboring grid points. In this way one ends up with one algebraic equation per differential equation per grid point. For large numbers of grid points the final number of algebraic equations can be huge, particularly in 3D, which explains the need for powerful computers.

When dealing with finite difference approximations there are three main concerns: consistency, convergence and stability. Consistency means that an approximation reduces to the original differential equation when the grid intervals go to zero. This can usually be verified by inspection, with the possible exception of places where the coordinate system is singular. Consistency is a local condition, but in practice one is interested

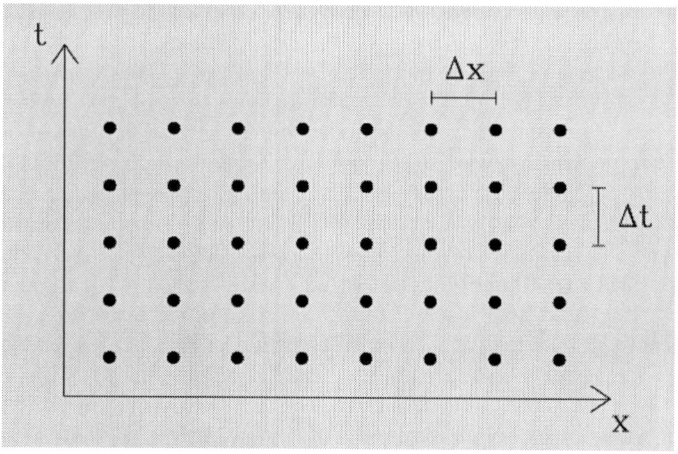

FIGURE 3. Discretization of space-time for finite differencing.

in a more global property. One looks for approximations that improve after a *finite time* as the grid is refined, a property known as "convergence". A consistent approximation is not necessarily convergent, as can be easily seen if we think that as the time interval goes to zero, we need an infinite number of steps to reach a finite time. This means that even if the error in each step is infinitesimal, the total integrated error could well be finite.

A third important property is that of stability. It turns out that finite difference approximations can easily produce solutions that grow exponentially even when the solution to the exact equations is well behaved. An unstable approximation is useless in practice, as the errors grow so fast that the true solution is quickly lost. A fundamental result in the theory of finite difference approximations is Lax theorem, that states that for a consistent approximation, stability and convergence are equivalent [54].

There is a simple condition that usually turns out to be necessary, and often also sufficient, for stability. This is known as the Courant-Friedrich-Lewy (CFL) condition, and states that the time interval Δt must be smaller than the spatial interval Δx. The condition has a clear interpretation: the numerical domain of dependence must be larger than the domain of dependence of the differential equation (see Fig. 4). If this wasn't the case, the numerical solution would not converge, as no matter the size of the grid there would always be information that is left out of the numerical domain of dependence. And as we have mentioned, if the approximation is not convergent it will be unstable.

6. ASTROPHYSICAL APPLICATIONS

On of the main applications of numerical relativity is the study of astrophysical systems that involve strong and highly dynamical gravitational fields, such as, for example, black holes and neutrons stars. In particular, binaries of compact objects have been identified as the most promising candidates for early detection by gravitational wave observatories.

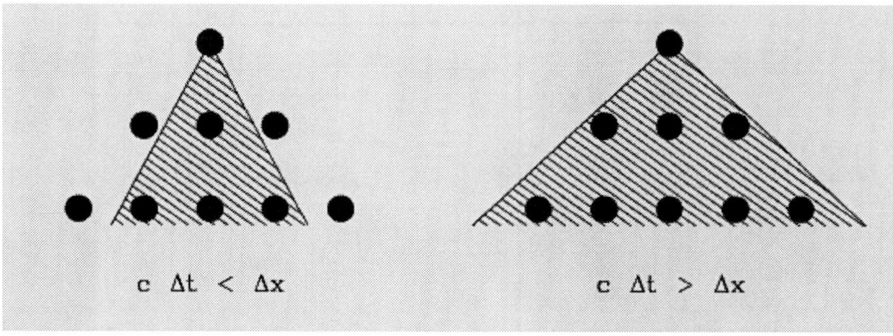

FIGURE 4. CFL stability condition. The situation on the left is stable, the one on the right is unstable.

As these systems orbit they emit gravitational radiation and thus loose energy. The orbits therefore continually shrink until the objects collide. In the final moments before a collision a burst of gravitational waves is expected that can carry as much as 5% of the total mass-energy of the system.

6.1. Black hole spacetimes

The simulation of black holes spacetimes has been a major theme in numerical relativity since its beginnings. The pioneering work of Hahn and Lindquist in the 1960's [1], as well as the work of Smarr and Eppley in the 1970's [2, 3], was precisely on simulations of the head-on collision of black holes. Black hole spacetimes are in vacuum, so the simulation of these systems has the advantage of not having to deal with hydrodynamics. On the other hand, black holes have a singularity, and dealing with this singularity is far from trivial. The traditional technique to evolve black holes has been the use of singularity avoiding slicing conditions such as maximal slicing, but this has the weakness of leading to slice stretching which eventually becomes a serious problem on its own. Figure 5, for example, shows the time evolution of the lapse function α as a function of the radial coordinate when evolving the initial data for a single Schwarzschild black hole found in the previous section, using maximal slicing. The figure clearly shows the lapse collapsing in the central regions to avoid the singularity. In Fig. 6 one can see the evolution of the radial metric γ_{rr} for the same system. Notice how the radial metric grows rapidly due to slice stretching.

An alternative approach is known as "singularity excision". The original idea has been attributed to Unruh by Thornburg [59, 60], and consists on placing a boundary inside the black hole, thus excising its interior from the computational domain, and also using a non-zero shift vector that keeps the horizon roughly in the same coordinate location during the evolution. Since no information can leave the interior of the black hole, excision should have no effect on the physics outside. Black hole excision was first attempted successfully by Seidel and Suen in spherical symmetry [61].

Issues related with singularity excision, gauge conditions and stability of the evolution

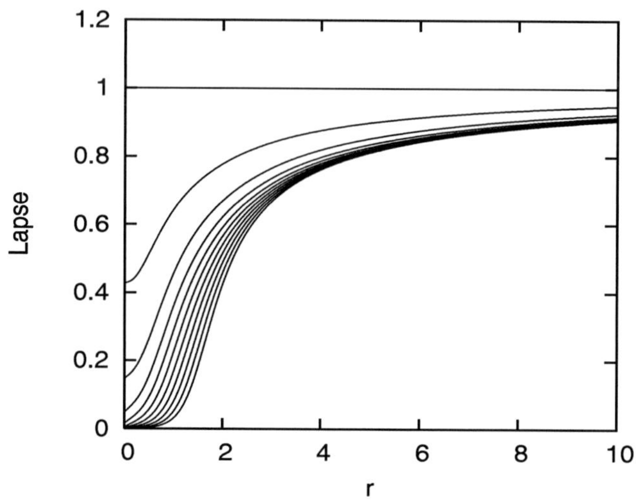

FIGURE 5. Collapse of the lapse for a simulation of a Schwarzschild black hole using maximal slicing.

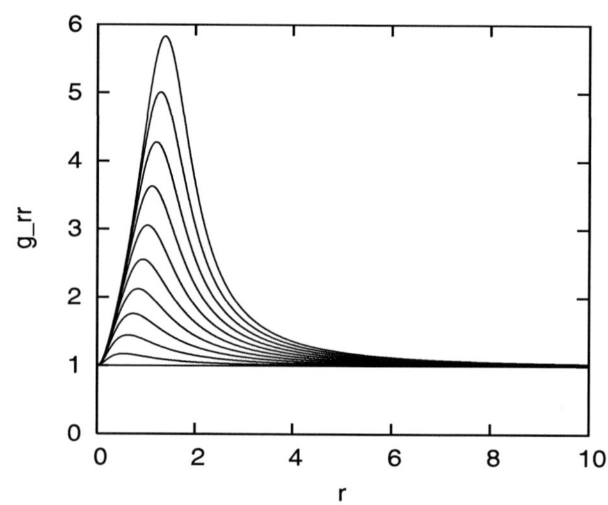

FIGURE 6. Slice stretching can be seen as a rapid growth of the radial metric component.

equations make the simulation of 3D black hole spacetimes and extremely hard problem. The 3D simulations of black holes done to this day include highly distorted single black holes [62], as well as more advanced versions of the head-on collision [63, 64]. In 2000 the first simulations of the grazing collision of two black holes, *i.e.* with a non-zero but small impact parameter, were done by Brandt *et al.* [65] and Alcubierre *et al.* [66]. Recently several important developments have taken place. On the one hand, excision techniques have improved to the point where one can now have a single black hole moving in an oscillating and even circular trajectory (controlled by gauge choices) in a stable way for essentially unlimited time [67]. Also, an approach based on using a co-rotating frame to keep the black holes fixed in coordinate space has finally led to the first ever simulation of a full orbit of two black holes by Bruegmann *et al.* [44]. And more recently, using the same approach but an independent code, Alcubierre *et al.* have studied sequences of black holes in quasi-circular orbits of the Cook/Baumgarte type [45]. Still, the simulation of the several orbits required in order to extract reliable gravitational waveforms remains an elusive goal.

6.2. Relativistic hydrodynamics and neutron stars

Apart from black hole spacetimes, there is also a lot of interest in systems with matter. In particular, there has been considerable work on fully 3D general relativistic hydrodynamical codes. At this time, several codes exist that are capable of simulating core collapse, rotating stars and binary star mergers. Among the groups with advanced 3D hydro codes one can mention the Tokyo group [68, 69, 70], the Illinois group [71, 72], the EU-Network group [73] and the JPL-WashU group [74].

Systems with matter have both advantages and disadvantages. If we start from a situation where no black holes are present we can safely ignore the problem of singularities, though if the system undergoes gravitational collapse the singularities will have to be dealt with. On the other hand, hydrodynamical systems can develop shock waves, and one has to be very careful about how these are handled numerically. Advanced numerical methods known as "high resolution shock capturing" schemes do a beautiful work of correctly handling shocks, but they are complex and hard to code.

Among the most important results so far related to hydro-dynamical systems one can mention the studies of collapse of super-massive stars [72] and the studies of collapsing rapidly rotating neutron stars [73].

7. CONCLUSIONS

Numerical relativity has come a long way in the last three decades. We are gaining a deeper understanding of the fundamental theoretical issues that underlie the field, such as well posedness of the evolution equations, gauge conditions, boundary conditions and initial data. At the same time, several advanced 3D codes exist that can handle both black holes and hydrodynamics. These codes have complex analysis tools incorporated: apparent and event horizon finding, horizon studies (mass, angular momentum), gravitational wave extraction, etc. In the next few years we hope to be able to simulate interesting astrophysical systems and extract gravitational waveforms that can be compared with the observations coming from the large interferometric detectors.

ACKNOWLEDGMENTS

It is a pleasure to thank the organizers of the "2nd mexican meeting on mathematical and experimental physics" for their invitation to give a talk and write these notes. This work was supported in part by DGAPA-UNAM through grants IN112401 and IN122002.

REFERENCES

1. Hahn, S. G., and Lindquist, R. W., *Ann. Phys.*, **29**, 304–331 (1964).
2. Eppley, K., *The Numerical Evolution of the Collision of Two Black Holes*, Ph.D. thesis, Princeton University, Princeton, New Jersey (1975).
3. Smarr, L., Čadež, A., DeWitt, B., and Eppley, K., *Phys. Rev. D*, **14**, 2443–2452 (1976).
4. LIGO - http://www.ligo.caltech.edu/.
5. VIRGO - http://www.virgo.infn.it/.
6. GEO600 - http://www.geo600.uni-hannover.de/.
7. TAMA - http://tamago.mtk.nao.ac.jp/.
8. Bruhat, Y., *Acta Mathematica*, **88**, 141–225 (1952).
9. Winicour, J., *Living Reviews in Relativity*, **1**, 5 (1998), [Online article].
10. Friedrich, H., *Lect. Notes Phys.*, **604**, 1–50 (2002).
11. Frauendiener, J., *Lect. Notes Phys.*, **604**, 261–282 (2002).
12. Husa, S., *Lect. Notes Phys.*, **604**, 239–260 (2002).
13. Arnowitt, R., Deser, S., and Misner, C. W., "The Dynamics of General Relativity," in *Gravitation: An Introduction to Current Research*, edited by L. Witten, John Wiley, New York, 1962, pp. 227–265.
14. York, J. W., "Kinematics and Dynamics of General Relativity," in *Sources of Gravitational Radiation*, edited by L. L. Smarr, Cambridge University Press, Cambridge, UK, 1979, pp. 83–126, ISBN 0-521-22778-X.
15. Kreiss, H.-O., and Lorenz, J., *Initial-Boundary Value Problems and the Navier-Stokes Equations*, Academic Press, New York, 1989.
16. Choquet-Bruhat, Y., in *Gravitation: An Introduction to Current Research*, edited by L. Witten, John Wiley, New York, 1962.
17. Choquet-Bruhat, Y., and York, J. W., Jr., "The Cauchy problem," in *General Relativity and Gravitation*, edited by A. Held, Plenum, New York, 1980, vol. 1, pp. 99–172.
18. Choquet-Bruhat, Y., and Ruggeri, T., *Comm. Math. Phys*, **89**, 269–275 (1983).
19. Friedrich, H., *Comm. Math. Phys.*, **100**, 525–543 (1985).
20. Friedrich, H., and Rendall, A. D., *Lect. Notes Phys.*, **540**, 127–224 (2000).
21. Bona, C., and Massó, J., *Phys. Rev. D*, **40**, 1022–1026 (1989).
22. Bona, C., and Massó, J., *Phys. Rev. Lett.*, **68**, 1097 (1992).
23. Bona, C., Massó, J., Seidel, E., and Stela, J., *Phys. Rev. Lett.*, **75**, 600–603 (1995).
24. Frittelli, S., and Reula, O., *Phys. Rev. Lett.*, **76**, 4667–4670 (1996).
25. Abrahams, A., Anderson, A., Choquet-Bruhat, Y., and York, J., *C.R. Acad. Sci. Paris*, **t. 323, Serie II**, 835–841 (1996).
26. Friedrich, H., *Class. Quantum Grav.*, **13**, 1451–1469 (1996).
27. Anderson, A., and York, J. W., *Phy. Rev. Lett.*, **82**, 4384–4387 (1999).
28. Baumgarte, T. W., and Shapiro, S. L., *Phys. Rev. D*, **59**, 024007 (1999).
29. Nakamura, T., Oohara, K., and Kojima, Y., *Progress of Theoretical Physics Supplement*, **90**, 1–218 (1987).
30. Shibata, M., and Nakamura, T., *Phys. Rev. D*, **52**, 5428 (1995).
31. Alcubierre, M., Allen, G., Brügmann, B., Seidel, E., and Suen, W.-M., *Phys. Rev. D*, **62**, 124011 (2000).
32. Sarbach, O., Calabrese, G., Pullin, J., and Tiglio, M., *Phys. Rev. D*, **66**, 064002 (2002).
33. Beyer, H., and Sarbach, O., *Phys. Rev.*, **D70**, 104004 (2004).
34. Frittelli, S., and Gómez, R., *J. Math. Phys.*, **41**, 5535–5549 (2000).
35. Kidder, L. E., Scheel, M. A., and Teukolsky, S. A., *Phys. Rev. D*, **64**, 064017 (2001).

36. Sarbach, O., and Tiglio, M., *Phys. Rev. D*, **66**, 064023 (2002).
37. Smarr, L., and York, J., *Phys. Rev.D*, **17**, 1945 (1978).
38. Smarr, L., and York, J. W., Jr., *Phys. Rev. D*, **17**, 2529–2552 (1978).
39. Anderson, A., and York, J., James W., *Phys. Rev. Lett.*, **81**, 1154–1157 (1998).
40. Alcubierre, M., *Class. Quantum Grav.*, **20**, 607–624 (2003).
41. Alcubierre, M., Brügmann, B., Diener, P., Koppitz, M., Pollney, D., Seidel, E., and Takahashi, R., *Phys. Rev. D*, **67**, 084023 (2003).
42. Lindblom, L., and Scheel, M. A., *Phys. Rev.*, **D67**, 124005 (2003).
43. Bona, C., and Palenzuela, C., *Phys. Rev.*, **D69**, 104003 (2004).
44. Brügmann, B., Tichy, W., and Jansen, N., *Phys. Rev. Lett.*, **92**, 211101 (2004).
45. Alcubierre, M., Brügmann, B., Diener, P., Guzmán, F. S., Hawke, I., Hawley, S., Herrmann, F., Koppitz, M., Pollney, D., Seidel, E., and Thornburg, J., *submitted to Physical Review D* (2004).
46. Lichnerowicz, A., *J. Math Pures et Appl.*, **23**, 37 (1944).
47. Cook, G. B., *Phys. Rev. D*, **50**, 5025–5032 (1994).
48. Brandt, S., and Brügmann, B., *Phys. Rev. Lett.*, **78**, 3606–3609 (1997).
49. Baumgarte, T. W., *Phys. Rev. D*, **62**, 024018 (2000).
50. York, J. W., *Phys. Rev. Lett.*, **82**, 1350–1353 (1999).
51. Grandclément, P., Gourgoulhon, E., and Bonazzola, S., *Phys. Rev. D*, **65**, 044021 (2002).
52. Tichy, W., and Brügmann, B., *Phys. Rev. D*, **69**, 024006 (2004).
53. Yo, H.-J., Cook, J. N., Shapiro, S. L., and Baumgarte, T. W. (2004).
54. Richtmyer, R. D., and Morton, K., *Difference Methods for Initial Value Problems*, Interscience Publishers, New York, 1967.
55. Mitchell, A. R., *The finite element method in partial differential equations*, J. Wiley and Sons, U.S.A., 1977.
56. Bonazzola, S., and Marck, J.-A., "Pseudo-Spectral Methods Applied to Gravitational Collapse," in *Frontiers in Numerical Relativity*, edited by C. Evans, L. Finn, and D. Hobill, Cambridge University Press, Cambridge, England, 1989, pp. 239–253.
57. Kidder, L. E., and Finn, L. S., *Phys. Rev. D*, **62**, 084026 (2000).
58. Kidder, L. E., Scheel, M. A., Teukolsky, S. A., Carlson, E. D., and Cook, G. B., *Phys. Rev. D*, **62**, 084032 (2000).
59. Thornburg, J., *Class. Quantum Grav.*, **4**, 1119–1131 (1987).
60. Thornburg, J., *Numerical Relativity in Black Hole Spacetimes*, Ph.D. thesis, University of British Columbia, Vancouver, British Columbia (1993).
61. Seidel, E., and Suen, W.-M., *Phys. Rev. Lett.*, **69**, 1845–1848 (1992).
62. Anninos, P., Bernstein, D., Hobill, D., Seidel, E., Smarr, L., and Towns, J., "Gravitational Waves from Oscillating Black Holes," in *Computational Astrophysics: Gas Dynamics and Particle Methods*, edited by W. Benz, J. Barnes, E. Muller, and M. Norman, Springer-Verlag, New York, 1997, in press.
63. Anninos, P., Hobill, D., Seidel, E., Smarr, L., and Suen, W.-M., *Phys. Rev. Lett.*, **71**, 2851–2854 (1993).
64. Anninos, P., Hobill, D., Seidel, E., Smarr, L., and Suen, W.-M., *Phys. Rev. D*, **52**, 2044–2058 (1995).
65. Brandt, S., Correll, R., Gómez, R., Huq, M. F., Laguna, P., Lehner, L., Marronetti, P., Matzner, R. A., Neilsen, D., Pullin, J., Schnetter, E., Shoemaker, D., and Winicour, J., *Phys. Rev. Lett.*, **85**, 5496–5499 (2000).
66. Alcubierre, M., Benger, W., Brügmann, B., Lanfermann, G., Nerger, L., Seidel, E., and Takahashi, R., *Phys. Rev. Lett.*, **87**, 271103 (2001).
67. Shoemaker, D., Smith, K. L., Sperhake, U., Laguna, P., Schnetter, E., and Fiske, D., *Class. Quantum Grav.*, **20**, 3729–3744 (2003).
68. Shibata, M., *Phys. Rev.*, **D67**, 024033 (2003).
69. Shibata, M., *Astrophys. J.*, **595**, 992 (2003).
70. Shibata, M., Taniguchi, K., and Uryu, K., *Phys. Rev.*, **D68**, 084020 (2003).
71. Duez, M. D., Shapiro, S. L., and Yo, H.-J., *Phys. Rev. D*, **69**, 104016 (2004).
72. Duez, M. D., Liu, Y. T., Shapiro, S. L., and Stephens, B. C., *Phys. Rev. D*, **69**, 104030 (2004).
73. Baiotti, L., Hawke, I., Montero, P. J., Löffler, F., Rezzolla, L., Stergioulas, N., Font, J. A., and Seidel, E., *Phys. Rev. D* (2004), in press; gr-qc/0403029.
74. Miller, M., Gressman, P., and Suen, W.-M., *Phys. Rev. D*, **69**, 064026 (2004).

Three-dimensional group manifold reductions of gravity

Román Linares

Departamento de Física, Universidad Autónoma Metropolitana Iztapalapa, San Rafael Atlixco 186, c.p. 09340, México D.F., México.

Abstract. We review the three-dimensional group manifold reductions of pure Einstein gravity and we exhibit a new consistent group manifold reduction of gravity when the compactification group manifold is S^3. The new reduction leads to a lower dimensional theory whose gauge group is $SU(2)$.

1. INTRODUCTION

Since Kaluza's pioneering paper [1], dimensional reductions of theories such as gravity, supergravities and extended objects, have been subject to an intensive research activity. The current status of the dimensional reductions has been discussed in [2, 3] and there it has been pointed out that the dimensional reductions can be divided into two types: *group manifold reductions* and *coset reductions*.

In this contribution we shall discuss the *group manifold reductions* [4, 5] of pure Einstein gravity on three-dimensional Euclidean group manifolds (G_3). These reductions are based on the assumption that the parametrization for the *metric* is invariant under a three-dimensional *simply transitive* acting group of isometries in the internal space which is given by the left action $(G_3)_L$ of the group manifold G_3. In [6], Bianchi classified the three-dimensional Lie algebras and he discussed how the algebras can be realized as isometries on the three-dimensional manifolds. It turns out that in three-dimensions there do exist eleven inequivalent Lie algebras. After apply the group manifold reduction to the $(D+3)$-dimensional pure Einstein gravity one obtains a coupled lower dimensional theory of gravity, non-Abelian gauged scalars and Yang-Mills fields in which the Lie algebra used for the reduction becomes the algebra of the gauged group. There are several contexts in which the G_3 group manifold reductions play a relevant role. For instance, in the Bianchi cosmologies [7], which can be considered as G_3 group manifold reductions of four dimensional pure Einstein gravity to one space-time dimension. One example of these reductions that we want to stress is the so-called $S^3 = SU(2)$ group manifold reduction [2]. In this case, the gauge group of the lower-dimensional theory is the left action of the three-dimensional isometry group $(G_3)_L = SU(2)$. More recently the group manifold reductions have also appeared in the construction of some massive/gauged supergravities in diverse dimensions, for instance, in the construction of the eleven different 8-dimensional supergravities which are obtained by dimensional reduction of the 11-dimensional supergravity on the eleven different three-dimensional group manifolds [8, 9, 10]. Also they appear in the Chamseddine and Volkov's obtention

CP758, *Gravitation and Cosmology: 2nd Mexican Meeting on Mathematical and Experimental Physics*
edited by A. Macías, C. Läemmerzahl and D. Nuñez

of an $SU(2) \times SU(2)$-gauged $N = 4$, $d = 4$ supergravity [11] and in the Chamseddine and Sabra's obtention of an $SU(2)$-gauged $N = 2$, $d = 7$ supergravity [12] from $N = 1$, $d = 10$ supergravity in both cases.

In this paper we exhibit a *new consistent* way to perform the group manifold reduction when the compactification group manifold is S^3. The basic idea is the following, gravitational theories in the vielbein formulation have two local symmetries: the general coordinate symmetry and the local Lorentz symmetry. Dimensional reduction of Einstein gravity on T^3 gives origin to a lower-dimensional theory with a $GL(3,\mathfrak{R})/SO(3)$ five-dimensional scalar coset where $GL(3,\mathfrak{R})$ is a global symmetry and $SO(3)$ a local one. Using the fact that there is a homomorphism that maps $SU(2)$ into the maximal compact subgroup of $GL(3,\mathfrak{R})$, i.e. into $SO(3)$, this group is gauged by introducing it into the dimensional reduction through the Maurer-Cartan 1-forms associated to the general coordinate symmetry. This group manifold reduction leads to a lower dimensional theory with $SU(2)$ gauge group [2]. We shall call to this reduction the standard group manifold reduction. The additional ingredient in the new reduction we are introducing is to perform a special gauge transformation in the tangent space of the vielbein, dictated by the adjoint representation of the $SU(2)$ Lie algebra. By doing it, we obtain a new consistent group manifold reduction that leads to an action and equations of motion in the lower dimension that are exactly the same as the ones obtained by apply the standard group manifold reduction [2], and therefore, the gauge group of the resulting theory is also $SU(2)$. However, we shall show that the new group manifold reduction introduces two differences with respect to the standard one. The first difference is in the covariant derivative of the internal "triad". It turns out that the gauge group of the derivative is $SU(2) \times AdSU(2)$. The $AdSU(2)$ group comes from the introduction of the special gauge transformation in the tangent space. The second difference appears in the spin connection, it turns out that the linear adjoint group generates a new term in the expression of the component of the spin connection with two internal indices.

2. GENERAL COORDINATE TRANSFORMATIONS

In the following discussion we assume a $(D+3)$ split of the $(D+3)$ space-time coordinates $x^{\hat{\mu}} = (x^{\mu}, z^{\alpha})$ where $\mu = \{0,1,\ldots,D-4\}$ are the indices of the D-dimensional space-time and $\alpha = \{1,2,3\}$ are the indices of the internal coordinates. The corresponding flat indices of the tangent space are denoted by $\hat{a} = (a,m)$. The group indices are also denoted with the letters $m,n,\ldots$,. We work in the conventions of [10].

In the vielbein formalism, the $(D+3)$-dimensional Einstein-Hilbert action

$$S = \int d^{D+3}\hat{x}\, \hat{e}\, \hat{\mathscr{R}}(\hat{\omega}) , \tag{1}$$

is invariant under the general coordinate transformations

$$\delta_{\hat{\mathbf{K}}} \hat{e}_{\hat{\mu}}{}^{\hat{a}} = \mathscr{L}_{\hat{\mathbf{K}}} \hat{e}_{\hat{\mu}}{}^{\hat{a}} = \hat{K}^{\hat{\nu}}\, \partial_{\hat{\nu}} \hat{e}_{\hat{\mu}}{}^{\hat{a}} + \partial_{\hat{\mu}} \hat{K}^{\hat{\nu}}\, \hat{e}_{\hat{\nu}}{}^{\hat{a}} . \tag{2}$$

As usual, $\hat{e}$ is the determinant of the vielbein, $\hat{\mathscr{R}}$ the Ricci scalar, $\hat{\omega}$ the spin connection and $\mathscr{L}_{\hat{\mathbf{K}}}$ denotes the Lie derivative along the infinitesimal vector field parameters $\hat{\mathbf{K}}$.

As it has been pointed out in [5], the group manifold reduction is specified by choosing the internal coordinate dependence of the parameters $\hat{K}^{\hat{\mu}}(x,z)$. If they are taken as

$$\hat{K}^{\mu}(x,z) = K^{\mu}(x), \qquad \hat{K}^{\alpha}(x,z) = K^{m}(x)(U^{-1}(z))_{m}{}^{\alpha}, \tag{3}$$

where $U_{\alpha}{}^{m}(z)$ are $GL(3,\Re)$ matrices which can be interpreted as the components of the left invariant Maurer-Cartan 1-forms $\sigma^m \equiv dz^{\alpha} U_{\alpha}{}^{m}(z)$, an arbitrary three-dimensional Lie algebra can be extracted out of the group of general coordinate transformations in $(D+3)$-dimensions. The algebra of general coordinate transformations

$$[\delta_{\hat{K}_1}, \delta_{\hat{K}_2}] = \delta_{\hat{K}_3} \qquad \text{where} \qquad \hat{K}_3^{\hat{\mu}}(x,z) = 2\hat{K}_{[2}^{\hat{\nu}}(x,z)\,\partial_{\hat{\nu}}\hat{K}_{1]}^{\hat{\mu}}(x,z), \tag{4}$$

gives origin to three different possibilities in D-dimensions. First, the algebra of two space-time transformations with parameters $K_1^{\mu}(x)$ and $K_2^{\mu}(x)$ gives a new space-time transformation with parameter $K_3^{\mu}(x) = 2K_{[2}^{\nu}(x)\,\partial_{\nu}K_{1]}^{\mu}(x)$ indicating that the theory has general coordinate transformations in the D-dimensional space-time. Second, the commutator of a space-time transformation with parameter $K_1^{\mu}(x)$ and an internal transformation with parameter $K_2^{m}(x)$ gives a new internal transformation with parameter $K_3^{m}(x) = K_1^{\mu}(x)\partial_{\mu}K_2^{m}(x)$ which means that the parameters of an internal transformation are space-time scalars. Finally, the commutator of two internal transformations with parameters $K_1^{m}(x)$ and $K_2^{m}(x)$ produces a new internal transformation with parameter $K_3^{p}(x) = f_{mn}{}^{p}K_1^{m}(x)K_2^{n}(x)$ where

$$f_{mn}{}^{p} = -2(U^{-1}(z))_{m}{}^{\alpha}(U^{-1}(z))_{n}{}^{\beta}\,\partial_{[\alpha}U_{\beta]}{}^{p}(z), \tag{5}$$

are the structure constants of the three-dimensional Lie group G_3, whose Lie algebra g_3 is given by

$$[\mathbf{K}_m, \mathbf{K}_n] = f_{mn}{}^{p}\mathbf{K}_p, \tag{6}$$

and the $f_{mn}{}^{p}$'s satisfy the Jacobi identity $f_{[mn}{}^{q}f_{p]q}{}^{r} = 0$.

After apply the standard group manifold reduction [4, 5] the *simply transitive* three-dimensional Lie algebra (6) becomes the algebra of the gauged group in the lower-dimensional theory. It turns out that in three dimensions there do exist eleven different ways to choose the structure constants [6, 13]. By using the class of group manifolds known as class A, is possible to construct six different lower dimensional theories with gauge group $SO(3)$, $SO(2,1)$, $ISO(2)$, $ISO(1,1)$, $Heis_3$ and $U(1)^3$. Here $Heis_3$ denotes the three-dimensional Heisenberg group (with generators corresponding to position, momentum and identity). The theory with gauge group $U(1)^3$ is obtained from a reduction over a three-torus T^3 and is referred as the ungauged theory, since there are no fields that carry any of the $U(1)$ charges. All groups mentioned above are related to $SO(3)$ by group contraction and/or analytic continuation. By using another class of manifolds, called class B, five more lower dimensional theories can be constructed whose gauge groups can be seen as extensions of $ISO(2)$, $ISO(1,1)$, $Heis_3$ and $U(1)^3$. An important characteristic of these class B reductions is that they are performed on-shell, i.e. at the level of the equations of motion. Metrics that have either an $SO(3)$ or $SU(2)$ isometry group acting transitively on 3-surfaces are called Bianchi IX metrics.

In [14] it has been discussed that these two topologically different isometry groups of the Bianchi IX metrics, are related to the two different definitions of homogeneity. One definition of homogeneity is related in a natural way to the isometry group $SO(3)$ (the maximal compact subgroup of $SL(3,\Re)$) which is topologically RP^3 i.e. S^3 with antipodal points identified. This definition of homogeneity involves only the Lie algebra (6). The second definition of homogeneity is related to the isometry group $SU(2)$ which is topologically S^3. This second definition of homogeneity involves besides the Lie algebra (6) another Lie algebra that generates physical rotations of the fields about the axes defined by the vielbein. Therefore, if we want to consider S^3 as the internal group manifold, the dimensional reduction must know about the extra three-dimensional Lie algebra. We indicate in the next section how this goal is achieved.

3. PARAMETRIZATION OF THE VIELBEIN

The first step in the dimensional reduction procedure is to make a suitable parametrization of the group-invariant vielbein in terms of lower-dimensional fields. The parametrization includes internal coordinates dependence dictated by the symmetries of the theory. In the standard group manifold reduction, the general coordinates symmetry is the only symmetry used in the parametrization. However for the new group manifold reduction there does exists an additional symmetry introduced as a rotation in the local tangent space. The parametrization for the latter possibility is

$$\hat{e}_{\hat{\mu}}{}^{\hat{a}}(x,z) = \begin{pmatrix} e^{c_1\varphi(x)} e_\mu{}^a(x) & e^{c_2\varphi(x)} A_\mu{}^\alpha(x,z)\, L_\alpha{}^p(x,z) \\ 0 & e^{c_2\varphi(x)} L_\alpha{}^p(x,z) \end{pmatrix}, \tag{7}$$

where c_1 and c_2 are constants whose values are $c_1 = -\sqrt{3}/\sqrt{2(D+1)(D-2)}$ and $c_2 = -c_1(D-2)/3$ [1]. The A_μ's are gauge fields and $L_\alpha{}^p(x,z)$ is a 3×3 matrix whose internal coordinate dependence is given by

$$\begin{aligned} A_\mu{}^\alpha(x,z) &= A_\mu{}^m(x)\,(U^{-1}(z))_m{}^\alpha\,, &(8)\\ L_\alpha{}^p(x,z) &= U_\alpha{}^m(z)\,L_m{}^n(x)\,\Lambda_n{}^p(z)\,. &(9)\end{aligned}$$

The novel ingredient in the parametrization (9) is the introduction of the *orthogonal* matrix $\Lambda(z)$ which is taken in the *adjoint representation* of the three-dimensional Lie algebra g_3 of the previous section [15, 16]. The property of orthogonality indicates that $\Lambda(z)$ is indeed a rotation in the internal tangent space. The matrix Λ satisfies the equation

$$(R_m)_n{}^p = (U^{-1}(z))_m{}^\alpha (\Lambda^{-1}(z))_n{}^q \partial_\alpha \Lambda_q{}^p(z)\,, \tag{10}$$

where the matrices R_m are the generators of $gl(3,\Re)$ and are given by the *adjoint representation* of the parameters of the internal transformations, $R_m = f_{mn}{}^p \mathbf{e}_p{}^n = ad_{\mathbf{K}}(\mathbf{K}_m)$

[1] The values of c_1 and c_2 ensure that the reduction of the Einstein-Hilbert action yields a pure Einstein-Hilbert term in D-dimensions, with no pre-factor involving the scalar φ, and that φ has a canonically normalized kinetic term in D-dimensions.

[16]. They satisfy the $SO(3)$ Lie algebra

$$[R_m, R_n] = f_{mn}{}^p R_p. \qquad (11)$$

The vielbein parametrization of the standard group manifold reduction differs from (9) in the matrix $\Lambda(z)$.

Using the vielbein parametrization (7)-(9) we can rewrite the $(D+3)$-dimensional interval in the way

$$ds^2 = e^{2c_1\varphi} g_{\mu\nu} dx^\mu dx^\nu - e^{2c_2\varphi} \mathscr{M}_{mn}(dx^\mu A_\mu{}^m + \sigma^m)(dx^\nu A_\nu{}^n + \sigma^n), \qquad (12)$$

where

$$\mathscr{M}_{mn}(x,z) \equiv -L_m{}^r(x,z) L_n{}^s(x,z)\, \eta_{rs}. \qquad (13)$$

Geometrically $L_m{}^n(x,z)$ describes the $GL(3,\Re)/SO(3)$ scalar coset of the internal space and can be interpreted as the internal "triad". It transforms under a global $GL(3,\Re)$ acting from the left and a local $SO(3)$ symmetry acting from the right. In contrast $\mathscr{M}_{mn}(x,z) = -L_m{}^p(x) L_n{}^q(x)\, \eta_{pq} = \mathscr{M}_{mn}(x)$ is the $SO(3)$ invariant metric of the internal manifold and it is parameterized by the same scalars.

4. THE D-DIMENSIONAL ACTION

Once discussed the general characteristics of the vielbein parametrization, we proceed to dimensional reduce the $(D+3)$-dimensional Einstein-Hilbert action.

As we have argued, it is important to perform the dimensional reduction in the vielbein formalism. The important quantities in this case are the components of the spin connection $\hat\omega_{\hat a \hat b}$. By using the parametrization (7)-(9), the $(D+3)$-dimensional spin connection is upon dimensional reduction

$$\begin{aligned}
\hat\omega_{ab} &= \omega_{ab} - 2c_1 e^{-c_1\varphi} \hat e_{[a}\partial_{b]}\varphi + \frac{1}{2} e^{(c_2-2c_1)\varphi} F_{ab}{}^m L_{mn} e^n, \qquad (14)\\
\hat\omega_{am} &= (\Lambda^{-1})_m{}^n \left[e^{-c_1\varphi} e^p \left(c_2 \partial_a \varphi \eta_{pn} + (L^{-1})_{(p}{}^q \mathscr{D}_a L_{q|n)} \right) + \frac{1}{2} e^{(c_2-2c_1)\varphi} F_{ab}{}^p L_{pn} \hat e^b \right],\\
\hat\omega_{mn} &= (\Lambda^{-1})_m{}^p (\Lambda^{-1})_n{}^q \left[-\hat e^a e^{-c_1\varphi} (L^{-1})_{[p}{}^r \mathscr{D}_a L_{r|q]} \right.\\
&\qquad \left. + e^r e^{-c_2\varphi} \left(\mathscr{F}_{r[pq]} - \frac{1}{2}\mathscr{F}_{pqr} + (\mathscr{R}_r)_{pq} \right) \right].
\end{aligned}$$

In these expressions $F^m = 2\partial A^m - f_{np}{}^m A^n A^p$ is the gauge vector field strength, the scalar functions $\mathscr{F}_{mnp}$ and $(\mathscr{R}_p)_{mn}$ are defined as $\mathscr{F}_{mnp} \equiv (L^{-1})_m{}^q (L^{-1})_n{}^r L_{sp} f_{qr}{}^s$, $(\mathscr{R}_p)_{mn} \equiv (L^{-1})_p{}^r (R_r)_{mn}$, whereas the covariant derivative of the scalar coset is given by

$$\mathscr{D}_\mu L_m{}^n = \partial_\mu L_m{}^n - A_\mu{}^p L_q{}^n f_{mp}{}^q + A_\mu{}^p L_m{}^q f_{qp}{}^n. \qquad (15)$$

Notice that as anticipated, the covariant derivative of the scalar coset reflects the gauging of the two Lie algebras under consideration. The second term corresponds to the standard

$SU(2)$ gauging of the internal coordinate symmetry whereas the third one corresponds to the $SO(3)$ gauging of the adjoint representation of the $SU(2)$ Lie algebra.

Using the reduced spin connection it turns out that the reduction of the $(D+3)$-dimensional action is

$$S = C\int d^D x\sqrt{|g|}\left[\mathscr{R} + \tfrac{1}{4}\mathrm{Tr}\left(\mathscr{D}\mathscr{M}\mathscr{M}^{-1}\right)^2 + \tfrac{1}{2}(\partial\varphi)^2 - \tfrac{1}{4}e^{-\frac{2c_1}{3}(D+1)\varphi}F^m\mathscr{M}_{mn}F^n - \mathscr{V}\right], \tag{16}$$

where $\mathscr{V}$ is the scalar potential

$$\mathscr{V} = \tfrac{1}{4}e^{\frac{2c_1}{3}(D+1)\varphi}\left[2\mathscr{M}^{mn}f_{mp}{}^q f_{nq}{}^p + \mathscr{M}^{mn}\mathscr{M}^{pq}\mathscr{M}_{rs}f_{mp}{}^r f_{nq}{}^s\right], \tag{17}$$

and C the group volume defined by $C(SU(2)) = \int d^3z \det(U_\alpha{}^m) = 16\pi^2$. In this expression of C we have used the property that $\det\Lambda = 1$. From the covariant derivative of the scalar coset (15), it is direct to compute the covariant derivative of the internal metric $\mathscr{M}$

$$\mathscr{D}\mathscr{M}_{mn} = \partial\mathscr{M}_{mn} + 2f_{q(m}{}^p A^q \mathscr{M}_{n)p}, \tag{18}$$

which reflects its $SO(3)$ invariant character.

In conclusion, the two differences produced by apply the new group manifold reduction with respect to the standard one are reflected in the term $(\mathscr{R}_p)_{mn}$ of the spin connection ω_{mn} and in the extra $SO(3)$ gauging of the covariant derivative of the scalar coset (15).

These differences are not manifest in the reduced action and therefore in the equations of motion neither. The reduced Lagrangian has the same functional form independently of the dimensional reduction used (either standard group manifold or the new group manifold reduction). The result is quite logic because the difference in the parametrization of the vielbein is a local rotation in the tangent space and the metric is invariant under such rotation.

5. CONCLUSIONS

In this paper we have introduced a new consistent group manifold reduction of Einstein pure gravity on S^3 and we have discussed its relation with the standard group manifold reduction. The new group manifold reduction considers explicitly two three-dimensional simply transitive groups of motions and we have showed that the resulting action and therefore the equations of motion of the resulting theory in the lower-dimension are exactly the same as in the standard group manifold reduction. However, we have discussed that there are two differences among the two different group manifold reductions. The first one is a new term in the covariant derivative of the internal triad and the second one a new term in the components of the spin connection with two internal indices. These differences are the main results of this contribution. It is possible to show that these differences have some important consequences [17]. For instance, because the equations of motion are the same in both cases, the domain wall type solutions coincide as well. However from the first order differential equations point of view, the solutions are divided into two disjoint sets, one of these sets is given by the metrics that solve the BGPP system [18] and the another one by the metrics that solve the Atiyah-Hitchin system [19]. If we reduce applying the standard group manifold reduction the domain walls that solve the BGPP system are self-dual in both the curvature and the spin connection whereas the metrics in the another set of solutions are self-dual only in the curvature. If instead we reduce applying the new group manifold reduction the conclusion is the opposite, i.e. now the domain walls that solve the Atiyah-Hitchin system are self-dual in both the curvature and the spin connection whereas metrics in the another set of solutions are self-dual only in the curvature. From the $(D+3)$-dimensional point of view the domain wall solutions are of the form $\Re^{D-2,1} \times M_4$. Therefore it turns out that by performing the new dimensional reduction the D-dimensional theory admits a self-dual (in the spin connection) domain wall solution which upon uplifting to $(D+3)$-dimensions leads to the Kaluza-Klein monopole [20, 21].

ACKNOWLEDGMENTS

We wish to thank the organizers of the Second Mexican Meeting in El Colegio Nacional in 2004 for the organization of the very nice meeting and the opportunity to give a presentation. We are also grateful to Eric Bergshoeff, Ulf Gran, Hugo Morales, Mikkel Nielsen, Tomás Ortín and Diederik Roest for interesting discussions at different stages of this work. This research was supported in part by the project CONACyT-40745-F.

REFERENCES

1. Kaluza, T., *Sitzungsber. Preuss. Akad. Wiss. Berlin (Math. Phys.)*, **K1**, 966 (1921).
2. Cvetic, M., Gibbons, G. W., Lu, H., and Pope, C. N., *Class. Quant. Grav.*, **20**, 5161 (2003).
3. Gibbons, G. W., and Pope, C. N., *Nucl. Phys.*, **B697**, 225 (2004).
4. DeWitt, B. S., *in Relativity, groups and topology, Les Houches 1963 (Gordon and Breach)*, p. 585 (1964).
5. Scherk, J., and Schwarz, J. H., *Nucl. Phys.*, **B153**, 61 (1979).
6. Bianchi, L., *Tomo XI*, p. 267 (1898), for an English translation see: On the three-dimensional spaces which admit a continuous group of motions, Gen. Rel. and Grav., **33**, 2171 (2001).
7. Taub, A. H., *Annals Math.*, **53**, 472 (1951).
8. Salam, A., and Sezgin, E., *Nucl. Phys.*, **B258**, 284 (1985).
9. Alonso, N., Bergshoeff, E., Gran, U., Linares, R., Ortín, T., and Roest, D., *JHEP*, **06**, 1 (2003).
10. Bergshoeff, E., Gran, U., Linares, R., Nielsen, M., Ortín, T., and Roest, D., *Class. Quant. Grav.*, **20**, 3997 (2003).
11. Chamseddine, A. H., and Volkov, M. S., *Phys. Rev. Lett.*, **79**, 3343 (1997).
12. Chamseddine, A. H., and Sabra, W. A., *Phys. Lett.*, **B476**, 415 (2000).
13. Estabrook, F. B., Wahlquist, H. D., and Behr, C. G., *J. Math. Phys.*, **9**, 497 (1968).
14. Graham, R., and Luckock, H., *Phys. Rev.*, **D49**, 4981 (1994).
15. Ellis, G. F. R., and MacCallum, M. A. H., *Commun. Math. Phys.*, **12**, 108 (1969).
16. Jantzen, R. T., *Commun. Math. Phys.*, **64**, 211 (1979).
17. Linares, R. (2004).
18. Belinskii, V., Gibbons, G., Page, D., and Pope, C., *Phys. Lett.*, **B76**, 433 (1978).
19. Atiyah, M. F., and Hitchin, N. J., *Phys. Lett.*, **A107**, 21 (1985).
20. Sorkin, R., *Phys. Rev. Lett.*, **51**, 87 (1983).
21. Gross, D. J., and Perry, M. J., *Nucl. Phys.*, **B226**, 29 (1983).

Mathematical Theories of Infinite and Finite Galactic Disks[1]

A. A. García and V. S. Manko

Departamento de Física, Centro de Investigación y de Estudios Avanzados del IPN, A.P. 14-740, 07000 México D.F., Mexico

Abstract. Theories of infinite and finite self–gravitating thin galactic disks recently developed by N. Sibgatullin *et al* within the classical Newtonian framework are reviewed. Special emphasis is made on showing the interrelation existing between the mathematical descriptions of the two types of disks.

1. INTRODUCTION

In the papers [1, 2] two efficient analytical methods have been developed allowing the reconstruction of the surface–density distributions in self–gravitating thin galactic disks from the known rotation curves (angular or circular velocity distributions of matter in the disks). The first method [1] deals with infinite disks possessing a central black hole, while the second one [2] treats the case of the disks of finite extension. These methods are based on a new representation of the gravitational potential Φ representing the disk and turn out to be mathematically interrelated. In Section 2 we shall consider the theory of infinite disks, and in Section 3 it will be shown how the results obtained in the infinite disks case can be used for the development of the theory of finite disks. In the last Section we will discuss some interesting applications of the methods to concrete rotation curves.

2. INFINITE DISKS WITH A CENTRAL BLACK HOLE

As was illustrated in [1], a direct use of the usual gravitational potential of a disk (involving the surface–density function $\sigma(\rho)$) in the condition determining balance of the gravitational and centrifugal forces for the disk matter leads to a singular integral equation of the non–Fredholm type from which the surface density cannot be readily obtained as the function of the angular–velocity distribution. This forces to look for different approaches. A possibility suggested in [1] explores the following representation

[1] We dedicate this contribution to the memory of Nail Sibgatullin, our colleague and dear friend.

CP758, *Gravitation and Cosmology: 2nd Mexican Meeting on Mathematical and Experimental Physics*
edited by A. Macías, C. Läemmerzahl and D. Nuñez

of the gravitational potential describing an infinite disk with a central point mass:

$$\Phi(\rho,z)=\frac{1}{2\pi^2}\int_0^{\pi}d\theta\int_1^{\infty}\ln[(s-\rho\cos\theta)^2+z^2]\alpha(s)ds+\frac{GM}{r},\quad r=\sqrt{\rho^2+z^2}. \tag{1}$$

Here $\alpha(s)$ is a real function which has the meaning of a source density, M is the mass of the central body, and G is Newton's gravitational constant.

This potential satisfies the Laplace equation everywhere except in the plane $z=0$, with the disk matter being distributed outside the circle $\rho<1$ (without loss of generality the inner radius of the disk can be set equal to 1). Besides, it possesses a remarkable property, namely,

$$\left.\frac{\partial\Phi}{\partial z}\right|_{z\to+0}=\begin{cases}0, & 0<\rho<1,\\ \dfrac{1}{\pi}\displaystyle\int_1^{\rho}\frac{\alpha(s_0)ds_0}{\sqrt{\rho^2-s_0^2}}, & \rho\geq 1,\end{cases} \tag{2}$$

that will permit us later on (see our formula (10) below) to relate the function $\alpha(s)$ to the genuine surface density $\sigma(\rho)$. The proof of (2) is given in [1] with the aid of the Sokhotsky–Plemelj formula for the Cauchy–type integrals.

On the other hand, the derivative of Φ with respect to ρ in the limit $z\to 0$ can be shown to have the form

$$\left.\frac{\partial\Phi}{\partial\rho}\right|_{z\to0}=-\frac{GM}{\rho^2}+\frac{1}{\pi\rho}\left(\int_1^{\infty}\alpha(s)ds-\int_{\rho}^{\infty}\frac{\alpha(s)sds}{\sqrt{s^2-\rho^2}}\right), \tag{3}$$

and its being substituted into the condition

$$\omega^2\rho+\left.\frac{\partial\phi}{\partial\rho}\right|_{z\to0}=0 \tag{4}$$

defining equilibrium of the gravitational and centrifugal forces, yields the equation

$$\omega^2\rho-\frac{GM}{\rho^2}=\frac{1}{\pi\rho}\int_{\rho}^{\infty}\frac{\alpha(s)sds}{\sqrt{s^2-\rho^2}} \tag{5}$$

(during the derivation of this formula one has to take into account that

$$\int_1^{\infty}\alpha(s)ds=0 \tag{6}$$

by virtue of the finiteness of the disk mass.)

After the introduction of the new variables $t = 1/s^2$ and $x = 1/\rho^2$, equation (5) can be solved with respect to α using the standard formula for the Abel–type equations (cf. Ref. [3], p. 318), the result being

$$\frac{\alpha(x)}{2x^{3/2}} = \frac{d}{dx}\int_0^x \frac{dt}{\sqrt{x-t}}\left(\frac{\omega^2}{t^{3/2}} - GM\right). \tag{7}$$

The substitution of (7) into (6) gives a formula for the mass of the central body

$$M = \frac{1}{2G}\int_0^1 \frac{\omega^2 dt}{t^{3/2}\sqrt{1-t}}, \tag{8}$$

which means that the function α can be expressed exclusively in terms of the known angular–velocity distribution:

$$\frac{\alpha(x)}{2x^{3/2}} = \frac{d}{dx}\int_0^x \frac{\omega^2 dt}{t^{3/2}\sqrt{x-t}} - \frac{1}{2\sqrt{x}}\int_0^1 \frac{\omega^2 dt}{t^{3/2}\sqrt{1-t}}. \tag{9}$$

The last step is to relate the surface density σ to the function α. This can be done via Poisson's equation, yielding

$$-4\pi G\sigma(\rho) = 2\left.\frac{\partial\phi}{\partial z}\right|_{z\to+0} = \frac{2}{\pi}\int_1^\rho \frac{\alpha(s_0)ds_0}{\sqrt{\rho^2 - s_0^2}}, \tag{10}$$

or in the variables t and s:

$$-4\pi G\sigma(x) = \frac{\sqrt{x}}{\pi}\int_x^1 \frac{\alpha(t)dt}{t\sqrt{t-x}}. \tag{11}$$

Therefore, in the case of infinite disks with a central black hole the reconstruction of the surface–density distribution from a known rotation curve reduces to carrying out the following two steps:

- calculation of the function α from the integral formula (9), and
- the subsequent estimation of σ with the aid of the formula (11).

3. THIN GALACTIC DISKS OF FINITE EXTENSION

It is clear that the inversion transformation applied to the potential (1), with M set equal to zero, must convert an infinite disk with the inner edge of radius 1 to a finite disk of radius 1. Therefore, in order to develop a general self–consistent theory of finite thin disks, one has to apply the well–known Thomson's transformation[2]

$$\Phi(\rho,z)=\sqrt{\tilde{\rho}^2+\tilde{z}^2}\,\tilde{\Phi}(\tilde{\rho},\tilde{z}) \tag{12}$$

for the solutions of Laplace's equation to the potential

$$\tilde{\Phi}(\tilde{\rho},\tilde{z})=\frac{1}{2\pi^2}\int\limits_0^{\pi}d\theta\int\limits_1^{\infty}\ln[(s-\tilde{\rho}\cos\theta)^2+\tilde{z}^2]\alpha(s)ds, \tag{13}$$

which is the potential (1), with $M=0$, written in the "old" cylindrical coordinates marked here with tildes, and where

$$\tilde{\rho}=\rho/r^2,\quad \tilde{z}=z/r^2,\quad r^2\equiv\rho^2+z^2. \tag{14}$$

Like in the previous Section, it is now necessary to relate, firstly, the function α to the angular–velocity distribution ω and, secondly, the surface density σ to α. This apparently should lead us to the integral relations which will differ from (9) and (11), and it should be pointed out that the Thomson transformation says nothing about how these relations can be derived directly from the formulas (9), (11).

Introducing the new variables

$$t=1/s^2,\quad x=\rho^2, \tag{15}$$

we obtain from the balance condition (4), after some manipulations, the equation

$$-\frac{1}{2\pi}\int\limits_0^x\frac{\alpha(t)dt}{t^{3/2}\sqrt{x-t}}=\int\limits_0^x(\omega^2t^{3/2})'_t\frac{dt}{\sqrt{t}}+A_0, \tag{16}$$

where the constant A_0 must be found from the condition that disk's mass is finite:

$$\int\limits_0^1\frac{\alpha(t)dt}{t^{3/2}}=0, \tag{17}$$

(see Ref. [2] for details).

The solution of (16) with respect to $\alpha(x)$ has the form [3]

$$\begin{aligned}\frac{\alpha(x)}{2x^{3/2}} &= -\frac{d}{dx}\int\limits_0^x\frac{dt}{\sqrt{x-t}}\left(A_0+\int\limits_0^t(\omega^2s^{3/2})'_s\frac{ds}{\sqrt{s}}\right)\\ &= -\frac{A_0}{\sqrt{x}}-\int\limits_0^x\frac{(\omega^2t^{3/2})'_tdt}{\sqrt{t}\sqrt{x-t}},\end{aligned} \tag{18}$$

---

[2] See, e.g., Ref. [4] where this transformation was used for obtaining a specific disk potential from the already known one.

and condition (17) yields for A_0 the following expression:

$$A_0 = -\frac{1}{2}\int_0^1 \frac{\omega^2(t)dt}{\sqrt{1-t}}. \tag{19}$$

To establish the relation between σ and α, we observe that

$$\left.\frac{\partial\Phi}{\partial z}\right|_{z\to+0} = -2\pi G\sigma(\rho) = \begin{cases} \dfrac{1}{\pi\rho^2}\displaystyle\int_1^{1/\rho} \frac{\alpha(s)ds}{\sqrt{1-\rho^2 s^2}}, & 0 \le \rho < 1, \\ 0, & \rho > 1, \end{cases} \tag{20}$$

whence the desired formula for σ written in the t and x variables is readily obtainable:

$$\sigma(x) = -\frac{1}{4\pi^2 Gx}\int_x^1 \frac{\alpha(t)dt}{t\sqrt{t-x}}. \tag{21}$$

Formulas (18), (19) and (21) provide one with a general recipe for the reconstruction of the surface–density distribution in the disks of finite extension from the known rotation curves.

To conclude this Section, it is worthwhile pointing out that the substitution of (18) and (19) into (21) leads to the formula for $\sigma(\rho)$ [2, 5] involving only one integration:

$$\begin{aligned} \sigma(\rho) &= \frac{1}{\pi^2 G}\left[\int_0^\rho \frac{d(\omega^2\rho'^2)}{d\rho'} F\left(\mu, \frac{\rho'}{\rho}\right)\frac{d\rho'}{\rho} + \int_\rho^R \frac{d(\omega^2\rho'^2)}{d\rho'} F\left(\mu', \frac{\rho}{\rho'}\right)\frac{d\rho'}{\rho'}\right], \\ \mu &\equiv \arcsin\sqrt{\frac{R^2-\rho^2}{R^2-\rho'^2}}, \quad \mu' \equiv \arcsin\sqrt{\frac{R^2-\rho'^2}{R^2-\rho^2}}, \end{aligned} \tag{22}$$

where $F(\phi,k) = \int_0^\phi (1-k^2\sin^2\alpha)^{-1/2}d\alpha$ is the incomplete elliptic integral of the first kind, and R is the radius of the disk (this can be easily introduced into the general formulas). Formula (22) generalizes the well–known Toomre's surface–density integral [6] obtained for infinite disks and resulting from (22) as the limit $R \to \infty$.

The mass of the disk is defined by the following simple formula:

$$M = \frac{2}{\pi G}\int_0^R \frac{\omega^2\rho^3 d\rho}{\sqrt{R^2-\rho^2}}. \tag{23}$$

4. SOME APPLICATIONS

The theories of infinite and finite disks were applied to a wide class of rotation curves [1, 2, 5]. The main results can be briefly summarized as follows.

(i) In the case of infinite disks with a central black hole and rotation curves of the polynomial form and Keplerian asymptotics

$$\omega^2\rho^3 = a_0 + \sum_{k=2}^{n} a_k/\rho^k, \tag{24}$$

where a_0 and a_k are some constant coefficients, the mass of the disk depends strongly on the mass of the central body; it turns out that the more terms are taken into account on the right of (24), the greater can be the disk mass compared to the mass of the black hole. The choice of the rotation curve in the form

$$\omega^2(\rho^2+b^2)^{3/2} = G(M+M_d), \quad b = \text{const} \tag{25}$$

leads to the following relation between the central–body mass M and the mass M_d of the disk:

$$M_d = bM, \tag{26}$$

which means that M_d in principle can exceed M arbitrarily many times.

(ii) The self–gravitating disks of finite extension corresponding to the superposition of the rotation curves[3]

$$\omega_n^2 = \frac{\omega_0^2 b^{2n+2}}{n!}\left(-\frac{\partial}{\partial b^2}\right)^{n-1}\frac{1}{b(b^2+\rho^2)^{3/2}}, \quad n = 1, 2, \ldots \tag{27}$$

were found to have the upper limit of the disk mass which imposes a natural restriction on the radius of the disk. The study of the flat rotation curve of the form [5]

$$v^2 = \begin{cases} \dfrac{v_0^2\rho^2}{\Delta^2}, & 0 \le \rho \le \Delta, \quad v_0 = \text{const}, \\ v_0^2, & \Delta < \rho \le R, \end{cases} \tag{28}$$

where Δ is the radius of the circle inside of which the rigid–body rotation takes place, and v_0 is the constant circular velocity at the flat region of the rotation curve, reveals that interestingly the mass of the galactic disk, with growing R, grows too due to the increase of the mass in the outer, not central, part of the disk. This clearly demonstrates that observations of only the central luminous part of a galaxy are not enough for determining the galactic mass.

---

[3] This family of rotation curves was suggested by Brandt [7] in the context of infinite disks possessing the Keplerian asymptotic behavior.

ACKNOWLEDGMENTS

This work has been supported by Projects 38495–E and 45946–E from CONACyT of Mexico.

REFERENCES

1. Sibgatullin, N. R., García, A. A., and Manko, V. S. Density distribution in massive galactic disks with a central black hole, *Astron. Lett.* **29** (2003) 70–76.
2. Sibgatullin, N. R., García, A. A., and Manko, V. S. Rotation curves and mass distributions in flat self–gravitating disks, *Astron. Lett.* **29** (2003) 825–830.
3. Sneddon I. N. The Use of Integral Transforms (McGraw–Hill Book Company: New York, 1972).
4. Lemos, J. P. S., and Letelier, P. S. Exact general relativistic thin disks around black holes, *Phys. Rev. D* **49** (1994) 5135–5143.
5. García, A. A., Manko, V. S., and Sibgatullin, N. R. New formulation of the theory of finite galactic disks (to be published in *Gen. Relat. Grav.*).
6. Toomre, A. On the distribution of matter within highly flattened galaxies *Astrophys. J.* **138** (1963) 385–392.
7. Brandt, J. C. On the distribution of mass in galaxies. I. The large–scale structure of ordinary spirals with applications to M31, *Astrophys. J.* **131** (1960) 293–303.

Generalized Bertotti-Robinson solution to the Einstein–Heisenberg–Euler theory

Alfredo Macías and Enrique Lozano

Departamento de Física, Universidad Autónoma Metropolitana–Iztapalapa, Apartado Postal 55–534, C.P. 09340, México, D.F., México.

Abstract. In this paper we present a Bertotti–Robinson–like analytical solution to the Einstein–Heisenberg–Euler system of field equations, endowed with cosmological constant. The solution is type D in Petrov's classification, possesses constant curvature and the electromagnetic field is static and homogeneous throughout the space–time. For vanishing Heisenberg–Euler non–linearities of the electromagnetic field the standard Bertotti–Robinson results are recovered.

1. INTRODUCTION

The interest in non–linear electrodynamics began 1912 when Mie [1] put forward the first model for non–linear electrodynamics. Between 1932 and 1935 Born and Infeld [2] proposed their non–linear theory, which represents a classical generalization of the Maxwell–Lorentz theory for accommodating stable solutions for the description of electrons. Because of the non–linearity, the field of a point charge turns out to be finite at $r = 0$, in contrast to the well known $1/r^2$ singularity of the Coulomb field in Maxwell–Lorentz electrodynamics. Moreover, the characteristic surface, the light cone, depends on the field strength, and the superposition principle for the electromagnetic field does not hold any longer. Black hole solutions to the Born–Infeld non–linear electrodynamics have been found by Salazar et al. [3]. Then, Plebański [4] postulated a more general non–linear electrodynamics, which contains the Born–Infeld theory as special case. Moreover, parity violating terms could emerge in Plebański non–linear electrodynamics [5]. A regular black hole solution to this theory has been obtained by Ayón et al. [6]. Additionally, Bretón [7] studied the trajectories of test particles in a geometry that is the Born–Infeld non–linear electromagnetic generalization of the Reissner–Nordstrom solution.

Quantum electrodynamical vacuum corrections to the Maxwell–Lorentz theory can be accounted for by an effective non–linear theory derived by Heisenberg and Euler [8, 9]. The vacuum is treated as a specific type of medium, the polarizability and magnetizability properties of which are determined by the clouds of virtual charges surrounding the real currents and charges [10]. Recently, Bordin et al. [11] proposed a possible direct measurement of the Heisenberg–Euler effect. This theory is a valid physical theory [12].

Nowadays, there exist a revival of interest in non–linear electrodynamics since the effective theory arising from superstrings is an electrodynamics of the Born–Infeld type [13, 14, 15, 16]. Besides, much attention has been deserved to the interpretation of the

CP758, *Gravitation and Cosmology: 2nd Mexican Meeting on Mathematical and Experimental Physics*
edited by A. Macías, C. Läemmerzahl and D. Nuñez

solutions to the Born–Infeld equations as states of D–branes [17].

In this paper we consider the Einstein–Heisenberg–Euler system of field equations for a static gravitational field and homogeneous electromagnetic field. The resulting solution is a non–linear electromagnetic generalization of the well known type D Bertotti–Robinson space–times [18, 19, 20, 21].

The plan of the paper is as follows: In Sec.2 the action and the field equations of the Einstein–Heisenberg–Euler system are briefly discussed. In Sec.3 the Einstein–Maxwell Bertotti–Robinson solution is revisited. In Sec.4 the generalized Bertotti–Robinson solution is presented. In Sec.5 the canonic form of the metric is displayed. In Sec.6 the results are discussed. In Sec.7 the main conclusion of this work is presented. In Sec.8 we summarize the results.

2. EINSTEIN–HEISENBERG–EULER FIELD EQUATIONS

The Einstein–Heisenberg–Euler action with cosmological constant reads

$$S=\frac{1}{16\pi}\int_{M_4} d^4x\sqrt{-g}(R-P+aP^2+bQ^2-2\Lambda), \tag{1}$$

with $a=\alpha^2/45m^4$ and $b=7\alpha^2/180m^4$, where α is the fine structure constant and m is the electron mass. We work in the units where the gravitational constant G and the velocity of light in vacuum c are equal to one. R is the Ricci scalar, Λ is the cosmological constant. $P\equiv F^{\mu\nu}F_{\mu\nu}$ and $Q\equiv\frac{1}{2}\varepsilon_{\mu\nu\sigma\rho}F^{\mu\nu}F^{\sigma\rho}$ are the two electromagnetic invariants. $\varepsilon_{\mu\nu\sigma\rho}$ is the completely antisymmetric tensor that satisfies $\varepsilon_{\mu\nu\sigma\rho}\varepsilon^{\mu\nu\sigma\rho}=-4!$.

In order to obtain the field equations we perform first the variation of the action (1) with respect to the metric $g_{\mu\nu}$. Hence, the Einstein equations are the following

$$G^{\mu}{}_{\nu}=T^{\mu}{}_{\nu}+\Lambda\delta^{\mu}{}_{\nu}, \tag{2}$$

with $T^{\mu}{}_{\nu}$ the energy–momentum tensor for the Heisenberg–Euler non–linear electromagnetic field given by

$$T^{\mu}{}_{\nu}=F^{\mu}{}_{\rho}F_{\nu}{}^{\rho}(1-2aP)-4bQ\,{}^{*}F^{\mu}{}_{\rho}F_{\nu}{}^{\rho}+\frac{1}{4}\delta^{\mu}{}_{\nu}(-P+aP^2+bQ^2)\,. \tag{3}$$

Setting $a=b=0$, Eq. (3) reduces to the standard linear Maxwell energy–momentum tensor. ${}^{*}F_{\mu\nu}$ is the dual of the Faraday tensor $F_{\mu\nu}=A_{\mu;\nu}-A_{\nu;\mu}$, and it is defined as usual ${}^{*}F_{\mu\nu}=\frac{1}{2}\varepsilon_{\mu\nu\sigma\rho}F^{\sigma\rho}$. Therefore, from (2) it is straightforward to obtain

$$R=-4\Lambda-T\,, \tag{4}$$

where T is the trace of the energy–momentum tensor (3). By means of (4) one can write the traceless Ricci tensor $S_{\mu\nu}$ in terms of the Ricci scalar and the trace of the non–linear energy–momentum tensor as

$$S_{\mu\nu}=R_{\mu\nu}-\frac{1}{4}g_{\mu\nu}R=T_{\mu\nu}-\frac{1}{4}g_{\mu\nu}T\,. \tag{5}$$

Performing the variation with respect to the electromagnetic potential A_μ we obtain the following equation for the Heisenberg–Euler non–linear electromagnetic field

$$\frac{\partial}{\partial x^\mu}\left[\sqrt{-g}\left((1-2aP)F^{\mu\nu}-2bQ^*F^{\mu\nu}\right)\right]=0. \qquad (6)$$

Once again , for $a=b=0$, Eq. (6) reduces to the standard Maxwell equations.

3. BERTOTTI–ROBINSON METRIC

The most general metric type D in Petrov classification can be written as [20, 21]

$$g=\frac{1}{H^2}\left[\frac{\Delta}{\Sigma}dx^2+\frac{\Sigma}{\Delta}(d\tau+Nd\sigma)^2+\frac{\Delta}{\Xi}dy^2-\frac{\Xi}{\Delta}(d\tau+Md\sigma)^2\right], \qquad (7)$$

where $H=H(x,y)$, $\Sigma=\Sigma(x)$, $\Xi=\Xi(y)$, $N=N(y)$, $M=M(x)$ and $\Delta=M-N$. Our coordinates are $\{x^\mu\}=\{x,y,\tau,\sigma\}$, their range is from $-\infty$ to ∞. Since we are assuming stationarity for ∂_τ, an axisymmetry for ∂_σ, one has to impose the signature requirements $\Delta/\Sigma>0$ and $\Delta/\Xi>0$. Regularity and elementary flatness on the rotation axis restricts the range of the coordinate σ. The metric (7) contains the well known Plabański–Demiański classes of solutions [22], as well as their Kerr–Newman subclasses [20, 21].

The coframe corresponding to the metric (7) reads

$$e^1 = \frac{1}{H\sqrt{2}}\left[\sqrt{\frac{\Delta}{\Sigma}}dx+i\sqrt{\frac{\Sigma}{\Delta}}(d\tau+Nd\sigma)\right], \qquad (8)$$

$$e^2 = \frac{1}{H\sqrt{2}}\left[\sqrt{\frac{\Delta}{\Sigma}}dx-i\sqrt{\frac{\Sigma}{\Delta}}(d\tau+Nd\sigma)\right], \qquad (9)$$

$$e^3 = \frac{1}{H\sqrt{2}}\left[\sqrt{\frac{\Delta}{\Xi}}dy+\sqrt{\frac{\Xi}{\Delta}}(d\tau+Md\sigma)\right], \qquad (10)$$

$$e^4 = \frac{1}{H\sqrt{2}}\left[-\sqrt{\frac{\Delta}{\Xi}}dy+\sqrt{\frac{\Xi}{\Delta}}(d\tau+Md\sigma)\right]. \qquad (11)$$

We assume, the electromagnetic field strength is aligned in the e^3 and e^4 principal directions. Therefore, only F_{12} and F_{34} are non–vanishing. Thus the electromagnetic two–form acquires the simple form

$$\begin{aligned}\omega &= (F_{12}+F_{34})\left(e^1\wedge e^2-e^3\wedge e^4\right)\\ &= \frac{1}{H^2}(\mathscr{E}+i\mathscr{B})\left[(d\tau+Md\sigma)\wedge dy+i(d\tau+Nd\sigma)\wedge dx\right],\end{aligned} \qquad (12)$$

where $\mathscr{E}=F_{34}$, $i\mathscr{B}=F_{12}$.

We look for a family of static solutions with $N=-1/2$, $M=1/2$, i.e., $\Delta=1$. Thus, the principal null directions e^3 y e^4 are geodesic, shear free, and rotation free. For

$H = H(x)$, the principal null congruences are expansion free. For $H = H(y)$, with $H_y \neq 0$ the principal null congruences posses non–vanishing expansion.

Introducing the coordinates $\phi = \tau - \frac{1}{2}\sigma$ and $t = \tau + \frac{1}{2}\sigma$ the metric (7) becomes

$$g = \frac{1}{H^2}\left[\frac{dx^2}{\Sigma} + \Sigma d\phi^2 + \frac{dy^2}{\Xi} - \Xi dt^2\right]. \tag{13}$$

The simplest solution corresponds to a spacetime filled with a homogeneous electromagnetic field, in this case the metric (13) splits into two 2–surfaces of constant curvature. By setting $H = 1$ (one can always achieve this simplification by accomplishing scaling transformations), the field equations for the structure constants Σ and Ξ assume the form

$$\Sigma_{xx} + \Xi_{yy} + 4\Lambda = 0, \tag{14}$$

$$\Sigma_{xx} - \Xi_{yy} + 4(E^2 + B^2) = 0. \tag{15}$$

Thus, subtracting (14) and (15), and integrating one obtains

$$\Sigma = \sigma_0 + \sigma_1 x - (\Lambda + E^2 + B^2)x^2, \tag{16}$$

$$\Xi = \xi_0 + \xi_1 y - (\Lambda - E^2 - B^2)y^2, \tag{17}$$

where σ_i and ξ_i are constants. As mentioned above, the metric can be written in the form of two 2–surfaces of constant curvature:

$$g = g_1 + g_2 \tag{18}$$

$$g_1 = \frac{dx^2}{\Sigma} + \Sigma d\phi^2, \quad \Sigma = \sigma_0 + \sigma_1 x + \sigma_2 x^2, \tag{19}$$

$$g_2 = \frac{dy^2}{\Xi} - \Xi dt^2, \quad \Xi = \xi_0 + \xi_1 y + \xi_2 y^2. \tag{20}$$

For g_1 and g_2 the only non–vanishing component of the Riemann tensor are given by

$$R_{1212} = \frac{\Sigma_{xx}}{4} = \frac{1}{2}\sigma_2 = const., \quad R_{3434} = \frac{\Xi_{yy}}{4} = \frac{1}{2}\xi_2 = const., \tag{21}$$

respectively. The class of Bertotti–Robinson solutions is characterized by

$$\Psi_2 = -\frac{1}{3}\Lambda, \qquad R = -4\Lambda, \qquad F_{34} = \mathscr{E} = E, \quad \text{and} \quad F_{12} = \mathscr{B} = B, \tag{22}$$

with E and B constants.

If Λ vanishes the Bertotti–Robinson metric becomes conformally flat. For vanishing electromagnetic field one arrives at the de Sitter and anti–de Sitter spaces.

4. BERTOTTI–ROBINSON–LIKE SOLUTION TO THE EEH FIELD EQUATIONS

We consider the Bertotti–Robinson metric (13), written as follows

$$g = \left[\frac{dx^2}{\Sigma} + \Sigma d\phi^2 + \frac{dy^2}{\Xi} - \Xi dt^2\right]. \tag{23}$$

Our ansatz for the electromagnetic potential is the following

$$A_\mu = -Eydt + Bxd\phi, \tag{24}$$

with E and B constants, corresponding to the homogeneous electric and magnetic fields, respectively. Replacing (24) into (4) and (5), they become

$$\left[\frac{d^2}{dx^2}\Sigma(x)\right] + \left[\frac{d^2}{dy^2}\Xi(y)\right] + 4\Lambda - 8a\left(E^4 + B^4\right) - 76aB^2E^2 = 0, \tag{25}$$

$$\left[\frac{d^2}{dx^2}\Sigma(x)\right] - \left[\frac{d^2}{dy^2}\Xi(y)\right] + 4\left(E^2 + B^2\right) = 0, \tag{26}$$

where we use the fact that $b = \frac{7}{4}a$. Subtracting (25) from (26) and solving leads to

$$\Sigma(x) = \sigma_0 + \sigma_1 x - [\Lambda + E^2 + B^2 - 2a(B^4 + E^4) - 19aB^2E^2]x^2, \tag{27}$$

$$\Xi(y) = \xi_0 + \xi_1 y - [\Lambda - E^2 - B^2 - 2a(B^4 + E^4) - 19aE^2B^2]y^2, \tag{28}$$

where σ_i and ξ_i are constants.

5. CANONIC FORM OF THE METRIC

The Bertotti–Robinson metric can be split into two bidimensional surfaces with constant curvature, i.e.,

$$g = g_1 + g_2, \tag{29}$$

where g_1 and g_2 are defined as follows

$$g_1 = \frac{dx^2}{\Sigma} + \Sigma d\phi^2 \quad \text{with} \quad \Sigma = \sigma_0 + \sigma_1 x + \sigma_2 x^2, \tag{30}$$

$$g_2 = \frac{dy^2}{\Xi} - \Xi dt^2 \quad \text{with} \quad \Xi = \xi_0 + \xi_1 y + \xi_2 y^2. \tag{31}$$

For g_1 the only no–vanishing component of the Riemann tensor reads

$$R_{1212} = \frac{\Sigma_{xx}}{4} = \frac{\sigma_2}{2} = const. \tag{32}$$

Whereas for g_2 the only no–vanishing component of the Riemann tensor is given by

$$R_{3434} = \frac{\Xi_{yy}}{4} = \frac{\xi_2}{2} = const. \tag{33}$$

By means of shifting and scaling transformations, one arrives at the well known canonical form of the metric, in terms of more suitable coordinates it reads

$$\begin{aligned} \sigma_2 = -\alpha^2 : \; g_1 &= \frac{dx^2}{1-\alpha^2x^2} + (1-\alpha^2x^2)d\phi^2, \\ &= \frac{1}{\alpha^2}(d\theta^2 + \sin^2\theta d\phi^2), \quad \alpha x = \cos\theta; \end{aligned} \tag{34}$$

$$\sigma_2=0: \; g_1 = \frac{4}{\sigma_1^2}[(d\sqrt{\sigma_1 x})^2]+(\sqrt{\sigma_1 x})^2 d\phi^2 \,,$$
$$= \frac{4}{\sigma_1^2}(d\theta^2+\theta^2 d\phi^2)\,, \quad \sigma_1 x=\theta^2 \,; \tag{35}$$

$$\sigma_2=\alpha^2 : \begin{cases} \sigma_0-\frac{\sigma_1^2}{4\alpha^2}>0 \quad g_1=\frac{dx^2}{1+\alpha^2x^2}+(1+\alpha^2x^2)d\phi^2=\frac{1}{\alpha^2}\left(d\theta^2+\cosh^2\theta d\phi^2\right), \\ \alpha x=\sinh\theta\,; \\ \sigma_0-\frac{\sigma_1^2}{4\alpha^2}<0 \quad g_1=\frac{dx^2}{\alpha^2x^2-1}+(\alpha^2x^2-1)d\phi^2=\frac{1}{\alpha^2}\left(d\theta^2+\sinh^2\theta d\phi^2\right), \\ \alpha x=\cosh\theta\,. \end{cases} \tag{36}$$

The canonical expressions for the metric g_1 can be gathered in the form

$$g_1=\frac{4d\zeta d\bar{\zeta}}{(1+\varepsilon\alpha^2\zeta\bar{\zeta})^2}\,, \qquad \varepsilon=-1,0,1\,, \tag{37}$$

the bar denotes complex conjugation. For the different values of ε one has

$$\varepsilon = 1: \qquad \alpha\zeta=\tan\frac{\theta}{2}\exp(i\phi), \tag{38}$$
$$\varepsilon = 0(=\alpha^2): \qquad \sigma_1\zeta=\theta\exp(i\phi), \tag{39}$$
$$\varepsilon = -1: \qquad \alpha\zeta=\tanh\frac{\theta}{2}\exp(i\phi)\,. \tag{40}$$

Analogously, for the metric g_2 one has

$$\xi_2=-\beta^2: \; g_2 = \frac{dy^2}{1-\beta^2y^2}-(1-\beta^2y^2)dt^2,$$
$$= \frac{1}{\beta^2}(dz^2-\sin^2 z dt^2)\,, \quad \beta y=\cos z\,; \tag{41}$$

$$\xi_2=0: \quad g_2 = \frac{4}{\xi_1^2}[(d\sqrt{\xi_1 y})^2]-(\sqrt{\xi_1 y})^2 dt^2 \,,$$
$$= \frac{4}{\xi_1^2}(dz^2-z^2dt^2)\,, \quad \xi_1 y=z^2\,; \tag{42}$$

$$\xi_2=\beta^2: \quad g_2 = \frac{dy^2}{\beta^2y^2\mp 1}-(\beta^2y^2\mp 1)dt^2$$
$$= \frac{1}{\beta^2}\left[dz^2-\begin{pmatrix}\sinh^2 z\\ \cosh^2 z\end{pmatrix}dt^2\right], \quad \begin{pmatrix}\beta y=\cosh z\\ \beta y=\sinh z\end{pmatrix}. \tag{43}$$

These last expressions can also be gathered in the form

$$g_2 = \frac{4dudv}{(1+\eta\beta^2 uv)^2}, \qquad \eta = -1, 0, 1, \tag{44}$$

with

$$\eta = 1: \quad \beta u = \tan\frac{z}{2}\exp(t), \quad \beta v = \tan\frac{z}{2}\exp(-t), \tag{45}$$

$$\eta = 0(=\beta^2): \quad \xi_1 u = z\exp(t), \quad \xi_1 v = z\exp(-t), \tag{46}$$

$$\eta = -1: \quad \beta u = \tanh\frac{z}{2}\exp(t), \quad \beta v = \tanh\frac{z}{2}\exp(-t). \tag{47}$$

Consequently, the metric (29) acquires the Bertotti–Robinson form [18, 19]

$$g = \frac{4d\zeta d\bar{\zeta}}{(1+\varepsilon\alpha^2\zeta\bar{\zeta})^2} + \frac{4dudv}{(1+\eta\beta^2 uv)^2}, \tag{48}$$

where $\varepsilon, \eta = -1, 0, 1$. Nevertheless, since α^2 and β^2 are polynomial of fourth order in the electric and magnetic fields, corresponding to the Heisenberg–Euler non–linearities of the electromagnetic theory, i.e.,

$$\begin{aligned} &\alpha^2 \quad (\varepsilon = \pm 1) = \pm[\Lambda + E^2 + B^2 - 2a(B^4 + E^4) - 19aB^2E^2], \\ &\alpha^2 \quad (=\varepsilon) = 0, \quad -\Lambda = E^2 + B^2 - 2a(B^4 + E^4) - 19aB^2E^2, \end{aligned} \tag{49}$$

$$\begin{aligned} &\beta^2 \quad (\eta = \pm 1) = \pm[\Lambda - E^2 - B^2 - 2a(B^4 + E^4) - 19aE^2B^2], \\ &\beta^2 \quad (=\eta) = 0, \quad \Lambda = E^2 + B^2 + 2a(B^4 + E^4) + 19aE^2B^2, \end{aligned} \tag{50}$$

one has to take care of the values over that Λ ranges.

6. DISCUSSION OF THE RANGE OF VALUES OF Λ

In order to analyze the linear case we set $a = 0$ and look at the nine cases contained in (48). Three cases lead to inconsistencies for the value of the cosmological constant, and one case becomes trivial, i.e., $\Lambda = E^2 = B^2 = 0$. The remaining five cases put forward restrictions to the value of the cosmological constant as follows

1. $g = g_1(\varepsilon = 1) + g_2(\eta = -1)$ with $|\Lambda| < E^2 + B^2$.
   For $\Lambda = 0$ one has $\alpha^2 = \beta^2 = E^2 + B^2$, this particular solution of the Bertotti–Robinson class is commonly written as

$$g = \frac{1}{\alpha^2}[d\theta^2 + \sin^2\theta d\phi^2 + dz^2 - \sinh^2 z dt^2], \tag{51}$$

   and it is the single homogeneous conformally flat solution to the Einstein–Maxwell system of equations with homogeneous general electromagnetic field [20].

2. $g = g_1(\varepsilon = 1) + g_2(\eta = 0)$, with $\Lambda = (E^2 + B^2)$.
3. $g = g_1(\varepsilon = 1) + g_2(\eta = 1)$, with $\Lambda > 0$.
4. $g = g_1(\varepsilon = 0) + g_2(\eta = -1)$, with $\Lambda = -(E^2 + B^2)$.
5. $g = g_1(\varepsilon = -1) + g_2(\eta = -1)$, with $\Lambda < 0$.

For vanishing Λ the Bertotti–Robinson metric becomes conformally flat. For vanishing electromagnetic field this solution reduces to de Sitter or anti–de Sitter space.

Let us consider now the non–linear case $a \neq 0$, the conditions on the values of the cosmological constant which should be fulfilled are the following

1. $g = g_1(\varepsilon = 1) + g_2(\eta = -1)$, with $|\Lambda| < E^2 + B^2 + 2a(B^4 + E^4) + 19aB^2E^2$.
2. $g = g_1(\varepsilon = 1) + g_2(\eta = 0)$, with $\Lambda = E^2 + B^2 + 2a(B^4 + E^4) + 19aB^2E^2$.
3. $g = g_1(\varepsilon = 1) + g_2(\eta = 1)$, with $\Lambda > 2a(B^4 + E^4) + 19aE^2B^2$.
4. $g = g_1(\varepsilon = 0) + g_2(\eta = -1)$, with $\Lambda = -(E^2 + B^2) + 2a(B^4 + E^4) + 19aB^2E^2$.
5. $g = g_1(\varepsilon = -1) + g_2(\eta = -1)$, with $\Lambda < 2a(B^4 + E^4) + 19aB^2E^2$.

As in the linear case, there exist only five consistent nontrivial cases. The first relevant difference is that if we set $\Lambda = 0$ in the first case this does not imply that $\alpha^2 = \beta^2$ as it happens in the linear case. Moreover, in all the cases the non–linear part of the electromagnetic field appear in the restrictions on the cosmological constant values, in the second and in the fifth cases it is only this part which restricts the possible values of Λ. For vanishing a the standard Eintein–Maxwell results on the allowed values of Λ are recovered.

7. CONCLUSION

The Bertotti–Robinson metric can be represented as a direct sum of two 2–spaces of constant curvature and possesses a Ricci tensor of the type 2S — 2T, being 2S two spacelike eigenvectors while 2T two timelike eigenvectors.

Since any electromagnetic field allows an energy–momentum tensor of the type 2S — 2T, the Bertotti–Robinson metric is thus compatible with all those sources. If one considers any electromagnetic field as described by $F_{\mu\nu}$, the field has to be homogeneous, i.e., constant and hence $dF_{\mu\nu} = 0$.

8. SUMMARY

In this paper we considered the Einstein–Heisenberg–Euler theory, endowed with a cosmological constant term. Starting from the most general type D metric in the Petrov classification scheme we particularize to constant curvature and to homogeneous non–linear electromagnetic field. The resulting solution is a generalization to the Einstein–Heisenberg–Euler system of differential equations of the well known Bertotti–Robinson metric. This solution reduces to the standard Bertotti–Robinson solution for vanishing non–linearities of the electromagnetic field. More general exact solutions like black hole ones can be also obtained and will be reported elsewhere.

ACKNOWLEDGMENTS

This research was supported by CONACYT Grant 42191–F.

REFERENCES

1. G. Mie: *Grundlagen einer Theorie der Materie. Annalen der Physik* **37** (1912) 511–534.
2. M. Born, L. Infeld: *Foudation of the new field theory. Proc. Roy. Soc.* **A144** (1934) 425–451.
3. H. Salazar, A. Garcia, and J. Plebański: *Duality rotations and type D solutions to Einstein equations with nonlinear electromagnetic sources. J. Math. Phys.* **28** (1987) 2171–2181.
4. J.F. Plebański, *Lectures on Nonlinear Electrodynamics.* (Nordita, Copenhagen 1970).
5. A. Garcia, H. Salazar, and J. Plebański: *Type–D Solutions of the Einstein and Born–Infeld Nonlinear–Electrodynamics Equations. Il. N. Cimento* **84** (1984) 65–90.
6. E. Ayón-Beato and A. García: *Regular Black Hole in General Relativity Coupled with Nonlinear Electrodynamics. Phys. Rev. Lett.* **80** (1998) 5056–5059.
7. N. Bretón: *Geodesic structure of the Born–Infeld black hole. Class. Quantum Grav.* **19** (2002) 601–612. Eprint Archive gr-qc/0109022.
8. W. Heisenberg y H. Euler: *Folgerungen aus der Diracschen Theorie des Positrons. Z. Phys* **98** (1936) 714–732.
9. S. Weinberg, *The Quantum Theory of Fields.* (Cambridge University Press, 1995).
10. F.W. Hehl and Yu.N. Obukhov, *Foundations of Classical Electrodynamics.* (Birkhäuser Boston, USA 2003).
11. G. Bordin, M. Marklund, and L. Stenflo: *Proposal for Detection of QED Vacuum Nonlinearities in Maxwell's Equations by the Use of Waveguides. Phys. Rev. Lett.* **87** (2001) 171801 (3 pages).
12. G. Boillat: *Nonlinear Electrodynamics: Lagrangians and Equations of Motion. J. Math. Phys.* **11** (1970) 941–951 .
13. E. Fradkin and A. Tseytlin: *Non–linear Electrodynamics from Quantized Strings. Phys. Lett.* **B163**(1985) 123–130.
14. A. Tseytlin: *Vextor field effective action in the open superstring theory. Nucl. Phys.* **B276** (1986) 391–428. Erratum –ibid **B291** (1987) 876.
15. R. G. Leigh: *Dirac–Born–Infeld action from Dirichlet σ–model. Mod. Phys. Lett.* **A4** (1989) 2767–2772.
16. L. de Fosse, P. Koerber and A. Servin: *The Uniqueness of the Abelian Born–Infeld Action. Nucl. Phys.* **B603** (2001) 413–426. Eprint Archive hep-th/0103015.
17. G.W. Gibbons: *Born–Infeld particles and Dirichlet p–branes. Nucl. Phys.* **B514**(1998) 603–639.
18. B. Bertotti: *Uniform Electromagnetic Field in the Theory of General Relativity. Phys. Rev.* **116** (1959) 1331–1333.
19. I. Robinson: *A solution of the Einstein–Maxwell equations*. Bull. Acad. Polon. Sci., Ser. Math. Astr. *Phys.* **7** (1959) 351–352.
20. H. Stephani, D. Kramer, M. MacCallum, C. Hoenselaers, E. Herlt: *Exact Solution of Einstein's Field Equations.* (Cambridge University Press, Cambridge, UK, 2003).
21. A. García and A. Macías, in: *Black Holes: Theory and Observations.* F. W. Hehl, C. Kiefer and R. J. K. Metzler, eds. Lectures Notes in Physics Vol. 514 (Springer Verlag 1998) pp. 203–223.
22. J. Plebański and M. Demiański: *Rotating, Charged, and Uniformly Accelerating Mass in General Relativity. Ann. Phys.* (N.Y.) **98** (1976) 98–127.

Scalar-hairy Black Holes and Solitons: their fate in asymptotically flat spacetimes

Miguel Alcubierre*, José A. González†, Ulises Nucamendi** and Marcelo Salgado*

*Instituto de Ciencias Nucleares, Universidad Nacional Autonóma de México
A.P. 70-545, México 04510 D.F., México
†Institute for Gravitational Physics and Geometry, Penn State University
University Park, PA 16802
**Instituto de Física y Matemáticas, Universidad Michoacana de San Nicolás de Hidalgo,
Edif. C-3, Ciudad Universitaria, Morelia, Michoacán, C.P. 58040, México

Abstract. In previous investigations we had analyzed the existence of static scalar hairy black holes and their corresponding solitons in asymptotically flat spacetimes, and found that the solitons can evolve to a hairless black hole or expand in a domain-wall like form. Here we review the main ideas of such findings and provide new features of the solutions.

1. INTRODUCTION

The uniqueness theorems of black holes (BH) in the Einstein-Maxwell system support the Wheeler's no-hair conjecture (NHC) which establishes that the stationary BH are characterized by its global quantities such as the mass (M), the charge (Q) and the angular momentum (J). These theorems basically state that the only asymptotically flat (AF) and stationary BH are those contained within the Kerr-Newman family. Moreover, the so-called no-hair theorems (NHT) in theories coupling different classical fields to Einstein gravity establish that there are no quantities (i.e., hair) other than M, Q and J required to characterized a stationary AFBH in those theories. Among these NHT's are worth mentioning the ones concerning the Einstein-Higgs system which is relevant for the present case (see below).

In view of the uniqueness and no-hair theorems, it was then quite surprising the discovery of solitons within the Einstein-Yang-Mills (EYM) theory by Bartnick and McKinnon [1], followed by the discovery of the corresponding BH solutions by Bizon [2]. These solutions were in fact the first counterexamples to the no-hair conjecture. The EYMBH solutions (termed as *colored BH*) are characterized not only by their mass (and null global charges) but also by an *integer* that label the number of nodes in the no-trivial YM field. Ultimately the colored BH turned out to be linearly unstable and therefore were considered as a *weak* counterexample to the NHC. Furthermore, new counterexamples of the NHC were found in more complicated theories that couple non-Abelian fields with scalar fields.

In the case of matter composed by a single scalar field minimally coupled to gravity, Sudarsky [3], proved a very simple no-hair theorem (see also [4] and [5]). Among other

CP758, *Gravitation and Cosmology: 2nd Mexican Meeting on Mathematical and Experimental Physics*
edited by A. Macías, C. Läemmerzahl and D. Nuñez

assumptions, the validity of that theorem relies on the fulfillment of the weak energy condition (WEC) which constrains the scalar potential to be positive semi-definite.

In the case of nonminimally coupled scalar fields, there are only a few no-hair theorems (see for instance Refs. [6, 7, 8, 9]). The AF hairy black hole solution corresponding to a scalar field conformally coupled to gravity (the Bronnikov-Melnikov-Bocharova-Beckenstein solution [10, 11]) which is often used as a counterexample for the failure of the no-scalar-hair conjecture, has been criticized as a non genuine black hole solution in that the Einstein field equations are not satisfied at the horizon [12], and moreover, in that it is not a solution at all if one demands a bounded scalar field throughout the static region [6].

Recently, in the context of asymptotically anti-de Sitter (AdS) spacetimes, scalar-hairy black holes (SHBH) were found for some scalar-field potentials $V(\phi)$ [13]. In Ref. [13] it was first suggested that by adjusting the parameters of $V(\phi)$, the effective cosmological constant could be cancel out so that asymptotically flat SHBH might exist. A further analysis along this line [14], emphasized this possibility and established a no-hair theorem in the AdS context.

Finally, we mention that motivated by the findings of Refs. [13, 14], some of us [15] found AF scalar-hairy black holes and their corresponding solitons by using a non-positive semidefinite asymmetric scalar field potential (see below). Another feature of the potential for the AF solutions to exist is that the potential's local minimum coincides with one of its roots. The solutions found were such that the non-trivial scalar-field reached the local minimum asymptotically preventing the existence of an effective cosmological constant. The family of SHBH are parametrized by their horizon radius r_h (or equivalently by the horizon area $\mathscr{A} = 4\pi r_h^2$), satisfying a non-trivial relation between the Arnowitt-Deser-Misner (ADM) mass $M_{\rm ADM}$ and r_h which differs from the Schwarzschild relation $M_{\rm ADM} = r_h/2$. It then turns out that given a spherically symmetric AFBH with fixed mass $M_{\rm ADM}$ there are two possible black holes in the Einstein-Higgs theory analyzed in Ref. [15]: one is the hairy BH with a non-trivial scalar field, and the second one is the plain Schwarzschild BH. It is in this sense that the NHC fails within the Einstein-Higgs theory analyzed in Ref. [15].

A further analysis also showed that both the solitons and BH were linearly unstable [15]. While the fate of such static configurations were uncertain, a full-non-linear evolution [16] revealed that the soliton can have two possible fates: a) it collapses to a stationary Schwarzschild BH or b) it ends in a non-stationary situation that resembles to a boosted domain-wall. We believe that something similar will happen when the initial conditions correspond to a hairy BH rather than to a soliton. In that case, the evolution is more complicated from the numerical point of view due to the presence of the horizon initially where the metric can become singular. The detailed evolution of the hairy BH is currently under investigation [17].

In the following, we summarize the main aspects of the SHBH and solitons and their evolution.

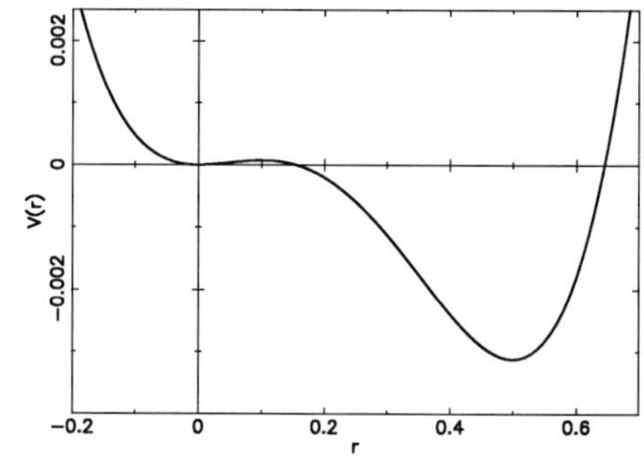

FIGURE 1. Scalar potential $V(\phi)$ corresponding to Eq. (2) with parameters $\sigma = 1$, $\eta_1 = 0.5$, $\eta_2 = 0.1$ and $a = 0$.

2. THE MODEL

We will consider a model of a scalar field minimally coupled to gravity and with a non-trivial self-interaction potential. The model is described by the Lagrangian (we will use units such that $G = c = 1$):

$$\mathscr{L} = \sqrt{-g}\left[\frac{1}{16\pi}R - \frac{1}{2}\nabla_\alpha\phi\nabla^\alpha\phi - V(\phi)\right] . \tag{1}$$

We choose the following asymmetric scalar-field potential leading to the desired asymptotically flat solutions:

$$V(\phi) = \frac{\sigma}{4}(\phi - a)^2\left[(\phi - a)^2 - \frac{4(\eta_1 + \eta_2)}{3}(\phi - a) + 2\eta_1\eta_2\right] , \tag{2}$$

with σ, η_1, η_2 and a constant parameters. For this class of potentials one can easily show that, if $\eta_1 > 2\eta_2 > 0$, $\phi = a$ corresponds to a local minimum, $\phi = a + \eta_1$ to the global minimum and $\phi = a + \eta_2$ to a local maximum. The key feature of this potential for the asymptotically flat and static solutions to exist is that the local minimum at $\phi = a$ is also a zero of $V(\phi)$ [15]. The factor σ in front of the potential fixes the scale, so one can always take $\sigma = 1$ and just re-scale everything for a different σ afterward. For the simulations discussed here we will take the following values for the parameters: $\sigma = 1$, $\eta_1 = 0.5$, $\eta_2 = 0.1$ and $a = 0$ (see Figure 1).

The field equations following from the Lagrangian (1) are the Einstein's field equations and the the Klein-Gordon (KG) equation:

$$G_{\mu\nu} = 8\pi T_{\mu\nu} , \qquad \Box\phi = \frac{\partial V(\phi)}{\partial\phi} , \tag{3}$$

The energy-momentum tensor for the scalar field is

$$T_{\mu\nu} = \nabla_\mu\phi\nabla_\nu\phi - g_{\mu\nu}\left[\frac{1}{2}\nabla_\alpha\phi\nabla^\alpha\phi + V(\phi)\right] . \tag{4}$$

We shall focus on a spherically symmetric space-time described by the metric

$$ds^2 = -\alpha^2 dt^2 + A dr^2 + B r^2 \left(d\theta^2 + \sin^2\theta d\varphi^2\right) \ . \tag{5}$$

The spherical symmetry implies that all dynamical functions depend only on r and t.

3. STATIC SOLUTIONS

For static solutions (specially for black holes) a convenient reparametrization is the following $\alpha^2 = [1-2m(r)/2]e^{2\delta(r)}$, $A = [1-2m(r)/2]^{-1}$, and taking $B \equiv 1$.

The first step consists in constructing static configurations with suitable boundary conditions. These configurations can serve then as initial conditions for analyzing the evolution and fate of black holes and solitons (although the BH evolution will not be discussed here).

Boundary conditions. For black hole configurations we demand regularity on the event horizon r_h and for solitons regularity at the origin $r=0$. This implies the following conditions for the fields (hereafter, the subscripts 'h' stand for a quantity to be evaluated at r_h):

$$m_h = \frac{r_h}{2} \ , \ \delta(r_h) = \delta_h \ , \ \phi(r_h) = \phi_h \ , \ (\partial_r \phi)_h = \frac{r_h(\partial_\phi V)_h}{1-8\pi r_h^2 V_h} \ , \tag{6}$$

The value δ_h is fixed so that the desired asymptotic behavior (see below) is obtained. For the soliton case, regularity at the origin $r=0$ results by taking the limit $r_h \to 0$ in the above regularity conditions. In addition to the regularity conditions, we impose asymptotically flat conditions on the spacetime (for black holes and solitons). These imply the following conditions on fields when $r \to \infty$,

$$m(\infty) = M_{\rm ADM} \ , \ \delta(\infty) = 0 \ , \ \phi(\infty) = \phi_\infty \ . \tag{7}$$

Here $M_{\rm ADM}$, is the ADM-mass associated with a given configuration. Actually, the value ϕ_∞ will correspond to the local minimum (which is also a root) of $V(\phi)$.

For a given scalar field potential (i.e., for a given theory), the family of SHBH configurations will be parametrized by the arbitrary free parameter r_h which specifies the location of the black hole horizon. The value ϕ_h is a shooting parameter rather than an arbitrary boundary value, and its value is determined so that the above AF conditions are fulfilled. Therefore for SHBH, $M_{\rm ADM} = M_{\rm ADM}(r_h)$. As mentioned above, the soliton will correspond to the case $r_h \to 0$ (in this limit, ϕ_0, the value of ϕ at $r=0$, is the shooting parameter), and the corresponding configuration is characterized by a unique $M_{\rm ADM}$.

Figures 2 and 3 depict the static solutions for BH and solitons for the potential shown in Fig. 1, obtained by integrating the Einstein and scalar field equations with the above boundary and asymptotic conditions (see Ref. [15] for more details).

As we mentioned above, the hairy BH's and the soliton turn to be unstable with respect to radial linear perturbations [15]. The unstable nature of such objects can also be exhibited from a heuristic energetic analysis similar to that of Sudarsky & Wald [18] in the EYM case. Such an analysis considers that the most stable configurations with fixed boundary conditions (same horizon area for BH or a regular origin for solitons and

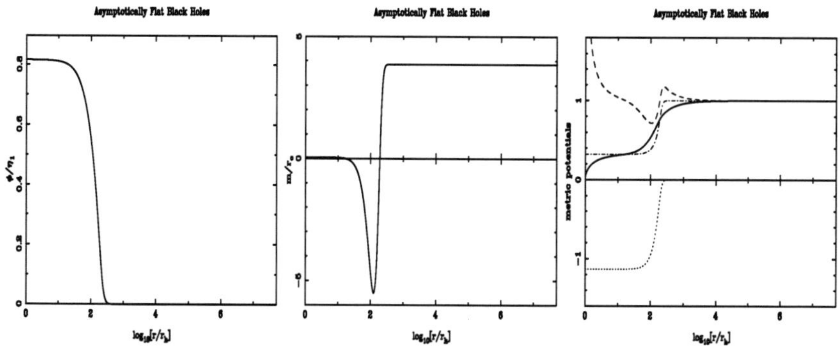

FIGURE 2. Black hole configuration constructed with $V(\phi)$ as given by Eq. (2) with parameters $\eta_1 = 0.5$, $\eta_2 = 0.1$, $a = 0$, and $r_h = 0.1/\sqrt{\sigma}$, $\phi_h \sim 0.40786$. The left and middle panels depict the scalar field and the mass function respectively. The latter converges to $M_{\rm ADM} \sim 3.843/\sqrt{\sigma}$. The right panel depicts the metric potentials $\sqrt{-g_{tt}}$ (solid line), $\sqrt{g_{rr}}$ (dashed line), e^{δ} (dash-dotted line) and δ (dotted line).

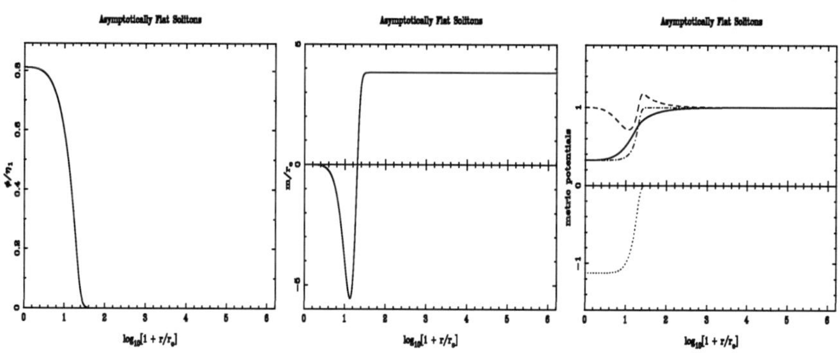

FIGURE 3. Same as Fig.1 for the soliton case. Here $\phi_0 \sim 0.40594$ and $M_{\rm ADM} \sim 3.827/\sqrt{\sigma}$. Note that the solutions are globally regular, notably at the origin.

same asymptotic conditions) are those that minimize the total energy (the ADM mass). Figure 4 (solid line) depicts the ADM mass of the hairy BH (including the soliton mass for $r_h = 0$) for different horizon radii. One appreciates that the ADM mass of the hairy configurations is always larger than the corresponding mass of the plain Schwarzschild solution (dashed line). In fact, for large hairy BH, one has $M_{\rm ADM} \approx 0.62 r_h$.

4. EVOLUTION OF SOLITONS

In view of the stability analysis it was interesting to investigate about the possible fates of the hairy objects. In the present contribution we shall summary the main features of the evolution of the soliton [16].

In order to perform a numerical analysis for the evolution, we used 3+1 approach

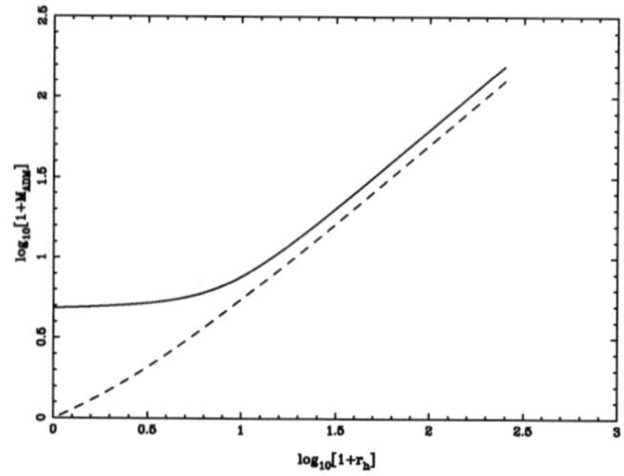

FIGURE 4. ADM masses of the hairy BH (solid line) and the Schwarzschild BH (dashed line) for different horizon radii. For the hairy BH, $M_{\rm ADM} \approx 0.62 r_h$ (for large r_h) while for the Schwarzschild BH $M_{\rm ADM} = 0.5 r_h$.

based on the standard ADM equations [19, 20]. The details can be consulted in Ref. [16], and suffice is to say that the system of evolution equations are strongly hyperbolic and regularized at the origin ($r = 0$). The outer boundary conditions imposed are of Sommerfeld type (outgoing waves). The numerical errors were controlled by using the Hamiltonian constraint, which was also employed to check the convergence of the numerical algorithm (finite differences).

For the case of a collapsing soliton into a BH, we used an apparent-horizon (AH) finder consisting in the location of the zero-expansion null surfaces. The expansion is given in 3+1 form as follows

$$H = \mathscr{D}_i s^i + K_{ij} s^i s^j - K = \frac{1}{A^{1/2}} \left(\frac{2}{r} + \frac{\partial_r B}{B} \right) - 2K^\theta_\theta \ , \tag{8}$$

with s^i the unit outward pointing normal vector to the surface, $\mathscr{D}_i$ is the covariant derivative compatible with the 3-metric and K_{ij} and K stand for the extrinsic curvature and its trace, respectively. An AH is then the outer-most closed surface such that $H = 0$ everywhere on the surface. One can associate a mass to the AH located at $r = r_{AH}$, through the formula

$$M_{AH} = \sqrt{\mathscr{A}_{AH}/(16\pi)} = \frac{r_{AH}\sqrt{B_{AH}}}{2} \ . \tag{9}$$

where $\mathscr{A}_{AH} = 4\pi r^2_{AH} B_{AH}$ is the area of the AH. This mass corresponds to the Bondi mass at late times of the evolution. From Fig.7 we appreciate that the BH mass almost equals the initial ADM-mass, the difference corresponding to the radiated energy. For low resolution, the AH-mass tends to become larger than the ADM-mass (dash-dotted line of Fig.7). However, this problem is overcome by increasing the resolution (solid line of Fig.7).

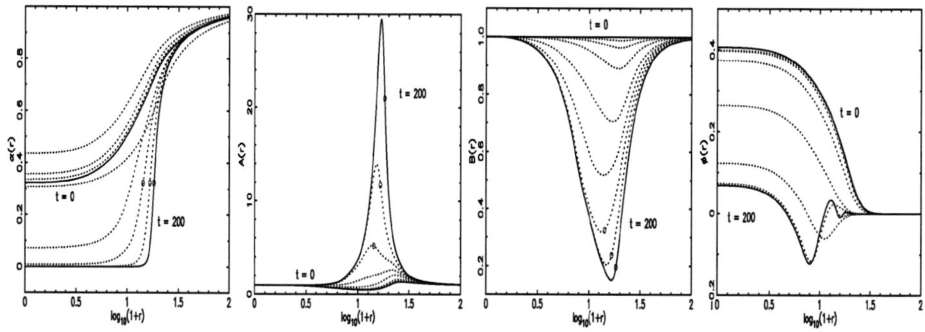

FIGURE 5. Evolution towards a BH formation of the spacetime generated by the soliton of Fig.3. The instability is triggered by a small inward Gaussian perturbation. The panels depict (from left to right) different snapshots of the evolution of the shift α, the metric potentials A, B and the scalar field ϕ, respectively. The small circles denote the position of the apparent horizon.

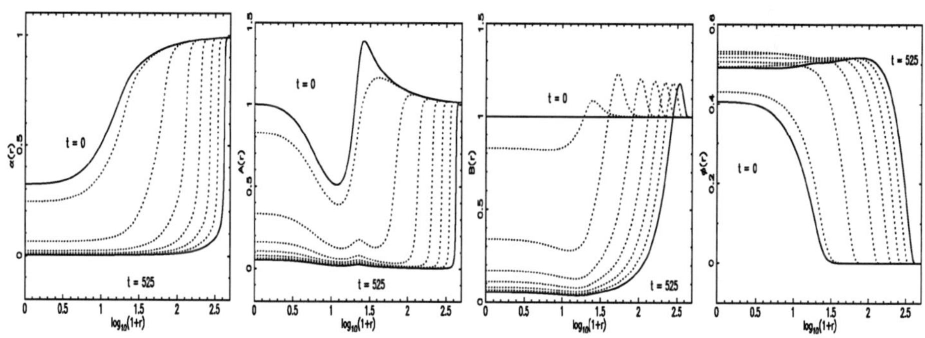

FIGURE 6. Evolution towards a domain-wall like expansion of the spacetime generated by the soliton of Fig.3. The instability is triggered by a small outward Gaussian perturbation. The panels depict (from left to right) different snapshots of the evolution of the shift α, the metric potentials A, B and the scalar field ϕ, respectively.

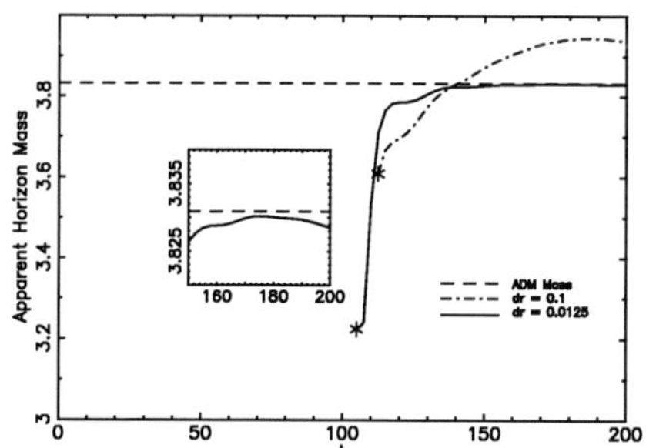

FIGURE 7. Apparent horizon mass as a function of time. An AH first appears at $t \sim 115$. The dashed line indicates the initial ADM mass of the spacetime, and the solid and dash-dotted lines the horizon mass for the highest and lowest resolutions respectively. The asterisks mark the first appearance of the apparent horizon.

5. DISCUSSION

At first glance and looking to Fig.1, one should expect that the lowest energy configuration would correspond to the one where the scalar-field settles at the global minimum. However, such configuration corresponds to an asymptotically anti-de Sitter spacetime with an effective cosmological constant provided by $V(\phi)$ at the global minimum, rather than to an asymptotically flat spacetime. The asymptotic flatness therefore forces the scalar-field to be "tied" in the local minimum. The instability and the low energy configuration on the other hand "pushes" the scalar-field towards the global minimum. The result of both expectations is then a non-stationary configuration (domain-wall) that tends to the global minimum but this is never reached in a final stationary situation due to the asymptotically flat condition (see Fig.6).

The other possibility that satisfy the energetic expectations without violating the asymptotic flatness condition is the one where the scalar-field evolves towards the local minimum. In that case, the final configuration is a stationary one which corresponds to the Minkowski spacetime or the Schwarzschild solution. However, the former is never reached but rather the latter due to the concentration of a large mass into a small sphere (Schwarzschild radius) when the scalar field wave travels towards the center of the soliton. As can be noted from Fig.5 the scalar-field outside the BH region has reached the trivial value $\phi = 0$ at the end of the evolution. It is very likely that for small mass solitons (obtained when changing the parameters of the scalar-field potential) the soliton disperses without forming a BH but leaving "behind" the globally regular Minkowski spacetime. The analysis of this possibility is currently under investigation [17].

ACKNOWLEDGMENTS

This work was partially supported by grants DGAPA-UNAM IN112401, IN122002, 4.8 CIC-UMSNH, PROMEP PTC-61, CONACYT 42949-F and NSF PHY-02-18750.

REFERENCES

1. Bartnik, R., and McKinnon, J., *Phys. Rev. Lett.*, **61**, 141 (1988).
2. Bizon, P., *Phys. Rev. Lett.*, **64**, 2844 (1990).
3. Sudarsky, D., *Class. Quantum Grav.*, **12**, 579 (1995).
4. Bekenstein, J., *Phys. Rev. D*, **51**, R6608 (1995).
5. Heusler, M., *J. Math. Phys.*, **12**, 3497 (1992).
6. Zannias, T., *J. Math. Phys.*, **36**, 6970 (1995).
7. Saa, A., *Phys. Rev. D*, **53**, 7377 (1996).
8. Mayo, A., and Bekenstein, J., *Phys. Rev. D*, **54**, 5059 (1996).
9. Ayon-Beato, E., *Class. Quantum Grav.*, **19**, 5465 (2002).
10. Bekenstein, J., *Ann. Phys. (N.Y.)*, **82**, 535 (1974).
11. Bocharova, N., Bronikov, K., and Melnikov, V., *Vest. Mosk. univ. Fiz., Astron.*, **6**, 706 (1970).
12. Sudarsky, D., and Zannias, T., *Phys. Rev. D*, **58**, 087502 (1998).
13. Torii, T., Maeda, K., and Narita, M., *Phys. Rev. D*, **64**, 044007 (2001).
14. Sudarsky, D., and González, J., *Phys. Rev. D*, **67**, 024038 (2003).
15. Nucamendi, U., and Salgado, M., *Phys. Rev. D*, **68**, 044026 (2003).

16. Alcubierre, M., González, J., and Salgado, M., *Phys. Rev. D*, **70**, 064016 (2004).
17. Alcubierre, M., González, J., and Salgado, M., *In preparation* (2005).
18. Sudarsky, D., and Wald, R., *Phys. Rev. D*, **46**, 1453 (1992).
19. Arnowitt, R., Deser, S., and Minser, C., "The Dynamics of General Relativity," in *Gravitation: An Introduction to Current Research*, edited by L. Witten, Wiley, New York, 1962, p. 227.
20. York, J., "Kinematics and Dynamics of General Relativity," in *Sources of Gravitational Radiation*, edited by L. Smarr, Cambridge University Press, Cambridge, 1979.

n–dimensional Friedmann-Robertson-Walker cosmologies

Alberto A. García*, Alberto García–Quiroz*, Mauricio Cataldo† and Sergio del Campo**

*Departamento de Física.
Centro de Investigación y de Estudios Avanzados del IPN.
Apdo. Postal 14-740, 07000 México DF, MEXICO.
†Departamento de Física, Facultad de Ciencias, Universidad del Bío–Bío,
Avenida Collao 1202, Casilla 5-C, Concepción, Chile.
**Instituto de Física, Facultad de Ciencias Básicas y Matemáticas, Pontificia Universidad Católica de Valparaíso, Avenida Brasil 2950, Valparaíso, Chile.

Abstract. In the framework of the n-dimensional Friedmann-Robertson-Walker cosmology, defined for a single scalar field minimally coupled to gravity in the presence of a perfect fluid, the n–dimensional spatially flat Lucchin–Mataresse, Barrow–Burd–Landcaster–Madsen, and Barrow-Saich cosmological solutions are determined.

1. INTRODUCTION

In the past few years there have been developed many approaches to deal with unified field theory based on the idea of multidimensional spacetime. Specially, inflationary universe models are candidates to play a significative role at the early epoch of the universe. Consequently, a fundamental theory such as string theory may tell us how inflation could be produced, and thus, it will tell us which of the many inflationary scenarios that exist nowadays is more adequate for describing our early universe[1]. In this respect, it is generally believed that higher dimensional theories must play a crucial role in the physics of the early universe. Based mostly on string theory ideas the brane word scenario has arose, yielding dramatic changes in our view of the universe[2]. Actually, if the brane world scenario becomes a reality, so that the string scale is close to the Planck scale, the main place to look at its possible implications will be cosmology. Also, much effort is devoted to understand experimentally the physical implications of higher dimensions at accessible energies. The cosmic microwave background (CMB) experiments have provided a lot of information on scalar density perturbations, among other things. On the light of these new approaches, it is worthwhile to consider higher dimensional theories and its solutions.

The aim of this contribution is the derivation of an n–dimensional generalization of the spatially flat Lucchin–Mataresse (LM)[3], Barrow–Burd–Landcaster–Madsen (BBLM)[4], and Barrow-Saich (BS)[5] cosmological solutions for a single scalar field coupled minimally to gravity in the presence of a perfect fluid. The derived solution possesses the property, similarly as the BS solution, of proportionality between potential and

CP758, *Gravitation and Cosmology: 2nd Mexican Meeting on Mathematical and Experimental Physics*
edited by A. Macías, C. Läemmerzahl and D. Nuñez

kinetic energies.

The spacetime to be considered is described by the n–dimensional spherically symmetric Friedmann–Robertson–Walker (FRW) metric

$$ds^2 = -dt^2 + a(t)^2\left(\frac{dr^2}{1-kr^2} + r^2 d\Omega_{n-2}^2\right), \tag{1}$$

where the spherical sector, related to $n-2$ angular variables θ_i, with i running from 1 to $(n-2)$, is determined as $d\Omega_{n-2}^2 = d\theta_1^2 + \sin^2\theta_1 d\theta_2^2 + \ldots + \sin^2\theta_1 \ldots \sin^2\theta_{n-3} d\theta_{n-2}^2$, for $n \geq 3$. The Einstein equations, $G_{\alpha\beta} = R_{\alpha\beta} - \frac{1}{2}Rg_{\alpha\beta} = \kappa_n T_{\alpha\beta}$, where Greek indices run from 1 to n, and κ_n stands for the n–dimensional gravitational constant, for a single scalar field ϕ_n associated to a potential $V(\phi_n)$ minimally coupled to gravity , a perfect fluid characterized by the matter perfect fluid energy density ρ_n and the isotropic pressure p_n, in the presence of a cosmological constant Λ_n, if any, amount to

$$\begin{aligned}
\frac{(n-1)(n-2)}{2}\left(\frac{\dot{a}^2+k}{a^2}\right) &= \Lambda_n + \kappa_n\left(\rho_n + \frac{1}{2}\dot{\phi_n}^2 + V_n\right), &(2)\\
\frac{\ddot{a}}{a} + \frac{(n-2)(n-3)}{2}\frac{\dot{a}^2+k}{a^2} &= \Lambda_n - \kappa_n\left(p_n + \frac{1}{2}\dot{\phi_n}^2 - V_n\right), &(3)
\end{aligned}$$

where dots denote derivatives with respect to t, and for short V_n denotes the potential $V(\phi_n)$.

The evolution equation of the ϕ_n–field is given by

$$\ddot{\phi}_n + (n-1)\frac{\dot{a}}{a}\dot{\phi}_n + \frac{d}{d\phi_n}V_n = 0. \tag{4}$$

Eq. (3) represents the matter energy conservation; in fact, substituting $(\dot{a}^2+k)/a^2$ from Eq. (2) into the derivative of Eq. (3),

$$\ddot{a} \equiv \frac{1}{2}\left(\frac{a^2}{\dot{a}}\right)\frac{d}{dt}\left(\frac{\dot{a}^2+k}{a^2}\right) + \frac{\dot{a}^2+k}{a} \tag{5}$$

and taking into account Eq. (4), one arrives at the conservation equation

$$\dot{\rho}_n + (n-1)\frac{\dot{a}}{a}(\rho_n + p_n) = 0, \tag{6}$$

which is also derivable from the conservation equation of the whole energy–momentum tensor $T^{\alpha\beta}{}_{;\beta} = 0$. Commonly as the set of independent dynamical equations is accepted

$$\begin{aligned}
\frac{(n-1)(n-2)}{2}\left(\frac{\dot{a}^2+k}{a^2}\right) &= \Lambda_n + \kappa_n\left(\rho_n + \frac{1}{2}\dot{\phi_n}^2 + V_n\right), &(7a)\\
\dot{\rho}_n + (n-1)\frac{\dot{a}}{a}(\rho_n + p_n) &= 0, &(7b)\\
\frac{1}{2}\frac{d}{dt}\dot{\phi_n}^2 + (n-1)\frac{\dot{a}}{a}\dot{\phi_n}^2 + \frac{d}{dt}V_n &= 0. &(7c)
\end{aligned}$$

The conventional perfect fluid formulation of the scalar field is achieved by introducing the scalar field density and pressure through

$$\rho_{\phi_n} = \frac{1}{2}\dot{\phi_n}^2 + V_n, \; p_{\phi_n} = \frac{1}{2}\dot{\phi_n}^2 - V_n, \tag{8}$$

hence, the dynamical equation (7) becomes

$$\dot{\rho}_{\phi_n} + (n-1)\frac{\dot{a}}{a}(\rho_{\phi_n} + p_{\phi_n}) = 0. \tag{9}$$

The system to be integrated in the forthcoming sections for spatially flat solutions, $k = 0$, is

$$\dot{\rho}_{\phi_n} + (n-1)\frac{\dot{a}}{a}(\rho_{\phi_n} + p_{\phi_n}) = 0, \tag{10}$$

$$\frac{(n-1)(n-2)}{2}\left(\frac{\dot{a}}{a}\right)^2 = \Lambda_n + \kappa_n\left(\rho_n + \rho_{\phi_n}\right), \tag{11}$$

2. DERIVATION OF THE N–DIMENSIONAL LM POWER LAW SOLUTION

In order to show the uniqueness of the power law solution within the conventional perfect fluid formulation we start with the equation of state

$$\rho_{\phi_n} + p_{\phi_n} = \Gamma_n \rho_{\phi_n}, \tag{12}$$

where Γ_n is an arbitrary constant.

Substituting this expression into Eq. (10) one gets

$$\dot{\rho}_{\phi_n} + (n-1)\frac{\dot{a}}{a}\Gamma_n \rho_{\phi_n} = 0, \tag{13}$$

which could be integrated to give

$$\rho_{\phi_n} = \rho_{\phi_{n0}} a^{(1-n)\Gamma_n}. \tag{14}$$

Moreover, when Eq. (13) is replaced into Eq. (11) one may get $\dot{a}/a \equiv H$, the Hubble parameter, giving rise to a first order differential equation for the energy density $\rho_{\phi_n}(t)$

$$\frac{(n-1)(n-2)}{2}\left(\frac{\dot{\rho}_{\phi_n}}{(n-1)\Gamma_n \rho_{\phi_n}}\right)^2 = \kappa_n \rho_{\phi_n}. \tag{15}$$

Integrating, one arrives at

$$\rho_{\phi_n} = 2\frac{n-2}{n-1}\frac{1}{\Gamma_n^2 \kappa_n}(t-t_0)^{-2}. \tag{16}$$

Therefore, from Eq. (14) we obtain the expression for the scale factor, $a(t)$

$$a(t) = \left(\frac{\rho_{\phi_n}}{\rho_{\phi_{n0}}}\right)^{\frac{1}{(1-n)\Gamma_n}} = a_0(t-t_0)^{\frac{2}{(n-1)\Gamma_n}}, \tag{17}$$

where a_0 is given by

$$a_0 = \left(2\frac{n-2}{n-1}\frac{1}{\Gamma_n^2\kappa_n\rho_{\phi_{n0}}}\right)^{\frac{1}{(1-n)\Gamma_n}}, \tag{18}$$

which, by a shifting transformation of the t coordinate, can be brought to the standard power law form

$$a(t) = a_0 t^{\frac{2}{(n-1)\Gamma_n}} =: a_0 t^{\alpha}, \qquad \alpha := \frac{2}{(n-1)\Gamma_n}, \tag{19}$$

and thus the energy density follows the law $\rho \sim a^{-2/\alpha}$. Next, from Eq. (8) we have

$$\begin{aligned} \rho_{\phi_n} + p_{\phi_n} &= \dot{\phi_n}^2, \\ V_n &= \rho_{\phi_n} - \dot{\phi_n}^2. \end{aligned} \tag{20}$$

Therefore, using Eq. (8) we obtain

$$\dot{\phi_n} = \varepsilon_\phi\sqrt{\Gamma_n\rho_{\phi_n}} = \varepsilon_\phi\sqrt{\frac{2}{\Gamma_n}\frac{n-2}{n-1}\frac{1}{\kappa_n}}\,t^{-1}, \tag{21}$$

where ε_ϕ is $+1$ or -1. Integration of the above expression yields

$$\phi_n - \phi_{n0} = \varepsilon_\phi\sqrt{\Gamma_n\rho_{\phi_n}} = \varepsilon_\phi\sqrt{\frac{2}{\Gamma_n}\frac{n-2}{n-1}\frac{1}{\kappa_n}}\ln t. \tag{22}$$

On the other hand, taking into account Eq. (21), the potential V from Eq. (20) becomes

$$V(\phi_n) = \frac{1}{2}(2-\Gamma_n)\rho_{\phi_n} = (2-\Gamma_n)\frac{n-2}{n-1}\frac{1}{\kappa_n\Gamma_n^2}t^{-2}. \tag{23}$$

Expressing t in terms of ϕ_n

$$t = \exp\left(\varepsilon_\phi\sqrt{\frac{n-1}{n-2}\frac{\Gamma_n\kappa_n}{2}}(\phi_n-\phi_{n0})\right), \tag{24}$$

we get

$$V(\phi_n) = (2-\Gamma_n)\frac{n-2}{n-1}\frac{1}{\kappa_n\Gamma_n^2}\exp\left(-\varepsilon_\phi\sqrt{2\frac{n-1}{n-2}\Gamma_n\kappa_n}(\phi_n-\phi_{n0})\right). \tag{25}$$

In terms of the standard power α we have the n–dimensional power law slow roll inflationary solution:

$$
\begin{aligned}
a(t) &= a_0 t^{\alpha}, \\
\phi_n &= \varepsilon_\phi \sqrt{\frac{\alpha(n-2)}{\kappa_n}} \ln t, \\
V_{\phi_n} &= \frac{\alpha}{2\kappa_n}(n-2)\left[\alpha(n-1)-1\right]\exp\left(-2\varepsilon_\phi \sqrt{\frac{\kappa_n}{\alpha(n-2)}}\phi_n\right).
\end{aligned} \tag{26}
$$

where α is assumed to satisfy the inequality $\alpha > 1$ in order to have an accelerated solution.

The scale factor $a(t)$ allows us to determine the number N of e-folding during the period of inflation. By definition the number of e-folding is given by $N = \ln[a(t_f)/a(t_i)]$ where t_f and t_i are the time when inflation ends and begins, respectively. By choosing $t_i = 0$ and $t_f = \Delta t$, we get

$$
N(n) = \frac{4\ln(\Delta t)}{\gamma_n}\left(\frac{1}{n-1}\right). \tag{27}
$$

We observe that those universes with low dimension are favored.

3. DERIVATION OF THE N–DIMENSIONAL BBLM SOLUTION

The demonstration of uniqueness of the BBLM n–dimensional solution within the conventional perfect fluid formulation of a single scalar field starts with the equation of state

$$
\rho_{\phi_n} + p_{\phi_n} = \Gamma_n \rho_{\phi_n}{}^{1/2}. \tag{28}
$$

Substituting this expression into Eq. (10), one has

$$
\frac{d}{dt}\rho_{\phi_n}{}^{1/2} + (n-1)\frac{\Gamma_n}{2}\frac{1}{a}\frac{d}{dt}a = 0, \tag{29}
$$

with general integral equal to

$$
a = a_0 \exp\left(-\frac{2}{\Gamma_n(n-1)}\rho_{\phi_n}{}^{1/2}\right). \tag{30}
$$

Moreover, from Eq. (29) one may obtain $\dot{a}/a$ which, after substituting into Eq. (11), gives rise to a first order differential equation for $\rho_{\phi_n}(t)$

$$
\rho_{\phi_n}^{-1/2}\frac{d}{dt}\rho_{\phi_n} + \varepsilon_a \Gamma_n \sqrt{2\kappa_n \frac{n-1}{n-2}}\rho_{\phi_n}^{1/2} = 0. \tag{31}
$$

Integrating, one arrives at

$$
\rho_{\phi_n} = \rho_{\phi_{n0}} \exp\left(-\Gamma_n \varepsilon_a \sqrt{2\kappa_n \frac{n-1}{n-2}}\, t\right). \tag{32}
$$

Therefore, from Eq. (30) one gets the expression of the scale factor $a(t)$, namely

$$a(t) = a_0 \exp\left[-\frac{2}{\Gamma_n(n-1)}\rho_{\phi_{n0}}{}^{1/2}\exp\left(-\frac{1}{2}\Gamma_n\varepsilon_a\sqrt{2\kappa_n\frac{n-1}{n-2}}\,t\right)\right]. \tag{33}$$

Next, from Eq. (8) one has

$$\begin{aligned} \rho_{\phi_n} + p_{\phi_n} &= \dot{\phi_n}^2, \\ V_n &= \rho_{\phi_n} - \frac{1}{2}\dot{\phi_n}^2, \end{aligned} \tag{34}$$

therefore, using Eq. (8) one obtains

$$\dot{\phi_n} = \varepsilon_\phi\sqrt{\Gamma_n}\rho_{\phi_n}{}^{1/4} = \varepsilon_\phi\sqrt{\Gamma_n}\rho_{\phi_{n0}}{}^{1/4}\exp\left(-\frac{\Gamma_n}{4}\varepsilon_a\sqrt{2\kappa_n\frac{n-1}{n-2}}\,t\right) \tag{35}$$

thus, integrating, one arrives at

$$\phi_n - \phi_{n0} = -\varepsilon_\phi\varepsilon_a\rho_{\phi_{n0}}^{1/4}\frac{4}{\sqrt{\Gamma_n}}\sqrt{\frac{1}{2\kappa_n}\frac{n-2}{n-1}}\exp\left(-\frac{\Gamma_n}{4}\varepsilon_a\sqrt{2\kappa_n\frac{n-1}{n-2}}\,t\right) \tag{36}$$

$$t = -\frac{4\varepsilon_a}{\Gamma_n C_n}\ln\left(-\varepsilon_a\varepsilon_\phi\frac{\sqrt{\Gamma_n}}{4}C_n\left(\phi_n-\phi_{n0}\right)\rho_{\phi_{n0}}^{-1/4}\right), \tag{37}$$

where $C_n := \sqrt{2\kappa_n\left(\frac{n-1}{n-2}\right)}$.

On the other hand, taking into account Eq. (21), the potential V from Eq. (34) becomes

$$V(\phi_n) = \frac{\Gamma_n^2}{32}C_n{}^2\left(\frac{C_n^2}{8}(\phi_n-\phi_{n0})^4 - (\phi_n-\phi_{n0})^2\right) \tag{38}$$

Summarizing, with $\varepsilon_\phi = -1$, the n–dimensional generalization of the BBLM solution is determined by

$$\begin{aligned} a &= a_0\exp\left(-\frac{2}{(n-1)\Gamma_n}\rho_{\phi_{n0}}{}^{1/2}e^{-\varepsilon_a\Gamma_n\sqrt{\kappa_n}\sqrt{\frac{1}{2}\frac{n-1}{n-2}}\,t}\right), \\ \phi_n - \phi_{n0} &= 4\sqrt{\frac{n-2}{n-1}}\varepsilon_a\frac{1}{\sqrt{2\Gamma_n\kappa_n}}\rho_{\phi_{n0}}{}^{1/4}\exp\left(-\frac{1}{4}\varepsilon_a\Gamma_n\sqrt{\kappa_n}\sqrt{2\frac{n-1}{n-2}}\,t\right), \\ V_n(\phi_n) &= \frac{1}{16}\frac{n-1}{n-2}\Gamma_n^2\kappa_n\left[\frac{n-1}{n-2}\frac{\kappa_n}{4}(\phi_n-\phi_{n0})^4 - (\phi_n-\phi_{n0})^2\right], \end{aligned} \tag{39}$$

where $\varepsilon_a = \pm 1$, and $\rho_{\phi_{n0}}$ and ϕ_{n0} are arbitrary constants.

From Eq.(39) it becomes apparent that the four order term dominates at large values of n. The main features of these expanding ($\varepsilon_a = 1$) solutions consist in: as t tends to

$-\infty$ there is a singularity, $a \to 0$ and $\phi \to \infty$, as t tends to ∞, $\phi \to 0$ and $a \to a_0$. This is true for any value of n. At $t = 0$ the scale factor and the scalar field are finite, finally, the Mexican–hat-shaped potential V_n possesses minima at $\phi_n = \pm\frac{2}{\sqrt{\kappa_n}}\sqrt{\frac{n-2}{n-1}}$.

As before, the scale factor $a(t)$ from Eq. (39) allows us to determine the number N of e-folding during the period of inflation. By choosing $t_i = 0$ and $t_f = \Delta t$ this number amounts to

$$N(n) = \beta(n)\left[1 - e^{-\alpha(n)\Delta t}\right], \tag{40}$$

where $\alpha(n)$ and $\beta(n)$ are given by $\alpha(n) = \varepsilon_a \Gamma_n \sqrt{\kappa_n}\sqrt{\frac{1}{2}\frac{n-1}{n-2}}$ and $\beta(n) = \frac{2}{(n-1)\Gamma_n}\rho_{\phi_{n0}}{}^{1/2}$, respectively. In order to make sure that $N(n) > 0$ we need $\varepsilon_a = 1$. We note that $N(n)$ increases when n decreases. This could be interpreted as followed: those universes that have less number of dimension are favored in concern with the value of N. Thus, in this respect, universes with $n = 3$ and $n = 4$ dimensions should be chosen.

The Barrow, Burd and Lancaster representation of the (2+1) solution is achieved by identifying: $A_{BBL} = 2\varepsilon_a \rho_{\phi_{n0}}{}^{1/4}/\sqrt{\kappa_3 \Gamma_3}$ and $\mu_{BBL} = \Gamma_3^2 \kappa_3/8$. The four–dimensional Madsen solution, arising for $n = 4$, describes an unusual cosmological solution for $\lambda\phi^4$ theory with broken symmetry, as stated in the title of the Madsen's publication; it occurs to be a unique solution for the non-linear state equation $p_{\phi_4} + \rho_{\phi_4} = \Gamma_4\sqrt{\rho_{\phi_4}}$, see Ref. [6] for details.

4. DERIVATION OF THE N–DIMENSIONAL BS SOLUTION

For a perfect fluid fulfilling the equation of state

$$\rho_n + p_n = \gamma_n \rho_n, \tag{41}$$

where γ_n is the equation of state parameter, the Eq. (7b) becomes

$$\dot{\rho}_n + (n-1)\frac{\dot{a}}{a}\gamma_n \rho_n = 0. \tag{42}$$

Its general integral is given by

$$\rho_n = A_n a^{(1-n)\gamma_n}, \tag{43}$$

where A_n is an constant of integration.

Similarly as in the (3+1)–dimensional BS case, the scalar potential is assumed to be proportional to the kinetic energy , namely

$$V(\phi_n) = \frac{1}{2\alpha_n}\dot{\phi}_n^2. \tag{44}$$

In such a case, the scalar field energy conservation equation (7) amounts to

$$\frac{d}{dt}\left[\frac{1+\alpha_n}{2\alpha_n}\dot{\phi}_n^2\right] + (n-1)\frac{\dot{a}}{a}\dot{\phi}_n^2 = 0, \tag{45}$$

which possesses as first integral

$$\dot{\phi}_n = \Delta_n a^{(1-n)\frac{\alpha_n}{1+\alpha_n}}, \tag{46}$$

where Δ_n is a constant of integration.

Taking into account Eq. (43), Eq. (44), and Eq. (46), the equation (7a) becomes

$$\begin{aligned}\frac{(n-1)(n-2)}{2}\frac{\dot{a}^2}{a^2} &= \kappa_n\left(A_n a^{(1-n)\gamma_n} + \frac{1+\alpha_n}{2\alpha_n}\dot{\phi}_n^2\right)\\ &= \kappa_n\left(A_n a^{(1-n)\gamma_n} + \frac{1+\alpha_n}{2\alpha_n}\Delta_n^2 a^{2(1-n)\frac{\alpha_n}{1+\alpha_n}}\right).\end{aligned} \tag{47}$$

Introducing the auxiliary dependent variable z,

$$z := a^{(n-1)\gamma_n/2} \quad \rightarrow \quad \frac{d}{dt}z = \frac{n-1}{2}\gamma_n a^{(n-1)\gamma_n/2}\frac{\dot{a}}{a}, \tag{48}$$

the Eq. (47) rewrites as

$$\frac{2}{\gamma_n^2}\frac{n-2}{n-1}\left(\frac{dz}{dt}\right)^2 = \kappa_n\left(A_n + \frac{1+\alpha_n}{2\alpha_n}\Delta_n^2 z^{2(1-\frac{2\alpha_n}{(1+\alpha_n)\gamma_n})}\right). \tag{49}$$

The general solution of this equation, given in terms of a hypergeometric function, amounts to

$$\begin{aligned}\gamma_n\sqrt{\frac{\kappa_n}{2}}\sqrt{\frac{n-2}{n-1}}(t+t_0) &= \frac{z}{\sqrt{A_n}}F\left(\left[\frac{1}{2},\frac{1}{2}\frac{(1+\alpha_n)\gamma_n}{(1+\alpha_n)\gamma_n-2\alpha_n}\right],\left[1+\right.\right.\\ &\left.\left.\frac{1}{2}\frac{(1+\alpha_n)\gamma_n}{(1+\alpha_n)\gamma_n-2\alpha_n}\right],-\Delta_n^2\frac{1+\alpha_n}{2\alpha_n}z^{2(1-\frac{2\alpha_n}{(1+\alpha_n)\gamma_n})}\right).\end{aligned} \tag{50}$$

In order to have an integral in terms of elementary functions, following Barrow and Saich, we search for solutions under the condition

$$\frac{1+\alpha_n}{\alpha_n} = \frac{4}{\gamma_n}, \tag{51}$$

which establishes a relationship between perfect fluid and scalar field. Under this condition, the equation for z amounts to

$$\frac{dz}{\sqrt{A_n + \frac{2}{\gamma_n}\Delta_n^2 z}} = \varepsilon_z\sqrt{\frac{n-1}{n-2}}\sqrt{\frac{\kappa_n}{2}}\gamma_n dt, \tag{52}$$

where $\varepsilon_z = \pm$. Its integration yields

$$\sqrt{A_n + \frac{2}{\gamma_n}\Delta_n^2 z} = \varepsilon_z\sqrt{\frac{n-1}{n-2}}\sqrt{\frac{\kappa_n}{2}}\Delta_n^2(t+t_0), \tag{53}$$

where t_0 is an integration constant.

Since the metric (1) is invariant under the translation (shifting) of the time t–coordinate, $t \to t + T_0$, thus without loss of generality t_0 can be equated to zero. Nevertheless, by choosing t_0, one can determine the value of $a(t)$, for instance, at the initial time $t = 0$. Hence, z can be given as

$$z = a^{(n-1)\gamma_n/2} = \frac{n-1}{n-2}\frac{\kappa_n}{8}B_n^2\gamma_n^2(t+t_0)^2 - \frac{A_n}{B_n}, \tag{54}$$

where it has been introduced the constant B_n instead of Δ_n by means of

$$B_n := \frac{2\Delta_n^2}{\gamma_n}. \tag{55}$$

Now we can determine ϕ_n by integrating Eq. (46). Taking into account Eq. (51), and B_n from (55), the differential equation (46) rewrites as

$$\frac{d\phi_n}{dt} = \Delta_n a^{(1-n)\gamma_n/4} = \frac{\sqrt{\frac{1}{2}B_n\gamma_n}}{\sqrt{\frac{n-1}{n-2}\frac{\kappa_n}{8}B_n\gamma_n^2(t+t_0)^2 - A_n/B_n}}. \tag{56}$$

The integration yields

$$\begin{aligned}\phi_n &= \sqrt{\frac{n-2}{n-1}\frac{4}{\kappa_n\gamma_n}}\ln\left[C_{\phi_n}\sqrt{\frac{n-1}{n-2}\frac{\kappa_n}{8}}B_n\gamma_n(t+t_0)\right.\\ &\left.+ C_{\phi_n}\sqrt{\frac{n-1}{n-2}\frac{\kappa_n}{8}B_n\gamma_n^2(t+t_0)^2 - A_n/B_n}\right],\end{aligned} \tag{57}$$

where C_{ϕ_n} is a constant of integration, which we choose as $C_{\phi_n} = c_{\phi_n} + \exp\left(\phi_{n0}\sqrt{\frac{n-1}{n-2}\frac{\kappa_n\gamma_n}{4}}\right)$ in order to determine convenient representations of the solutions by assigning certain definite values to c_{ϕ_n}.

To express t as function of ϕ_n we write Eq. (57) as

$$\begin{aligned}c_{\phi_n}\sqrt{\frac{n-1}{n-2}\frac{\kappa_n}{8}}B_n\gamma_n(t+t_0) &+ c_{\phi_n}\sqrt{\frac{n-1}{n-2}\frac{\kappa_n}{8}B_n\gamma_n^2(t+t_0)^2 - A_n/B_n}\\ &= \exp\left((\phi_n - \phi_{n0})\sqrt{\frac{n-1}{n-2}\kappa_n\gamma_n}\right).\end{aligned} \tag{58}$$

Consequently

$$c_{\phi_n}\sqrt{\frac{n-1}{n-2}\frac{\kappa_n}{8}}B_n\gamma_n(t+t_0) = \frac{\exp\left((\phi_n-\phi_{n0})\sqrt{\frac{n-1}{n-2}\kappa_n\gamma_n}\right) + \frac{A_n}{B_n}c_{\phi_n}}{2\exp\left(\frac{1}{2}(\phi_n-\phi_{n0})\sqrt{\frac{n-1}{n-2}\kappa_n\gamma_n}\right)}. \tag{59}$$

Substituting this last expression into Eq. (54) we obtain a in terms of ϕ_n,

$$a^{(n-1)\gamma_n/2} = \frac{1}{4c_{\phi_n}^2} \frac{\left[\exp\left((\phi_n - \phi_{n0})\sqrt{\frac{n-1}{n-2}\kappa_n\gamma_n}\right) - \frac{A_n}{B_n}c_{\phi_n}^2\right]^2}{\exp\left((\phi_n - \phi_{n0})\sqrt{\frac{n-1}{n-2}\kappa_n\gamma_n}\right)}. \tag{60}$$

Moreover, we are able now to evaluate $V(\phi_n)$,

$$\begin{aligned} V(\phi_n) &= \frac{1}{4}(4-\gamma_n)Ba^{(1-n)\gamma_n/2} \\ &= (4-\gamma_n)B_n c_{\phi_n}^2 \frac{\exp\left((\phi_n - \phi_{n0})\sqrt{\frac{n-1}{n-2}\kappa_n\gamma_n}\right)}{\left[\exp\left((\phi_n - \phi_{n0})\sqrt{\frac{n-1}{n-2}\kappa_n\gamma_n}\right) - \frac{A_n}{B_n}c_{\phi_n}^2\right]^2}, \end{aligned} \tag{61}$$

Note that this potential goes to a simple exponential function for very large ϕ_n.

Summarizing, we have established that the n–dimensional generalization of the BS solution is given by the metric (1) and the structural functions:

$$a^{(n-1)\gamma_n/2} = \frac{n-1}{n-2}\frac{\kappa_n}{8}B_n\gamma_n^2(t+t_0)^2 - \frac{A_n}{B_n}, \tag{62a}$$

$$\begin{aligned} \phi_n - \phi_{n0} &= \sqrt{\frac{n-2}{n-1}\frac{4}{\kappa_n\gamma_n}} \ln\left[c_{\phi_n}\sqrt{\frac{n-1}{n-2}\frac{\kappa_n}{8}}B_n\gamma_n(t+t_0)\right. \\ &\quad \left. + c_{\phi_n}\sqrt{\frac{n-1}{n-2}\frac{\kappa_n}{8}B_n\gamma_n^2(t+t_0)^2 - A_n/B_n}\right], \end{aligned} \tag{62b}$$

and

$$V(\phi_n) = (4-\gamma_n)B_n c_{\phi_n}^2 \frac{\exp\left((\phi_n - \phi_{n0})\sqrt{\frac{n-1}{n-2}\kappa_n\gamma_n}\right)}{\left[\exp\left((\phi_n - \phi_{n0})\sqrt{\frac{n-1}{n-2}\kappa_n\gamma_n}\right) - \frac{A_n}{B_n}c_{\phi_n}^2\right]^2}. \tag{62c}$$

The perfect fluid fulfills the equation of state $p_n + \rho_n = \gamma_n\rho_n$ with energy density given by $\rho_n = A_n a^{(1-n)\gamma_n}$. In this case we could get also the number of e-folding, $N(n)$. From Eq. (62a) we get that

$$N(n) = \beta(n) ln\left[\frac{(\Delta t + t_0)^2 - T(n)^2}{t_0^2 - T^2(n)}\right], \tag{63}$$

where again we have taken $t_i = 0$ and $t_f = \Delta t$. Here, $T(n)$ is defined by means of $T(n) = \frac{8(n-2)A_n}{(n-1)\kappa_n B_n^2}$ and we assume that $T(n) < t_0$. Since $\Delta t \ll 1$ we may write

$$N(n) \sim \frac{2}{(n-1)\gamma_n} ln\left\{1 + \frac{2t_0}{t_0^2 - T^2(n)}\Delta t\right\}. \tag{64}$$

From this later expression we may conclude as before that those universes with low dimensionality will be favored, in the sense that they give a big number of e-folding.

An alternative form of the n–dimensional BS solution can be achieved via an alternative parameterization; choosing

$$t_0^2 = \frac{A_n}{B_n \gamma_n^2} \frac{n-2}{n-1} \frac{8}{\kappa_n} = \frac{A_n}{4B_n} c_{\phi_n}^2 \tag{65a}$$

and

$$c_{\phi_n} = \frac{2}{\gamma_n} \sqrt{8 \frac{n-2}{n-1} \frac{1}{\kappa_n B_n}} \tag{65b}$$

the field functions acquire the form:

$$a^{(n-1)\gamma_n/2} = \frac{n-1}{n-2} \frac{\kappa_n}{8} B_n {\gamma_n}^2 (t^2 + 2t_0 t) = \frac{4}{c_{\phi_n}^2} (t^2 + 2t_0 t), \tag{66a}$$

$$\phi_n - \phi_{n0} = \sqrt{\frac{n-2}{n-1} \frac{4}{\kappa_n \gamma_n}} \ln \left[2(t + t_0) + 2\sqrt{t^2 + 2t_0 t} \right], \tag{66b}$$

$$V(\phi_n) = (4 - \gamma_n) \frac{32}{\gamma_n^2} \frac{n-2}{n-1} \frac{1}{\kappa_n} \frac{\exp\left((\phi_n - \phi_{n0}) \sqrt{\frac{n-1}{n-2} \kappa_n \gamma_n} \right)}{\left[\exp\left((\phi_n - \phi_{n0}) \sqrt{\frac{n-1}{n-2} \kappa_n \gamma_n} \right) - 4t_0^2 \right]^2}. \tag{66c}$$

Notice that under this parameterization $a(t=0) = 0$. For $n = 4$, and $\kappa_4 = 1$, one arrives just at the structural functions of the proper BS solution [5], namely

$$a^{3\gamma_4/2} = \frac{3}{16} B_n {\gamma_4}^2 (t^2 + 2t_0 t), \tag{67a}$$

$$\phi_4 - \phi_{40} = \sqrt{\frac{8}{3\gamma_4}} \ln \left[2(t + t_0) + 2\sqrt{t^2 + 2t_0 t} \right], \tag{67b}$$

$$V(\phi_4) = \frac{64}{3} \frac{(4 - \gamma_4)}{\gamma_4^2} \frac{e^{(\phi_4 - \phi_{40})\sqrt{3\gamma_4/2}}}{\left[e^{(\phi_4 - \phi_{40})\sqrt{3\gamma_4/2}} - 4t_0^2 \right]^2}. \tag{67c}$$

It is worthwhile to mention that there is a misprint in equation (BS,2.14) of Ref. [5] for C_{BS}, the correct expression should be $C = C_{BS}/\sqrt{3}$ as it can derived from V_0, (BS,2.16).

Note that the potential presents a little change under a variation of the n parameters. We should notice also that this potential has the same properties that the simple inverse power law potential $V(\phi) \sim \phi^{-\alpha}$ used for describing quintessence scalar fields. Therefore, this sort of potential is suitable for describing accelerated universe models[7].

It is clear that the n-dimensional generalization of the BS solution must contain the n-dimensional generalization of the famous power law inflationary cosmology derived by Lucchin and Matarrese [3] for the case where the matter perfect fluid is absent. This has to be so since both solutions arise for scalar fields having a scalar potential energies, $V(\phi)$, proportional to the kinetic term $\dot{\phi}^2$.

5. CONCLUDING REMARKS

We have established the n-dimensional generalizations of the LM, BBLM and BS cosmological solutions in a FRW background coupled to a single scalar field. For each of these solutions the scalar potential, $V(\phi_n)$, is proportional to the kinetic energy, $\dot{\phi}_n^2$. As far as the behavior of these solutions is concerned, they exhibit the same properties as the proper BS solution, namely, at the origin $t = 0$, they present initial curvature singularity and ϕ_n rolls from an initial value $\phi_n(t = 0) = C_n \ln|2t_0|$ to the asymptotic expression $\phi_n(t) = C_n \ln|4t|$ as t tends to ∞. Moreover, the potential $V(\phi_n)$ tends to infinity at $\phi_n = C_n \ln|2t_0|$ and it splits into two branches; one with monotonically increasing slope from zero to infinity through positive values in the interval $\phi_n \in (-\infty, C_n \ln|2t_0|)$, and a second one with monotonically decreasing slope from infinity to zero through negative values in the interval $\phi_n \in (C_n \ln|2t_0|, \infty)$. Inflation takes place under the condition $\ddot{a}(t) > 0$, hence $\gamma_n < \frac{4}{n-1}$, which for $3+1$ gravity becomes $\gamma_4 < \frac{4}{3}$.

ACKNOWLEDGMENTS

MC and SdC would like to thank the Phys. Dept. of the CINVESTAV for hospitality where part of this work was carried out. We acknowledge also the partial support to this research by CONICYT through grants N^0 7010485 (MC and AG), FONDECYT N^0 1010485 (MC and SdC), N^0 1030469 (SdC and MC); CONACYT N^0 38495–E (AG), CONACYT N^0 32427–E (AG-Q), and CONICYT/CONACYT N^0 2001-5-02-159. SdC acknowledge also The Vice Rectoría de Investigación y Estudios Avanzados de la PUCV for support.

REFERENCES

1. N. Ohta, hep-th/0411230; A. Linde,hep-th/0402051; Sh. Kachru, R. Kallosh, A. Linde, J. Maldacena, L. McAllister and S. P. Trivedi, JCAP **0310**, 013 (2003); N. Ohta, Prog. Theor. Phys. **110**, 269 (2003); A. Linde, Int. J. Mod. Phys., **A17S1**, 89 (2002); R. Brustein and P.J. Steinhardt, Phys. Lett B, **302**, 196 (1993); R. Brandenberger and C. Vafa, Nuc. Phys. B **316**, 391 (1989).
2. R. Maartens, Living Rev. Rel. **7**, 1 (2004); Y. Himemoto, T. Tanaka and M. Sasaki, *ibid* **65**, 104020 (2002); Y. Himemoto and M. Sasaki, Phys. Rev. D **63**, 044015 (2001); R. Maartens, V. Sahni and T. D. Saini, Phys. Rev. D **63**, 063509 (2001); R. Kallosh and A. Linde, JHEP **0002**, 005 (2000); A. Lukas, B. Ovrut, K. Stelle and D. Waldram, Phys. Rev. D **59**, 086001 (1999);*ibid*, Nucl. Phys. **B 552**, 246 (1999); L. Randall and R. Sundrum, Phys. Rev. Lett. **83**, 3370 (1999).
3. F. Lucchin and S. Matarrese, Phys. Rev. D **32**, 1316 (1985).
4. J.D. Barrow, A.B. Burd and D. Lancaster, Class. Quantum Grav.**3**, 551 (1986); M. S. Madsen, Gen. Relativ. Gravit. **18**, 879 (1986).
5. J.D. Barrow and P. Saich, Class. Quantum Grav. **10**, 279 (1993).
6. A.A. García, M. Cataldo and S. del Campo, Phys. Rev. D **68**, 123022 (2003).
7. B. Ratra and P.J.E. Peebles, Phys. Rev. D **37**, 3406 (1988); I. Zlatev, L. Wang and P.J. Steinhardt, Phys. Rev. Lett. **80**, 1586 (1998).

Non-commutativity in classical and quantum cosmology

Luis O. Pimentel* and César Mora†

*Departamento de Física ,
Universidad Autónoma Metropolitana,
Apartado Postal 55-534,CP 09340 México D.F.,MEXICO.
†Departamento de Matemáticas,
UPIBI-Instituto Politécnico Nacional,
Av. Acueducto s/n Col. Barrio La Laguna Ticomán,
CP 07340 México DF, MEXICO.

Abstract. We consider the effects of noncommutativity in classical and quantum cosmology for the scalar–tensor theory of gravitation.

INTRODUCTION

The study of the noncommutativity for the space time coordinates goes back to Snyder [1]. The renewed interest on the subject come from two main avenues, from the mathematical point of view due to the work of Connes [2], and from the physical point of view in connection with the M and string theories [4]. In this work we will explore the consequences of noncommutativity among the gravitational and matter fields for a simple model in cosmology based on the scalar–tensor theories of gravitation as was done in the work of Garcia–Compean, Obregon and Ramirez [5]. We consider the noncommutativity in the quantum and classical cases.

THE BERGMANN-WAGONER MINISUPERSPACE MODEL

In order to construct the minisuperspace model, our starting point is the action for Bergmann-Wagoner scalar-tensor theory:

$$S=\frac{1}{l_p^2}\int_M\sqrt{-g}\left[\phi R^{(4)}-\frac{\omega(\phi)}{\phi}g^{\mu\nu}\phi_{,\mu}\phi_{,\nu}+2\phi\,\lambda(\phi)\right]d^4x+\frac{2}{l_p^2}\int_{\partial M}\sqrt{h}\phi h_{ij}K^{ij}\,d^3x, \quad (1)$$

where $g=det(g_{\mu,\nu}), R^{(4)}$ is the scalar curvature of the Friedmann-Robertson-Walker theory, $\phi(t)$ is the conventional real scalar gravitational field, l_p is the Planck length and $\lambda(\phi)$ is the cosmological term. The second integral is a surface term involving the induced metric h_{ij} and second fundamental form K^{ij} on the boundary, needed to cancel the second derivatives in $R^{(4)}$ when the action is varied with the metric and scalar field, but not their normal derivatives, fixed on the boundary. We are interested to study an

CP758, *Gravitation and Cosmology: 2nd Mexican Meeting on Mathematical and Experimental Physics*
edited by A. Macías, C. Läemmerzahl and D. Nuñez

homogeneous and isotropic cosmological model, consequently we used the FRW metric

$$ds^2 = \sigma^2 \left[N^2(t)dt^2 - a^2(t)d\Omega_3^2)\right], \tag{2}$$

where N is the lapse function, a is the scale factor of the Universe, $\sigma^2 = l_P^2/24\pi^2$, the metric on S^3 with the unit volume is

$$d\Omega_3^2 = d\chi^2 + \sin^2\chi(d\theta^2 + \sin^2\theta d\phi^2), \tag{3}$$

and the Ricci scalar

$$R^{(4)} = -\frac{6k}{a^2} - 6\frac{\dot{a}^2}{N^2a^2} - 6\frac{\ddot{a}}{N^2a} + 6\frac{\dot{a}\dot{N}}{N^3a}, \tag{4}$$

substituting (4) in (1) and integrating with respect to space coordinates, after simplifications we have

$$S = \frac{1}{2}\int \left[-Nka\phi + \frac{a\phi}{N}\dot{a}^2 + \frac{a^2}{N}\dot{a}\dot{\phi} - \frac{N\omega(\phi)}{6\phi}a^3\dot{\phi}^2 + \frac{N}{3}a^3\phi\lambda(\phi)\right]dt, \tag{5}$$

where dot denotes the time derivative with respect to the time t, now we introduce the following new variables

$$x = ln(a\phi^{\frac{1}{2}}), \quad y = \int \left(\frac{2\omega(\phi)+3}{12}\right)^{\frac{1}{2}} \frac{d\phi}{\phi}, \quad \Lambda(y) = \frac{\lambda(\phi)}{3\phi}, \quad d\tau = \phi^{\frac{1}{2}}dt, \tag{6}$$

then the BW action simplifies to

$$S_{BW} = \frac{1}{2}\int \left[e^{3x}\{x'^2 - y'^2 + \Lambda(y)\} - ke^x\right]d\tau, \tag{7}$$

prime denotes the time derivative with respect to the new time τ and we have chosen $N = 1$. The corresponding hamiltonian is

$$H = \frac{1}{2}\left[e^{-3x}\{\pi_x^2 - \pi_y^2\} + ke^x - e^{3x}\Lambda(y)\right]. \tag{8}$$

A very similar Hamiltonian was obtained for the case of General relativity with a scalar field with different variables [3].

QUANTUM COSMOLOGY

Commutative Quantum Cosmology

The Wheeler–DeWitt (WDW)equation is obtained by the canonical quantization,

$$e^{-3x}\left[\partial_x^2 - \partial_y^2 - ke^{4x} + e^{6x}\Lambda(y)\right]\psi(x,y) = 0. \tag{9}$$

Where we have used a simple factor ordering. It is clear that this equation can be separated and we will consider some simple cases in which an exact solution can be obtained.

For the flat space, $k = 0$ and a constant Λ the WDW equation is

$$e^{-3x}\left[\partial_x^2 - \partial_y^2 + e^{6x}\Lambda\right]\psi(x,y) = 0. \tag{10}$$

and the solution by separation of variables is

$$\psi^{\pm}(x,y) = e^{\pm i\omega y}\left[C_1 J_{i\omega/3}\left(1/3\sqrt{\Lambda}e^{3x}\right) + C_2 Y_{i\omega/3}\left(1/3\sqrt{\Lambda}e^{3x}\right)\right]. \tag{11}$$

It is also possible to separate the equation in the variables $x \pm y$ and the solution is

$$\psi_c(x+y, x-y) = e^{\frac{4c^2 e^{3x+3y} - \Lambda e^{3x-3y}}{12c}}. \tag{12}$$

Noncommutative Quantum Cosmology

For the noncommutative case, we assume, following [5], that the minisuperspace variables are noncommuting,

$$[x,y] = i\theta. \tag{13}$$

And it is well known that that the noncommutativity can be implemented by means of the substitution

$$\mathbf{x} \to \mathbf{x} - \frac{\tilde{\mathbf{p}}}{\mathbf{2}}, \tag{14}$$

here

$$\tilde{\pi}_i = \theta^{ij}\pi_j = \theta\varepsilon^{ij}\pi_j. \tag{15}$$

The WDW equation for the noncommutative case is now

$$\left[\partial_x^2 - \partial_y^2 + \Lambda e^{6(x - \frac{\theta\pi_y}{2})}\right]\psi(x,y) = 0. \tag{16}$$

The solutions are

$$\psi_\theta^{\pm}(x,y) = e^{\pm i\omega y}\left[C_1 J_{\frac{i\omega}{3}}\left(\frac{\sqrt{\Lambda}e^{3(x\mp\omega\theta/12)}}{3}\right) + C_2 Y_{\frac{i\omega}{3}}\left(\frac{\sqrt{\Lambda}e^{3(x\mp\omega\theta/12)}}{3}\right)\right]. \tag{17}$$

CLASSICAL COSMOLOGY

Commutative Classical Cosmology

The classical equations of motion follows from the hamiltonian or the Lagrangian, implicit in the action, they are,

$$x'' + \frac{3}{2}[x'^2 + y'^2 - \Lambda] = 0, \tag{18}$$

$$(e^{3x}y')' = 0, \tag{19}$$

and also the Hamiltonian constraint

$$-x'^2 + y'^2 + \Lambda = 0. \tag{20}$$

From the equation for y we obtain a first integral,

$$y' = \pi_{x0}e^{-3x}, \tag{21}$$

here π_{y0} is the constant value of π_y. After the substitution of this first integral into the constraint we obtain

$$-x'^2 + \pi_{y0}^2 e^{-6x} + \Lambda = 0. \tag{22}$$

The solution for x is given by

$$x = -\frac{1}{6}\ln\left(\frac{\Lambda}{\pi_{y0}{}^2}\right) + \frac{1}{3}\ln\left(\sinh\left(3\sqrt{\Lambda}\tau\right)\right). \tag{23}$$

Substituting into the Eq.(21) and integrating we obtain the solution for y.

$$y = y_0 - \frac{2}{3}\mathrm{arctanh}\left(e^{-3\tau\sqrt{\Lambda}}\right). \tag{24}$$

We notice that this solution gives for late τ a constant y (ϕ)that implies that $x = a\phi^{1/2}$ approaches a DeSitter phase for late τ with ϕ reaching a constant value.

Noncommutative Classical Cosmology

In classical mechanics the noncommutativity shows up in a deformation of the Poisson brackets,

$$\{x_i, x_j\} = \theta_{ij}, \qquad \{x_i, \pi_j\} = \delta_{ij}, \qquad \{\pi_i, \pi_j\} = 0. \tag{25}$$

Here, as in the quantum case, we can return to canonical variables by the substitution

$$\mathbf{x} \to \mathbf{x} - \frac{\tilde{\pi}}{\mathbf{2}}, \tag{26}$$

where

$$\tilde{\pi}_i = \theta^{ij}\pi_j = \theta\varepsilon^{ij}\pi_j. \tag{27}$$

If we apply this substitution to the the Hamiltonian for the classical cosmology Eq. (8), we obtain

$$H_\theta = \frac{1}{2}\left[e^{-3(x-\theta\pi_y/2)}\left(-\pi_x^2 + \pi_y^2\right) + \Lambda e^{3(x-\theta\pi_y/2)}\right]. \tag{28}$$

The equations of motion are now

$$x' = -\pi_x e^{-3(x-\theta\pi_y/2)}, \tag{29}$$

$$\pi_x{}' = \frac{3}{2}\left[e^{-3x+3/2\,\theta\,\pi_y}\left(-\pi_x{}^2+\pi_y{}^2\right) - \Lambda\, e^{3\,\alpha-3/2\,\theta\,\pi_y}\right], \tag{30}$$

$$y' = e^{-3x+3/2\,\theta\,\pi_y}\left[\frac{3\,\theta}{4}\left(-\pi_x{}^2+\pi_y{}^2\right)+\pi_y\right] - \frac{3\,\theta\,\Lambda}{4}\,e^{3\,\alpha-3/2\,\theta\,\pi_y}, \tag{31}$$

$$\pi_y{}' = 0. \tag{32}$$

We also have to take into account the Hamiltonian constraint

$$H_\theta = \frac{1}{2}\left[e^{-3(x-\theta\pi_y/2)}\left(-\pi_x^2+\pi_y^2\right)+\Lambda e^{3(\alpha-\theta\pi_y/2)}\right] = 0. \tag{33}$$

The equation for π_y implies that it is a constant , if we eliminate π_y in the last equation, after a little algebra we obtain

$$-x'^2 + \pi_{y0}{}^2 e^{-6(x-\theta\pi_{y0}/2)} + \Lambda = 0. \tag{34}$$

Here we have called π_{y0} the constant value of π_y If we compare Eq.(34)with Eq.(22), we notice that they are the same if in the last one we substitute π_{y0} by $\tilde{\pi}$ with

$$\tilde{\pi} = \pi_{y0} e^{\frac{3\theta\pi_{y0}}{2}}, \tag{35}$$

hence the solution for x in the noncommutative case is

$$x_\theta = -1/6\ln\left(\frac{\Lambda}{\tilde{\pi}^2}\right) + 1/3\ln\left(\sinh\left(3\sqrt{\Lambda}t\right)\right). \tag{36}$$

Then the effect of non commutativity on x coordinate is just a shift by a constant:

$$x(\tau) \to x_\theta = x(\tau) + \frac{\theta\tilde{\pi}}{2}, \tag{37}$$

and therefore we still have a DeSitter asymptotic solution. The effect on the variable y is more complicated, the corresponding equation is now

$$y' = \pi_{y0} e^{-3(x-\theta\pi_{y0}/2)} + \frac{3\theta}{4}\left[\pi_{y0}{}^2 e^{-3(\alpha-\theta\pi_{y0}/2)} - (x'^2+\Lambda)e^{3(x-\theta\pi_{y0}/2)}\right]. \tag{38}$$

Since we already know $x(\tau)$ the last equation can be solved:

$$y = y_0 - \left(\frac{2}{3\pi_{y0}} + \frac{\theta}{2}\right)\sqrt{\Lambda}\,\operatorname{arctanh}\left(e^{-3\sqrt{\Lambda}\tau}\right)$$

$$
\begin{aligned}
&+ \left(\frac{\pi_{y0}{}^6\theta}{\Lambda^{5/2}}\right)\left[48\left(\cosh\left(\sqrt{\Lambda}\tau\right)\right)^9 - 108\left(\cosh\left(\sqrt{\Lambda}\tau\right)\right)^7 + 81\left(\cosh\left(\sqrt{\Lambda}\tau\right)\right)^5\right.\\
&- \left.\frac{81}{4}\left(\cosh\left(\sqrt{\Lambda}\tau\right)\right)^3\right]\\
&+ \left(\frac{\pi_{y0}{}^2\theta}{\Lambda^{1/2}}\right)\left[\left(\cosh\left(\sqrt{\Lambda}\tau\right)\right)^3 - \frac{3}{4}\cosh\left(\sqrt{\Lambda}\tau\right)\right]. \qquad (39)
\end{aligned}
$$

We can see that for small values of θ the correction due to the noncommutativity is small, however, for large values of the parameter the character of the solution can be changed dramatically. In particular for the critical value $\theta_c = -4/3\pi_{y0}$ the original solution is no longer present. In the differential equation for the noncommutative case we have a bifurcation.

We have considered a toy model for which exact solution can be obtained in the classical an quantum cosmologies in the commuting and noncommuting cases. The implications for observational cosmology are under consideration. Some author have considered this conection in a different context [6].

ACKNOWLEDGMENTS

This work was supported in part by PROMEP.

REFERENCES

1. H. Snyder, Phys. Rev. **71**, 38 (1947).
2. A. Connes, *Noncommutative Geometry* (Academic Press, London, 1994).
3. L. O. Pimentel and O. Obregon, to be published in GRG Journal 2005.
4. N. Seiberg and E. Witten, J. High Energy Phys. **09**, 032 (1999).
5. H. García-Compeán, O. Obregón and C. Ramírez, Phys. Rev. Lett. **88**, 161301-1 (2002).
6. S. Tsujikawa, R. Maartens, R. Brandenberger, *Phys.Lett. B* **574** (2003) 141; Q-G. Huang, M. Li, Power Spectra in Spacetime Noncommutative Inflation, astro–ph/0311378 ;M. Fukuma, Y. Kono and A. Miwa, Nucl.Phys. B682 (2004) 377; ibid, Noncommutative inflation and the large-scale damping in the CMB anisotropy,hep–th/0401153; H. Kim, G. S. Lee, Y. S. Myung, Noncommutative spacetime effect on the slow-roll period of inflation, hep–th/0402019.

Author Index

T

U